DEEP TIME

Deep Time

A LITERARY HISTORY

NOAH HERINGMAN

PRINCETON UNIVERSITY PRESS
PRINCETON & OXFORD

Published by Princeton University Press
41 William Street, Princeton, New Jersey 08540
99 Banbury Road, Oxford OX2 6JX

press.princeton.edu

ISBN 978-0-691-23677-3
ISBN (pbk.) 978-0-691-23579-0
ISBN (e-book) 978-0-691-23580-6

British Library Cataloging-in-Publication Data is available

Editorial: Anne Savarese and James Collier
Production Editorial: Kathleen Cioffi
Cover Design: Chris Ferrante
Production: Erin Suydam
Publicity: Alyssa Sanford and Charlotte Coyne
Copyeditor: Martin Schneider

Cover image: Engraving by J. E. Geriare based on the illustration from *Versuch einer Geschichte von Flötz-Gebürgen* (*Attempt at a History of Stratified Mountains*), p. 162b (Tafel 7) O 2530 RES by Johann Gottlob Lehmann, 1756. Courtesy of the Heidelberg Historical Holdings for Geological Literature, Heidelberg University Library.

This book has been composed in Arno

10 9 8 7 6 5 4 3 2 1

If they watched their house
turn above trees silvered in ice,
its walls framed in chains of glass that
throw back light in periodic waves, they
could tell time.

—MIKE BARRETT, "BABYLON XIV"

CONTENTS

ILLUSTRATIONS

ACKNOWLEDGMENTS

MY FIRST THANKS are due to the students in my "Deep Time in Literature and Science" class (2012 and 2013) with whom I first explored the topics of this book. It was a privilege to work with these two successive Freshman Interest Groups in Mathematics and Life Sciences, who brought their scientific acumen, their intellectual curiosity, and at times their skepticism (which echoes in the afterword to this book) with unfailing energy. At the University of Missouri, the Research Board provided funding for research travel and the College of Arts and Science provided a semester of leave in Fall 2018 that aided in the completion of this book.

I am very grateful to the National Humanities Center, which awarded me a fellowship in 2014–15 that enabled me to begin writing the book. Among the colleagues in that year's splendid cohort, I would especially like to thank Josephine McDonagh, Jonathan Sachs, and Yasmin Solomonescu, who read the first drafts with friendly rigor. I would also like to thank the Center's excellent librarians, Brooke Andrade and Sarah Harris. I accrued a debt of thanks to the staff of several other libraries over the next six years, including staff at the rare books reading room at the central library of the Muséum national d'Histoire naturelle, Paris; the Linda Hall Library of Science and Technology, Kansas City; and Special Collections, Ellis Library, University of Missouri.

During these years, I had the good fortune to present material from the book at a number of institutions as it took shape. I would like to thank the audiences at those talks and especially the following individuals and institutions for hosting me and providing feedback: Stefan Waldschmidt and Phillip Stillman at Duke University (September 2013); Adam Sneed at the University of Michigan (March 2014); John Savarese at UC Berkeley (April 2014); Adriana Craciun and Mary Terrall at the William Andrews Clark Memorial Library (November 2014); Eva Horn and Peter Schnyder at the Internationales Forschungszentrum Kulturwissenschaften (IFK), Vienna (January 2015); Anne D. Wallace at the University of North Carolina-Greensboro (March 2015);

Tilottama Rajan and Joel Faflak at the University of Western Ontario (May 2015); Crystal B. Lake at Wright State University (April 2016); Philipp Erchinger at the Heinrich-Heine-Universität Düsseldorf (April 2016); Peter Schnyder at the Université de Neuchâtel (November 2017); Emily Patton Smith at Randolph College in Virginia (November 2018); and Erik Flesch at the Rollo Jamison and Mining Museums (February 2020). Thanks also to Tobias Boes and Kate Marshall for organizing the "Writing the Anthropocene" panel at the American Comparative Literature Association (April 2013), where some of my ideas for the book first began to take shape.

Two chapters of this book (Ch. 1 and Ch. 2) include material adapted from earlier publications. I would like to thank the University of California Press, Routledge, and the University of Toronto Press for allowing me to reprint material from the following articles and chapters in revised form:

"Deep Time at the Dawn of the Anthropocene," *Representations* 129 (Winter 2015): 56–85.

"Deep Time in the South Pacific," *Marking Time: Romanticism and Evolution*, ed. Joel Faflak (Toronto: University of Toronto Press, 2017), 95–121. ©2017 University of Toronto Press. Reprinted with permission of the publisher.

"Calabrian Hounds and Roasted Ivory, or, Swerving from Anthropocentrism," *Multispecies Archaeology*, ed. Suzanne Pilaar Birch (New York: Routledge, 2018), 10–25.

"Stadial Environmental History in the Voyage Narratives of George and John Reinhold Forster," *Curious Encounters: Voyaging, Collecting, and Making Knowledge in the Long Eighteenth Century*, ed. Adriana Craciun and Mary Terrall (Toronto: University of Toronto Press, 2019), 206–28. ©2019 Regents of the University of California. Reprinted by permission.

Many friends and colleagues provided helpful feedback on draft material while the book was in progress. My heartfelt thanks to Mike Barrett, Ian Duncan, Stefani Engelstein, Joshua David Gonsalves, Lily Gurton-Wachter, Eva Horn, Jon Klancher, Ted Koditschek, Crystal B. Lake, Tim Langen, Tobias Menely, Amit Prasad, Brian E. Rodriguez, Peter Schnyder, Emily Patton Smith, Carsten Strathausen, Jesse Oak Taylor, and Anne D. Wallace for their insightful comments. At a critical stage, Nancy West and Craig Kluever generously gave me the use of their cabin on the North Fork of the White River, the perfect writer's retreat. Thanks also to Anne Savarese, Martin Schneider, and the team at Princeton University Press who saw this project through with great professional gusto.

Very special thanks to Amy M. King, my most loyal and exacting reader since 1997; and to Sean Franzel, who read every page of this book in manuscript, many of them more than once. I would like to thank two new friends, Howard Lidsky and Emily Patton Smith, for lively conversation that provided two sets of new ideas during the final stages of writing. For the gift of music and inspiration, I thank Mike Barrett, Sean Franzel, Trudy Lewis, Ray Ronci, and my friends the Hellbenders. More inspiration came from my mother, Christiane B. Marks, who finished her first book while I was writing this one. Everything else I owe to my loving family: my mom, the Marks kids, Celia, Jake, Zan, Edie, and *especially* Elizabeth and Eli.

ABBREVIATIONS

E William Blake, *The Complete Poetry and Prose of William Blake*, ed. David Erdman (New York: Anchor/Doubleday, 1988).

EN Comte de Buffon (Georges-Louis Leclerc), *Les Époques de la Nature*, ed. Jacques Roger (Paris: Éditions du Muséum, 1988 [1962]).

FA William Smellie (trans.), "Facts and Arguments in Support of the Count de Buffon's Epochs of Nature," *Natural History, General and Particular*, second ed., 9 vols. (London: Strahan and Cadell, 1785), IX.258–410.

DM Charles Darwin, *The Descent of Man and Selection in Relation to Sex*, ed. James Desmond and Adrian Moore (Harmondsworth, UK: Penguin, 2004).

I Johann Gottfried Herder, *Ideen zur Philosophie der Geschichte der Menschheit*, ed. Martin Bollacher (Frankfurt: Deutscher Klassiker Verlag, 1989).

O John Reinhold Forster, *Observations Made During a Voyage Round the World*, ed. Nicholas Thomas, Harriet Guest, and Michael Dettelbach (Honolulu: U of Hawaii P, 1996).

PT John Lubbock, *Pre-Historic Times, as Illustrated by Ancient Remains and the Manners and Customs of Modern Savages*, 2nd ed. (London: Williams and Norgate, 1869).

R Georg Forster, *Reise um die Welt*, ed. Gerhard Steiner (Frankfurt: Insel, 1967).

V George Forster, *A Voyage Round the World*, ed. Nicholas Thomas and Oliver Berghof, 2 vols. (Honolulu: U of Hawaii P, 2000).

Z Comte de Buffon (Georges-Louis Leclerc), *The Epochs of Nature*, ed. and trans. Jan Zalasiewicz, Anne-Sophie Milon, and Mateusz Zalasiewicz (Chicago: U of Chicago P, 2018).

Translations from *EN*, *I*, and *R* are my own unless otherwise noted.

DEEP TIME

INTRODUCTION

Deep Time: A Counterhistory

Deep Time before Geology

Until recently, deep time was defined against human history. It was nearly synonymous with geological time, which was widely understood to measure processes of planetary formation so ancient, and so slow, that margins of error in the tens of thousands of years were the least that could be expected. Stephen Jay Gould, John McPhee, and other natural historians popularized deep time in the 1980s by showing that the thin glaze of human history was comparatively trivial, incommensurable with the grand narratives preserved in strata that were pushed down deep into the earth's crust, only to resurface tens or hundreds of millions of years later. It was often argued that geology established itself as a science by diverging from the field of history.[1] But then twenty-first-century earth system scientists, among others, began to entertain the possibility that human activity might be causing system-level changes that would endure in the geological record as an Anthropocene epoch. Earth system–level changes were known as revolutions by the naturalists who first began to expand the Geologic Time Scale (GTS) in eighteenth-century Europe, and they are some of the main actors in this counterhistory of deep time before geology. The history of deep time involves a paradox that is sharpened by the proliferation of scholarship on our newest revolutionary epoch. As Jeremy Davies puts it in *The Birth of the Anthropocene,* we are now "living in deep time."[2] At this moment, deep time comes into view as a concept with a history.

Deep time is supposed to be the time of the other, the not-human, outside history. But as a way of imagining the cosmos, it has a history much older than geology and broader than the field of European ideas. Geology has established the age of the earth as 4.6 billion years. In Hindu cosmology, this is roughly equivalent to one *kalpa,* a day in the life of Brahma; since Brahma is about fifty years old, this makes our current world much older, well over 150 trillion years.[3] Events on this scale remain inaccessible to history, but a conceptual history of deep time can accomplish two things: it can help to explain how human actors, who were insignificant by definition in the context of the earth's deep past, could suddenly become "players in geologic time," as the climatologist David Archer noted in 2009; and it can recover deep time as a field of imagination, providing a counterhistory to the twentieth-century narrative of a "time revolution" happening exclusively within modern empirical science.[4] Unlike this time revolution, which banished human actors from deep time, and unlike the Anthropocene, which restores them to it, the concept of deep time has many possible histories. By choosing the mid-eighteenth century as a starting point, I aim to show how deep time became associated with earth history in the first place, expanding its conceptual domain to include colonial natural history, oral tradition, and scientific romance—all frontiers of the expanded time horizon associated with modernity.[5]

One of the many unsung literary actors in this counterhistory is the science fiction writer J. G. Ballard. The central characters in Ballard's novel *The Drowned World* (1962), most of them scientists, find that their dreams are disturbed by transformations of the geobiosphere caused by a rapid rise in global temperatures: all major cities become flooded and giant reptiles reemerge within two human generations. The characters dream of a giant sun that triggers deep evolutionary memories, causing them to embark on a quest in their waking hours to merge once again with their ancient reptilian ancestors. One character who seems to be immune to this delusion asks the others, knowingly, "How are things in deep time?" Writing twenty years before McPhee supposedly coined the phrase "deep time"—according to Gould's partial account—Ballard uses it numerous times in this novel, which deliberately invokes geological and evolutionary time to establish temporal scale.[6] *The Drowned World* has rightly attracted attention since a revival of interest in global warming sparked the rise of climate change fiction (cli-fi) in the mid-2000s.

It is more surprising to find an early-nineteenth-century geologist marginalized by some historians for his catastrophism, Georges Cuvier, reemerging as a prophet in another book about the Anthropocene, Elizabeth Kolbert's *The Sixth Extinction*. Endorsing Cuvier's paradigm of occasional "revolutions on the surface of the earth," Kolbert concludes that some of his "most wild-sounding claims have turned out to be surprisingly accurate."[7] *Epochs of Nature* by Georges-Louis Leclerc, Comte de Buffon, is an even earlier work on the theory of the earth that is now experiencing a startling revival. Early in the book, originally published in 1778, Buffon declares his motive for outlining the Epochs of Nature: "How many revolutions have taken place beyond the reach of human memory!"[8] Buffon, too, was uncannily accurate in some of his predictions. Less thorough than Cuvier but more daring, he identified the six definitive revolutions that altered the earth on an epochal scale, leading up to the "seventh and last" epoch belonging to human beings (Fig. 1a). One of the co-translators of *Epochs of Nature*, Jan Zalasiewicz—who is also a stratigrapher and the leader of the Anthropocene Working Group—has written that Buffon could be regarded as the founder of the geochronological time scale.[9] Today's GTS (Fig. 1b) looks quite different, and the epoch is one of its smaller units of periodization, but it too correlates geological features with a series of discrete events in the planet's history. The Anthropocene epoch, which has not been ratified by the International Union of Geological Sciences, would appear as a microscopic line at the top of the Holocene in the upper left-hand corner of the current version.

Geologists no longer describe the events that mark period boundaries as revolutions. The word *revolution* now has a predominantly sociopolitical meaning, and that is why Cuvier's and Buffon's revolutions suddenly seem to make sense again: planetary change is now human-driven. Resistance to the Anthropocene among scientists grows, in part, out of a sense that this line between the natural and the social should remain intact.[10] My counterhistory of "the time revolution" seeks to recover these revolutions of nature along with other key terms from the eighteenth century—expressions like "primitive rocks" and "the abyss of time"—that illuminate the qualitative dimensions of deep time as an imaginative experience.[11] The goal of this book is to juxtapose the emerging geological past with new experiences of ethnographic time: with the disruption of European chronology through contact with other cultures and with the deep past of poetry as it came into view through the study of oral traditions. The book concludes with a look ahead to Charles Darwin's *The Descent of Man* (1871), taking note of the materials that Darwin used to situate human origins more securely in the context of geological time once the long time scale had gained wider acceptance. These materials range from recent findings in the new disciplines of archaeology and anthropology back to Darwin's own youthful experience as a voyaging naturalist in the 1830s and even further back to the Enlightenment narratives that he read aboard the *Beagle*—and from which he and his contemporaries still imbibed many of the narrative strategies and aesthetic tropes through which deep time was conceptualized, including spatialized images of "the dim recesses of time" (*DM* 188) and the framework of conjectural history.

'Epoch firſt,—When the earth and planets firſt aſſumed their proper form.

'Epoch ſecond,—When the fluid matter conſolidated, and formed the interior rock of the globe, as well as thoſe great vitrifiable maſſes which appear on its ſurface.

'Epoch third, When the waters covered all the continents.

'Epoch fourth,—When the waters retired and volcanos began to act.

'Epoch fifth,—When the elephants, and other animals of the ſouth, inhabited the northern regions.

'Epoch ſixth,—When the continents were ſeparated from each other.

'Epoch ſeventh, and laſt,—When the power of man aſſiſted the operations of nature.'

FIGURE 1. (a, above) Buffon's Table of Epochs. From a review of *Natural History, General and Particular*, in *The Critical Review, or, Annals of Literature* 61 (May 1786): 368. Courtesy of HathiTrust. (b, facing page) J. D. Walker et al. (2018), Geologic Time Scale v. 5.0: Geological Society of America, https://doi.org/10.1130/2018.CTS005R3C.

According to the historian Dipesh Chakrabarty, the Anthropocene marks the "collapse of the age-old distinction between natural history and human history."[12] To grapple effectively with the collapse of two scales of time, we must try to understand how the separation came about in the first place and how the geological understanding of deep time might be embedded in human experience both before and after this modern division. Ironically, one dominant interpretation of the break between human and prehuman time locates it at the point where geology first met human prehistory: in 1859, Joseph Prestwich published his report on the first human artifacts found together with extinct animal bones—proving that our species was old enough to have entered the fossil record—and in the same year Darwin published *The Origin of Species*, which set the evolution of all species within the scale of geological time.[13] In fact, explorers in the Pacific had located the intersection between geology and prehistory almost a century earlier, before these sciences were formalized. Even earlier, the Copernican revolution greatly expanded the scope of cosmic time, and scholars reflecting on the history of astronomy began to suspect that this science, and therefore humanity itself, must be much older than the written record.[14] By the eighteenth century, colonial researchers in Asia began to communicate what they were learning about indigenous traditions—such as the cycle of Great Time in the *Mahabharata*, which seemed and still seems improbably long to Western readers.[15] Last but not least, naturalists working in the field established by the 1750s that the primitive

CENOZOIC

Period		Epoch	Age	Picks (Ma)
QUATERNARY		HOLOCENE		0.012
		PLEISTOCENE*	CALABRIAN	1.8
			GELASIAN	2.58
TERTIARY	NEOGENE	PLIOCENE	PIACENZIAN	3.600
			ZANCLEAN	5.333
		MIOCENE	MESSINIAN	7.246
			TORTONIAN	11.63
			SERRAVALLIAN	13.82
			LANGHIAN	15.97
			BURDIGALIAN	20.44
			AQUITANIAN	23.03
	PALEOGENE	OLIGOCENE	CHATTIAN	27.82
			RUPELIAN	33.9
		EOCENE	PRIABONIAN	37.8
			BARTONIAN	41.2
			LUTETIAN	47.8
			YPRESIAN	56.0
		PALEOCENE	THANETIAN	59.2
			SELANDIAN	61.6
			DANIAN	66.0

MESOZOIC

Period	Epoch	Age	Picks (Ma)
CRETACEOUS			66.0
	LATE	MAASTRICHTIAN	72.1
		CAMPANIAN	83.6
		SANTONIAN	86.3
		CONIACIAN	89.8
		TURONIAN	93.9
		CENOMANIAN	100.5
	EARLY	ALBIAN	~113
		APTIAN	~125
		BARREMIAN	~129.4
		HAUTERIVIAN	~132.9
		VALANGINIAN	~139.8
		BERRIASIAN	~145.0
JURASSIC	LATE	TITHONIAN	~152.1
		KIMMERIDGIAN	~157.3
		OXFORDIAN	~163.5
	MIDDLE	CALLOVIAN	~166.1
		BATHONIAN	~168.3
		BAJOCIAN	~170.3
		AALENIAN	~174.1
	EARLY	TOARCIAN	~182.7
		PLIENSBACHIAN	~190.8
		SINEMURIAN	~199.3
		HETTANGIAN	~201.3
TRIASSIC	LATE	RHAETIAN	~208.5
		NORIAN	~227
		CARNIAN	~237
	MIDDLE	LADINIAN	~242
		ANISIAN	247.2
	EARLY	OLENEKIAN	251.2
		INDUAN	251.90

RAPID POLARITY CHANGES

FIGURE 1. (b, *continued*).

or unstratified rocks underlying the layers of sedimentary rock had to be more ancient than anyone had ever imagined—simply because they began to measure the time needed for all those fine-grained sedimentary strata to be laid down, all of which must have happened after the formation of the primitive masses, as they were then called.

Cuvier attempted to define modern geology by putting all but the last of these factors to one side. The positive evidence concerning the revolutions of nature, he said, would be found in the fossil-bearing sedimentary rocks. In attacking geological speculation concerning the primitive rocks, Cuvier explicitly cited the ethnographic content of the time of the other, recognizing the qualitative, composite nature of deep time in its early European form.[16] Deep time was increasingly reformulated as geological time in the course of the nineteenth century, and this association was strongly reinforced by Gould's misleading priority claim on behalf of McPhee, who merely used the expression in a sense that was consistent with Gould's own view of the history of geology. In the question posed by an enigmatic colonial adventurer named Strangman to Kerans, the atavistic hero of Ballard's *The Drowned World*—"How are things in deep time?"—deep time operates in biological and "archaeopsychic" as well as geological registers. Ballard's treatment is no more original than McPhee's, but the unquestioning restatement of Gould's claim by more recent scholars (myself included) is symptomatic of a larger displacement of scientific romance by science writing (a twentieth-century genre) as the authorized source of explanations concerning the meaning of deep time.[17] More recently, deep time has been applied metaphorically by scholars to media history and to literary history, which operate on a human scale. Although this usage is accurately characterized as an analogy derived from geological time, my claim is that deep time as a figural register predates geology, and therefore the use of "deep time" by Siegfried Zielinski, Wai-Chee Dimock, and others seeking to recalibrate the temporality of media or other histories is no less original than the dominant geological usage.[18]

The point of the joke in Ballard's novel is that humans aren't supposed to know "how things are in deep time." By acknowledging in this way that his protagonist suffers from a kind of delusion, Ballard implicitly recognizes an earlier set of priority claims around the discovery of deep time, those of James Hutton and the uniformitarian tradition. This is the tradition recognized by Gould in the vast prehuman time scale so vividly evoked by McPhee's narratives of geological processes. Hutton is central to Gould's history as the thinker who incorporated "time's arrow" into the cyclical model of earth history established by Thomas Burnet in the seventeenth century. Many other histories of geology and geological time, especially those written in English, begin with Hutton's *Theory of the Earth* (1788/1795), and Hutton is widely known as the "father of geology." Geology textbooks continue to feature brief treatments of Hutton as an ancestor figure and allude to his famous declaration that he saw "no vestige of a beginning, no prospect of an end" in the geological record, placing it beyond human reckoning.[19] My account begins instead with two projects from the 1770s that have remained marginal to the history of geology: first, the comparative data gathered by John Reinhold Forster and his son George on their voyage around the world (1772–75), and second, the revisionist geochronology put forward by Buffon in *Epochs of Nature*. Hutton's bold paradigm provides a useful touchstone, especially in cases when the history of geological time is used to contextualize another body of thought, whether in science or in cultural history. However, the emphasis on Hutton also has the potential to distort or oversimplify the state of earth science around 1800, as Martin Rudwick has suggested by calling Hutton's paradigm "eternalist" rather than "geohistorical." Hutton's commitment to geological continuity led him to argue against the "supposition of the primitive" that is a central concern of this book.[20]

Charles Lyell, too, plays a major role in many histories of geology as the thinker who incorporated Hutton's insights on the slow and continual agency of geological forces into a systematic uniformitarian theory. In Gould's account, Hutton's *Theory of the Earth* marks a midpoint between Burnet's *Sacred Theory of the Earth* (1684–91) and Lyell's *Principles of Geology* (1830–33), which achieves a synthesis between the cyclical and linear models of its two predecessors. Lyell is of interest here as the first geologist to devote a book-length study to the chronological place of the human species, *Geological Evidences of the Antiquity of Man* (1869). Lyell's work will enter the picture in the later chapters of this book, but here again I have chosen to emphasize other sources that offer a more holistic approach to matters of uncertain antiquity, geological and otherwise. In a chapter on the early history of the ballad revival, I consider a largely independent body of thought on human antiquity that is brought into dialogue with the natural history of the earth by the philosopher Johann Gottfried Herder. Rather than taking Lyell as a guide on the integration of geology and anthropology, my final chapter addresses two evolutionary thinkers, Darwin and John Lubbock. Both younger contemporaries of Lyell and deeply influenced by him, Darwin and Lubbock situate human behavior (including the primal song that fascinated the ballad revivalists) in evolutionary and geological time.

This book is not a history of geological time, though I am much indebted to Gould's, Rudwick's, and other excellent scholarship on that subject.[21] Deep time, in my view, is not identical with geological time as it is currently understood; the metaphor "deep time" has a wider purchase and a longer history than the modern-day GTS. Without rejecting existing accounts of geological time,

I seek to recover that longer human history and to focus on the imaginative act of distinguishing *any* kind of long-scale time outside recorded history (including sacred history). Geological time is incommensurable with historical time, yet the distinction appears less absolute in light of the Anthropocene proposal, which recalls the uncertainty historically associated with deep time. The topos of uncertain antiquity is common to many earlier inquiries about origins, including human origins. It plays a critical role in the voyage narratives of the Forsters, which look to natural history for the keys to human history, and it informs the much later evolutionary narratives of Darwin and Lubbock, who consider the history of species as integral to the history of the earth. The experimental chronology of Buffon and the conjectural histories associated with the ballad revival, including those of Herder and William Blake, also share the naturalists' interest in linking anthropological and geological data to account for human origins. Hutton's more canonical "abyss of time," by contrast, rules out the question of origins and severs this connection. I am casting a wider net here, drawing in a broad constellation of genres and traditions to recover the generative area of uncertainty between geological and human time that persisted even beyond the moment when Lubbock claimed it for a new discipline, "prehistory." Pratik Chakrabarti has offered another kind of revision by re-centering the history of deep time geographically and historically on nineteenth-century India, where the nexus of geology and prehistory produced a "naturalization of antiquity."[22]

The attachment to deep time as a scientific truth only intensifies with the Anthropocene awareness of "living in deep time" as reality rather than science fiction. But Ballard's remarkable account of a "descent into deep time" stands as part of a longer literary tradition informing the current literature on deep time and on more recent concepts relating to the fusion of human and natural history, including the Anthropocene, deep history, and the Long Now. Thomas Carlyle's reference to the "deep time" of cultural history, the earliest use recorded by the *OED*, operates within the scope of recorded and future history, so it makes sense to leave it out of the story of geological time, as Gould does. The "abyss of time," made famous by Hutton's popularizer John Playfair, has a longer history, which incorporates both anthropocentric usage and the macro-scale, prehuman sense of deep time foregrounded in McPhee's *Annals of the Former World*. Like "deep time" in the twentieth century, the "abyss of time" and related figures appeared in both literature and science; in the eighteenth century, these figures crossed over more readily between the disciplines and began to put pressure on conventional ideas of large or small "spaces of time."[23] Carlyle's reference does share with these early approximations of geological time the strongly relative conception of time as comprising units of vastly differing magnitudes, the scalar flexibility also captured in poetry by Blake's infinitely extendable "moment of time."[24] Ballard articulates these magnitudes in post-Darwinian terms when he positions his narrator as "marooned in a time sea, hemmed in by the shifting planes of dissonant realities millions of years apart." I will argue later that Ballard's account of a "descent into deep time" follows *The Descent of Man* closely.[25]

The first stage of our descent here will be the critical distinction between primitive and secondary rocks established in the mid-eighteenth century. The following chapters trace the ramifications of this sequential view of natural history in narratives of Cook's second voyage around the world (Ch. 1) and in Buffon's *Epochs of Nature* (Ch. 2). My third chapter emphasizes the ethnographic side of the primitive past as brought into focus by the study of oral tradition, initiated by the ballad revival and further refined in poetic and philosophical form (respectively) by Blake and Herder. The fourth chapter rereads Darwin's *Descent of Man* in light of these precursors. The book concludes with an envoi proposing evolutionary nostalgia as a characteristic form of deep-time narrative for the twenty-first century.

In this eclectic body of work, reflection on geological time is commonly prompted by the question of human origins. There was no fixed discipline or genre dedicated to the problem of locating human origins within the natural history of the earth. Fostered by secular tendencies in Enlightenment thought and by a disciplinary ecology that had no fixed rubric for archaeology or prehistory, many writers from the mid-eighteenth century onward arrived at a concept of deep time by asking, in their own terms, what developments in the history of the earth must have been required for the human species to become viable. The imaginative urgency of creating a naturalistic context for human origins gave currency to the idea of a long prehuman history. Beginning with the discussion of primitive rocks that follows, my argument draws upon the frequently overlapping ideas of naturalists, explorers, philosophers, and poets in order to present the full range of locations where geological and anthropological questions are brought into proximity. The concepts and narrative forms associated with deep time, including revolution, reversion, catastrophe, species memory, and the primitive, result from cross-pollination rather than disciplinary specialization. As against the linear chronology of specialization, a literary history of deep time allows for a synchronic emphasis on the persistence of personification, analogy, anagnorisis, and other formal techniques in deep time narrative across a century (1770–1870), and likewise on the persistence of genres, including voyage narrative, ballads, myth, and scientific (or philosophical) romance. These formal continuities reveal the full scope of the uncertainty associated with "living in deep time," a long-term result of the convergence between human and nonhuman elements in the imaginative space of deep time.

Primitive

The German *Neuzeit* is generally translated as "modernity," as English lacks a native term to designate this reflexive awareness of belonging to recent history. But the idea of "new time"—the diachronic sense of which is absent from "modern," a Latinate adjective meaning "up to the minute" or "of the present"—enters English usage through the neo-Greek terminology used by geologists, who designate our current era as Cenozoic, the era of new life. Whereas German historians, most prominently Reinhart Koselleck, have established *Neuzeit* as the central frame of reference for postmedieval European history, especially since the French Revolution, the newness of modern times is harder to pin down in English.[26] The lack of such a vocabulary for human history may help to explain the enormous popularity among English speakers of the Anthropocene, a new geological term that deliberately crosses over into history. Given the popularity of this new geological epoch, the "new time of man," it is indeed surprising that no renegade earth system humanist has yet claimed for the Anthropocene the dignity of an era, superseding the Cenozoic, or even of a period (these are both higher-order categories than "epoch" on the current GTS, Fig. 1b). Why should Man content himself with a mere Epoch?

From the Anthropocene perspective, the deep future appears depopulated, a scene of very long-term human impacts without human witnesses. For Hutton, Buffon, and other writers who helped to shape the concept of deep time, the future that it opened to view was more compatible with Enlightenment notions of progress. In his philosophical history of time, Hans Blumenberg characterizes the opening of a "world time" much vaster than "life time" as a strategy of buying time for civilizational progress. In this reading, establishing a comparatively young age even for "advanced" societies allows them to project their progress into an expanded future.[27] The historian Frederik Albritton Jonsson, however, has identified historical anxieties about resource exhaustion that preceded and helped to spur industrial progress, suggesting that the "shallow" future is likewise a product of modernity.[28] Thomas Malthus's projection of inevitable food scarcity is only the most obvious example. Darwin acknowledges the influence of Malthus explicitly in *The Origin of Species,* and Malthus seems an important source for the later Darwin's gloomy certainty concerning the naturalized "extinction" of indigenous peoples in Tasmania and elsewhere, a view of cultural evolution consistent with the thought of Lubbock, Alfred Russel Wallace, and other Victorian contemporaries. On the whole, however, the human future associated with deep time seems to expand rather than contract in the course of the nineteenth century. There is no real precedent for the attenuation of the future evoked by our present preoccupation with anthropogenic impacts, though traditional deep-time narratives provide the allegorical figure of a far-distant future reader (or witness) that is sometimes used to describe these impacts.[29] According to the principle of superposition, the abyss of time opens downward, into the past. Conceptions of the deep past will be my main focus, but these carried with them specific implications for the future that are of special interest in the Anthropocene context.

Unlike *Neuzeit, Tiefenzeit* is a recent import from English—in fact, *Tiefenzeit* (deep time) entered the German language via the 1990 translation of Gould's *Time's Arrow, Time's Cycle.*[30] The contrast between these two German words illuminates the dialectical relationship between deep time and the modern intellectual history that produced this concept of time as a natural precondition of life exceeding the scale of history by several orders of magnitude—a "metahistorical" given, in Koselleck's vocabulary. This book offers a conceptual history of deep time that is indebted to Koselleck's method, though it will necessarily trace a different path from the histories of sociopolitical concepts foregrounded in his account of the structural transformation of modern societies. Temporalization (*Verzeitlichung*) is the common factor uniting the concepts that define modernity in Koselleck's reading. In pointing out the expanded "temporal horizon" associated with vastly increased estimates of the age of the earth proposed by Enlightenment thinkers including Buffon and Immanuel Kant, Koselleck's approach to modern temporality somewhat resembles Michel Foucault's influential account of the historicity of nature in *The Order of Things*. Koselleck's self-conscious use of geological metaphors such as "sediments of time" (*Zeitschichten*), however, suggests that the conceptual history of deep time is not simply a product of the Foucauldian historical turn.[31] This modern concept should be understood, rather, as the product of a *pre*historic turn, the discovery of a synchronic space outside or prior to the historicity of nature and of nations. This negative or paradoxical relation between deep time and its dialectical other, modernity, bears witness to the continuities underlying geological eras and historical epochs.

Davies's vividly paradoxical image of "living in deep time" echoes Chakrabarty's influential thesis on the collapse of human into natural history. Historicizing deep time, however, also means recognizing the time of the other that is inscribed in it. The history of deep time as a concept is routinely neglected in these otherwise trenchant accounts of recent anthropogenic effects that will remain visible in the fossil record on the same scale as some landmarks from the first few billion years of the planet's history. Two moments of this conceptual history are essential for the eventual consolidation of "deep time" as a geological metaphor in the twentieth century: the older classification of "primitive rocks" and the nonlinear temporalities assigned by European savants to non-European or unlettered peoples. The idea of primitive rocks (also current in French, Italian, and German as *terrains primitifs*, *monti primari*, and *uranfängliche Gebirge*) was fundamental to all Western accounts of the earth's deep history between roughly 1750 and 1850. Spatially and conceptually, the idea of primitive cultures emerged in close proximity to this geological idea. This proximity is still apparent in *The Voyage of the* Beagle, where Darwin locates "savages of the lowest grade" among "immense fragments of primitive rocks."[32] "Primitive" takes on its characteristically modern pejorative connotations as it disappears from the geological vocabulary and becomes the province of newer social science disciplines. My contention is that what Johannes Fabian called "natural time" did not preexist and then shape the "time of the other" as it was constituted by anthropology; instead, the separation of natural and cultural time postdates and depends on the ambivalent Enlightenment articulation of "primitive rocks" to establish the early history of the earth.[33] In this sense, deep time was a composite of human and natural history to begin with.

Examples from either end of this prehistory of deep time will indicate briefly the broad purchase of "primitive" in this multivalent sense. Johann Gottlob Lehmann's 1756 *Essay toward a History of Secondary Rocks* (*Flötzgebürge*) was widely cited throughout the nineteenth century as the original source for the subdivision of rocks and mountains into primitive (or primary) and secondary groups. Lehmann's work became widely known via the French translation of Baron d'Holbach. Cuvier and Humphry Davy are among the nineteenth-century followers who relied on this translation, which renders Lehmann's term *uranfänglich* as "primitive." Davy maintains the importance of this fundamental distinction, arguing that geology in the nineteenth century still requires a boundary to mark the difference between formations with observable causes and "that matter of our globe as yet unchanged by any known natural operations."[34] Although nineteenth-century geologists typically rejected Lehmann's premise that primitive rocks remained unchanged since the creation, it was still the case that these rocks occurred far below the hundreds of strata formed by processes that were increasingly known and recognized, while the formation of unstratified masses remained largely unknown. As George Forster observed of the "so-called human races," a primitive rock is simply "a kind without a known origin" (*ein Stamm, dessen Herkunft unbekannt ist*).[35] Darwin's account, informed by his shipboard reading of George and Reinhold Forster and of Alexander von Humboldt—another traveler trained in geognosy—makes it especially clear that the apparently primitive opens a window into deep time. He returns in his late work to his encounter with "savages of the lowest grade" among the "immense fragments of primitive rocks" at Tierra del Fuego because this encounter in situ leads him to regard those peoples as "ancestors" of modern humans. By the time Darwin articulated this view in *The Descent of Man* (1871), "primitive cultures" belonged to the remit of anthropology, but Darwinian natural history still shows clearly the composition of deep time out of earth history and human history, reframed as prehistory.[36]

Darwin's evolutionary theory depends explicitly on the depth of geological time. His juxtaposition of human and geological history in the field marks an important step along the way to the mobilization of the fossil record for this theory in *The Origin of Species*. Evolutionary time and geological time are in this sense both aspects of the same injury to anthropocentrism that is often described in Freudian terms as the third *Kränkung* (following the first two "outrages" or decenterings of humanity by Copernicus and Kant). It is hard to date the onset of this "temporal marginalization" of humanity, as Peter Schnyder observes, but a consciousness of something like deep time would seem to be a prerequisite.[37] Lehmann's untimely and surprising role as a creator of deep time accords with Koselleck's paradigm of a *Sattelzeit*, a transitional period after 1750 when the premodern conceptual "legacy is transformed into our present."[38] Rudwick's history of geology situates Lehmann's work in the geognostic tradition of Abraham Gottlob Werner and other mid-eighteenth-century thinkers whom Rudwick regards as anticipating modern stratigraphy. He presents these geognosts as historical actors who practice a qualitative insight that precedes and prefigures the recognition of deep time as such.[39] After 1750 it became increasingly apparent that even the empirically verifiable deposition of sedimentary rocks exceeded the six-thousand-year span of sacred history. At the same time, naturalists who were interested in human origins, such as Buffon, recognized that the common chronology was inadequate even for the history of the species. Rudwick makes the point that these speculations, due in part to their qualitative aspect, were not initially received as threatening. The qualitative sense of an unsuspected time dimension thereby achieved the broad currency presupposed by ideas of a "time revolution" associated with Darwin or any other nineteenth-century thinker. The tradition of scientific voyaging that Darwin inherited contributed strongly to this popularization, as I shall argue in the first and last chapters of this book. Humboldt, one of the more celebrated voyagers, presented the recognition of deep time as a kind of *mise-en-abîme*: "May the eagerly curious human spirit be permitted, from time to time, to swerve from the present into the darkness of prehistory [*Vorzeit*], to intuit [*ahnen*] what may not yet be clearly recognized, and so to delight itself with the ancient myths of geognosy, which return in many forms."[40] As often in these encounters, deep time confronts the geologist in the form of human prehistory, a reflection of science in myth.

Lehmann's "fundamental distinction" (as Cuvier termed it) between primitive and secondary mountains should not be equated with deep time, but his prose and that of some of his contemporaries introduce key narrative structures that enable the formation of this modern concept. Lehmann's terminology is among the salient survivals from this early period in nineteenth-century geology, when the substantive part of his work had long been superseded. Lehmann devotes considerable energy to establishing the "endless depth" of the primitive masses, a dimension amplified in d'Holbach's translation.[41] The key feature of these masses (now known as igneous rocks) is that they are not stratified, and hence, in a certain sense, not temporal at all. Koselleck recognizes the pun on *Geschichte* (history) and *geschichtet* (stratified) in his explanation of *Zeitschichten* (literally "layers of time"). This is not merely a pun, he insists, because it denotes a real affinity between geological unconformity and historical discontinuity, the interaction of "strata [*Zeitebenen*] of varying duration and diverse origin, which are nonetheless simultaneously present." The materialist d'Holbach, in a lengthy editorial preface to Lehmann, points out that the depth of the secondary masses provides evidence of a colossal lapse of time postdating the formation of the primitive masses themselves.[42] As a translator, d'Holbach used language to exploit the abyssal, unlayered, almost atemporal qualities of Lehmann's paradigm.

These early experiments with figures of temporality have implications for the shape of time not only in modern geology but also in history, as Koselleck's heavily geological vocabulary shows. Koselleck develops this vocabulary in his late work on "sediments of time," which marks a departure from his earlier account of temporalization, or historicity as it was more widely understood in that poststructuralist moment. His recognition of Enlightenment natural history as laying the groundwork for the expanded "temporal horizon" of modernity inspired his own adoption of a geological vocabulary and attests to the deep synchronic dimension as well as to the dynamism of the natural history tradition, both of which the present study sets out to recover. This dynamism is opposed to Foucault's conception of a "static" natural history and to the idea of a historical turn that follows from it. Foucault's archaeology of knowledge illuminates the process of disciplinary change more broadly, but Koselleck's geological metaphors help to identify a geological dimension of history that precedes and destabilizes the historicity of nature. His concept of a *Sattelzeit*, the period of transition to modernity, derives from a geological model for the formation of mountain passes or *Bergsattel*, literally "mountain saddles."[43]

By virtue of its origins in this period after 1750, the geognostic distinction between primitive and secondary masses occupies a clear position in conceptual history, namely that of an antecedent that allows us to recognize deep time as an object of conceptual history. In this "determination of its meaning for us," deep time occupies a special place in relation to modernity, the onset of which is "witnessed," according to Koselleck, by conceptual history as a whole. This initial glimpse of deep time, then, promotes the reflection on structural transformation that defines modernity in his account. Deep time, however, continues to structure historical analysis in its role as the dialectical other of modernity, the "new time" that Koselleck explains by means of geological metaphors such as "temporal depth" and "deep diachronic structure."[44] He even cites the distinction between primitive and secondary rocks as a critical stage in the historiography of modernity itself, thereby casting doubt on his own conclusion that historicity is merely "transferred back" from nature into history: my contention here is that the figures informing the history of deep time as a concept ultimately exceed and determine its literal meaning. Temporalization in and of itself is neutral, indifferent to the disciplines of geology and history. It is therefore reductive to speak of deep time, or indeed of any temporal dynamism outside history, in strictly Foucauldian terms as a historicization of nature.[45]

Gould's history of geological time rightly draws attention to geological theories and their narrative techniques, too often marginalized in favor of empirically driven fieldwork. In mapping the wider literary field of deep time,

FIGURE 2. Engraving by J. E. Geriare from a drawing by Johann Gottlob Lehmann. Lehmann, *Versuch einer Geschichte von Flötz-Gebürgen* (1756), Plate 7. Universitätsbibliothek Heidelberg / O 2530 RES / 162.

I seek to attend equally to the geognostic fieldwork of Lehmann and his contemporaries and to theories of the earth, more particularly to the voyage narratives that synthesized both these dimensions of early geological thought, and ultimately to the narrative forms transmitted through all these approaches to writing about the earth and its temporality. One of Rudwick's signal interventions in this history is to recognize the conceptual weight of the early geognostic fieldwork, which produces "a widespread qualitative sense of the likely immensity of time" that precedes the quantification of deep time.[46] Citing precise cross-sections of stratigraphic sequences created by Lehmann, Giovanni Arduino, and Georg Christian Füchsel, Rudwick argues that these images made vivid the sheer diversity and depth of the strata as a wholly secular index of the lapse of time. One engraving in Lehmann's *Essay* (Fig. 2), based on his own carefully labeled drawing made in the field in Thuringia, identifies thirty-one strata in a single section. Siegfried Rein goes so far as to credit Füchsel with the "discovery of deep time."[47] The proliferation of these priority claims indicates the fundamental place of deep time in histories of geology, on the one hand, and the broadly diffused character of the ideas and materials associated with its discovery, on the other.

The Boundaries of Geology

These mid-eighteenth-century accounts of structural relationships within the earth's material are indispensable for a construction of deep time that becomes associated with geology. My account of Lehmann and the continental tradition in early geology also sets the stage for the first chapter of this book, which concentrates on voyage narratives by the Forsters (1778–82) in which these principles of geological succession are applied and transformed in the field, in conjunction with European ideas of the progress of society. The encounter with landscapes, and even more so with cultures that did not fit these paradigms, further disrupted the framework of chronology. My second chapter positions Buffon's metropolitan theory of the earth, synthesized from the mass of material available to him as the director of a national collection, against the Forsters' empirically driven narratives. In *Epochs of Nature* (1778), Buffon became the first to give empirically based absolute dates for geological periods, but he continued as well to emphasize the qualitative experience of long sequences of strata that emerged from fieldwork. In one evocative passage he observes a thousand-foot hill of shale in Normandy that is topped by a stratum of limestone. The glimpse of limestone, postdating the many strata of shale, prompts a minute analysis of geological processes so slow that human generations are inadequate to measure them: "our all-too-short existence," he declares, provides no adequate scale for "the number of centuries [*siècles*] that were needed to produce all of the animals and shells with which the Earth is replete; and then, the yet greater number of centuries which have passed for the transport and deposition of these shells and their detritus," all of which follow the equally vast periods required for the shale to accumulate at the rate of five inches a year, and then erode again, before the shells were deposited in the first place (Z 36).[48]

Though Cuvier and most later geologists consigned Buffon to the prehistory of geology, his narrative approach to temporality and scale is more timely for the early twenty-first century and brings a capacious genealogy of deep time into view. Reflecting on the formation of this Norman topography in a manner that anticipates Hutton and Darwin, Buffon enters on a philosophical digression, much of it deleted in revision, that problematizes temporal apperception on this scale: "why does the human spirit seem to lose itself in an extent of time, rather than one of space or in a consideration of measures, weights, numbers?" (Z 35–36). Darwin, confronting the inadequacy of numbers to express duration, directs his reader to "watch the sea at work, grinding down old rocks," thereby compressing eons of geological process into moments of visual perception.[49] This rhetoric of compression is a technique shared with poetry, as I argue in my third chapter, which compares these moments of geological retrospect to scenarios created around the deep past of poetry, from the conjectural history of the ballad revival to the eternal present constructed by Blake as the time of his epics. The fourth chapter on Darwin integrates the anthropological approach mooted by early students of ballads and folklore into an account of the larger synthesis of prehistory, geological time, and human evolution that Darwin achieved late in his career. The importance for this late work of his early encounters with the Yaghan and other native people of Tierra del Fuego also bridges the seeming distance between Enlightenment voyage narratives, such as that of the Forsters, and modern evolutionary thinking. Rather than a "time revolution," this gradual history brings out the importance of inherited tropes, and of cross-cultural encounters, for the sense of deep time.

Buffon's innovation in 1778 was to augment Lehmann's qualitative sense of the lapse of time with absolute dates for the epochs of nature, which he verified experimentally. These dates nevertheless proved elusive and arbitrary, as indicated by the heavily revised manuscript of *Epochs of Nature* (Fig. 3), in which values in the millions of years appear periodically only to be crossed out—sometimes more than once—and ultimately replaced with the chronology of 75,000 years that Buffon settled on in print. His experiments with white-hot balls of iron, intended to represent the young earth, relied upon the familiar concept of primitive mountains—in this version a solidified core of granite thrown off in molten form when a comet grazed the sun. In Buffon's chronology, the secondary formations are deposited in the third epoch of nature, and officially the human species appears only in the seventh and last Epoch. Just here, however, his chronology is disrupted by the anthropological moment of deep time. This account of human origins begins by attributing to "the first men" an inherited or evolutionary memory of the "great convulsions" of the fourth epoch, the global upheaval marked by widespread earthquakes and eruptions that predates this ostensible "time of man" by tens of thousands of years, according to Buffon's own reckoning. Reinhold Forster offers a similar kind of explanation for an indigenous creation myth that he encountered in the Pacific islands, which was that they were formed when the god Maui dragged a great net through the ocean. Forster explains this legend as a deep inherited memory of prehistoric "revolutions of nature," comparable to that attributed by Buffon to ancestral humans (*O* 112).

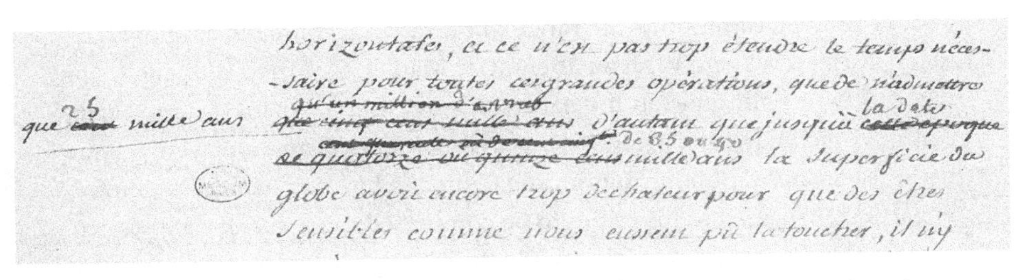
horizontales, et ce n'est pas trop étendre le temps néces-
-saire pour toutes ces grandes opérations, que de n'admettre
que 25 mille ans
d'autant que jusqu'à la date
de 85 ou 90
mille ans la superficie du
globe avait encore trop de chaleur pour que des êtres
sensibles comme nous eussent pu la toucher, il n'y

FIGURE 3. Georges-Louis Leclerc, Comte de Buffon, "4eme Epoque," fol. 24 (detail). MS 883, *Cahier* 3. © Muséum national d'histoire naturelle, Paris.

From a historical point of view, both these moments together—the geological engagement with primitive mountains and the ethnographic engagement with the time of the other—contribute in an essential way to the formation of deep time as a concept. Eighteenth- and nineteenth-century naturalists themselves reflected on this convergence, as we have seen in Humboldt's account of the myths of geognosy, and Schnyder has noted the role of wonder tales in popular geology as a way of reckoning with "the alterity of past epochs."[50] More often than not, however, naturalists met the mythical qualities of deep time narration with skepticism, as when d'Holbach, in his critical examination of Lehmann's source texts, rejects all accounts of a Biblical Deluge. Similarly, Jean-André Deluc defended the realm of fact against fable in his harshly critical review of Hutton's *Theory of the Earth*, and Reinhold Forster, among other readers, criticized Buffon's *Epochs of Nature* for indulging in romance.[51] The historiography of deep time in fact begins here, with these reflections within the nascent discipline of geology itself, which preserved the old consensus concerning primitive and secondary rocks not for its poetic merit but for its empirical efficacy (as suggested by Davy's and Cuvier's qualified praise of Lehmann in the early nineteenth century).

I have positioned Lehmann as a co-creator of the deep time concept because he played a key role in establishing this consensus and the chronology that follows from it. Similar claims might be adduced on behalf of much earlier natural histories of the earth. The paleontologist Rein, as I noted, effectively backdates the "discovery of deep time" by attributing it to Lehmann's contemporary Füchsel. Many geologists look back even further to the Danish polymath Nicolaus Steno (1638–86), not for discovering deep time per se but for establishing the principle of superposition that assigns a progressively younger age to each overlying stratum.[52] There is a more pressing reason, however, to look to seventeenth-century geological thought in a conceptual history of deep time: the wide range of narrative techniques employed by Steno, Burnet, William Whiston, Benoît de Maillet, Gottfried Wilhelm Leibniz, and their colleagues in the natural history and/or theory of the earth. Narrative technique and prose style, as well as empirical method, were at issue in the criticisms of Buffon, Hutton, and Lehmann by other naturalists; Buffon in particular seemed to his critics to be committed to an out-of-date system. Myth, fable, romance, and eventually even Scripture were deemed to fall outside the boundaries of "geology" beginning in the late eighteenth century (when the word was coined). Notably, Forster embraces indigenous myth as a geological source but condemns Buffon for writing a "beautiful romance" in the guise of a treatise on the earth.

One of Buffon's greatest failings, according to many of his critics, was to adopt a hypothesis very close to those put forward by seventeenth-century thinkers including Whiston and Leibniz: that the passage of a comet through the solar system could be a major agent of geological change (see Fig. 1.1 below). Buffon's interest in geogony (earth-formation) recalled an unfettered style of bold theorizing and grand narrative that had fallen into disrepute by the late eighteenth century, ridiculed by Playfair as "a species of mental derangement, in which the patient raved continually of comets, deluges, volcanos and earthquakes; or talked of reclaiming the great wastes of the chaos, and converting them into a terraqueous and habitable globe."[53] Writing in 1987, Gould reclaimed one of the "villains" of this progressive history of geology, Burnet, by arguing at length that he conceptualized geological time in the same cyclical manner as Steno, the ancestor generally considered to be a "hero" for his empirical rigor. Gould's reclamation of Burnet is the first step in his demonstration that "metaphors of time's arrow and time's cycle . . . proved as fundamental to the formulation of deep time as any observation about the natural world."[54] As Gould himself acknowledges, the conceptual field of deep time was shaped by a large number of actors in many branches of early modern inquiry; he limits his account heuristically to a small number of canonical actors and two specific metaphors. My account takes in a wider cross-section of all three—actors, disciplines, and narrative forms—but favors the period between the mid-eighteenth and mid-nineteenth centuries.

The boundary between the history of geology and its ancestral romances is not a stable one. Rudwick, for example, realigns Hutton with pre-paradigm geology by tracing his "eternalism" to de Maillet's *Telliamed* (published posthumously in 1748) and ultimately back to Aristotle. De Maillet posited a geological past of millions of years based on a vision of geological processes extending "to all eternity" and therefore lacking any meaningful historicity, Rudwick observes.[55] Rudwick's intervention has not stopped textbook authors from positioning Hutton as the "father of geology," though geology students today are getting a more nuanced account than I did as a first-year undergraduate in 1987.[56] Buffon, whose theory seemed romantic and amodern to his contemporaries, now seems more relevant to geochronology and may in time enter the textbooks again. It will be sufficient here to keep in mind the early modern antecedents ruled out of bounds by later geologists, noting that many of these antecedents have a renewed interest in the twenty-first century, as Lydia Barnett has argued.[57] In the eighteenth- and nineteenth-century context, it is striking that early modern European thought is often relegated to the same doubtful sphere as non-Western antiquity by those seeking to establish the history of the earth as the preserve of a separate discipline.

Cuvier's contributions to the history of geology reinforce the old distinction between primitive and secondary masses while also casting doubt on early modern and on some more recent approaches. His account is essential for my argument because it cites the ethnographic content of the time of the other explicitly in attacking geological speculation concerning the primitive. Cuvier's research agenda was to draw the attention of geologists away from the *terrains primitifs* and toward the *terrains secondaires,* represented as a vastly underexploited archive of clearly demarcated geohistorical events. His *Rapport historique* on the progress of the geological sciences (1810) begins by noting this "fundamental distinction in geology," the recognition (co-credited here to Lehmann and Guillaume-François Rouelle) that the horizontal strata, deposited along with their fossil content on the flanks of the primitive mountains, were necessarily younger in age than the latter.[58] In his famous preface to *Recherches sur les ossemens fossiles des quadrupèdes,* Cuvier adds that only the fossils contained in the *terrains secondaires* can "testify with certainty" that the crust of the earth has undergone any change at all and that even the basic distinction between primitive and secondary would be impossible without organic fossils. By analogy to these fossil archives, Cuvier infers that the primitive mountains too have undergone "revolutions" in the past, but he argues that Werner and H.-B. de Saussure have already established all that can be conjectured concerning these primitive revolutions.[59]

Compared with their work, he declares, research on the fossiliferous strata has only just begun. By dedicating his science to this new field, Cuvier seeks to redirect the young discipline of geology away from the domain of the primitive, which is too reliant on speculation and tends too readily to lapse into mythology and antiquarianism. In this context, Cuvier's famous self-description as "a new species of antiquary" refocuses disciplinary attention on the fossil archive of extinct species, which offers a sure path toward reconstructing the more recent "revolutions." He concedes that a study of the more shadowy revolutions of primitive times may promise to reveal the planet's origins but argues that the archival detail of the fossil record makes its interpretation a more suitable task for human beings, "to whom is permitted only a moment on this earth."[60] At the end of his preface, in spite of himself, Cuvier is very much concerned with old-style antiquarianism, because he feels compelled to refute antiquarian arguments for the high antiquity of some non-European cultures. While not literally true, the Biblical Deluge for Cuvier stood as a folkloric record of the last great revolution of nature; since he claimed that the species postdated this global revolution, non-European chronologies that claimed to predate the Deluge had to be wrong. Thus he rejects Chaldean and Egyptian claims to an antiquity of several thousand years on the basis of insufficient evidence and dismisses other chronologies with contempt. Cuvier denies India's high antiquity, for example, dismissing the *Mahabharata* as "nothing more than a poem" and insisting that all textual references to India's supposedly two-million-year-old astronomy postdate the Middle Ages.[61]

By banishing the human species entirely from deep time, Cuvier reduces the latter to purely "natural time," sharply distinguished from history. The vision of "nature's own history," often defended by scientists and historians of science today, may be seen in this way as a disciplinary construct, an intervention in the way deep time was conceived.[62] The grounds for reconsidering this division are widely apparent in our own geohistorical moment.

By contrast to Cuvier's self-conscious geology, the Forsters' Pacific travel narratives place natural history and geology at the service of a revisionist history of the human species. All but two of the islands they visited with Captain Cook between 1772 and 1775 were of volcanic origin. The elder Forster had published an introduction to mineralogy (1768), and his voyage on the *Resolution* offered the opportunity to deepen his knowledge in the field; his knowledge of volcanism proved especially serviceable. Both his account and George's cite the corresponding European scholarship, including material that appeared during their voyage, such as Johann Jakob Ferber's important letters on the Italian volcanoes (inspired in turn by the stratigraphy of Arduino). Volcanoes and their products differed from both primitive and secondary masses and remained a hotly contested category in the specialized debate around basalt as well as in broader geological texts such as Cuvier's *Rapport*—which counts volcanic materials as the most modern strata of all—or Davy's *Lectures on Geology*, in which volcanoes feature as "the most sublime of the phenomena belonging to our globe," whose unknown powers still hold great potential for the theory of the earth.[63] The Forsters prefer Ferber as a source because he delivers exact observations without committing to a theory. It was precisely the ambiguity of volcanoes—which acted in real time, on the one hand, but also left behind unstratified masses of indeterminate age, on the other—that allowed the Forsters to acknowledge their uncertainty concerning the history of the indigenous peoples without attempting to fit their development into the progressive scheme of the Scottish and French "armchair philosophers."[64]

Several points on the *Resolution*'s itinerary, especially in Melanesia, were sites of first contact between the Europeans and indigenous peoples. On Malekula, George Forster brought the large active volcano on the island to bear directly on the history of its indigenous people, who, he argues, must have arrived recently (*R* 699). In the two exceptional cases of sites that lacked any evidence of volcanism, New Caledonia and Dusky Sound, the absence of volcanoes plays a comparable role. Forster projects the primitive or "original, chaotic condition" of Dusky Sound onto the "patriarchal family" of Maori they encounter there (181; see Fig. 1.2). The geological puzzle of Grand Terre, New Caledonia, leads him to declare that the native Kanak are "a race of men very distinct from all we had hitherto seen" (*V* II.588). He qualifies this claim by showing the inadequacy of climate, diet, and other conventional explanations for understanding their distinctive characteristics. These could only have developed over "a series of many ages" from a multiplicity of largely unknown causes (II.592). George's own German rendering of "many ages" is *geraume Zeit* (*R* 827), literally "a spacious time," a figure very much akin to deep time in its self-conscious spatiality.

George Forster's account of another volcanic island, Tanna (Vanuatu), not only refers to Ferber and the European literature but also calls on indigenous knowledge concerning the volcano, just as Reinhold retells the native legend he claims to have heard concerning Maui's creation of the archipelago. The Forsters' empirical revisions of the "history of mankind" appear again and again as a motive for their geological investigations; conversely, geological uncertainty concerning the origin of volcanoes contributes substantially to the chronological openness that characterizes this revisionist history. The Forsters sometimes manage to steer clear of the fallacy of imagining Pacific Islanders as ancestral Europeans, and their occasional glimpses into the deep time of the species as a whole, by way of geology, ultimately lead to the unexpected conclusion that "the history of the human species in the South Seas cannot yet be unravelled" (*V* II.563).[65]

These human and natural histories together articulate the unknown origin of the primitive, reiterated in George's later definition of a "so-called human race" as simply "a kind without a known origin."[66] Thus, a significant moment of deep time is lost when Cuvier decides to shore up the boundaries of geology by marginalizing primitive mountains along with non-European peoples and their traditions. Nevertheless, this moment reappears throughout the nineteenth and twentieth centuries. Darwin, for example, even though he had a much more substantial geological time scale at his disposal than the Forsters, nonetheless entered into an ethnographic experience of deep time at Tierra del Fuego. From this point of view, evolutionary theory itself might be seen as an attempt to bridge the two time scales separated by geology. In his late work on human evolution, Darwin returns to this encounter on Tierra del Fuego because it allows him to produce the image of a prehistoric human ancestor. The Forsters' travel narratives do not explicitly challenge the common chronology of their milieu, but they attest clearly to the pressure to which this chronology was subjected by ethnographic and geological questions and by the relation of these questions to each other. Hence my argument in this book that the collision of geological and ethnographic temporalities produces a qualitative experience of deep time that predates and in some ways anticipates geological time.

This clash of temporalities reverberates through the critique of colonialism and through the modern struggle to liberate natural time from the constraints of sacred history. These ideological fault lines remain decisive for our present moment. The legacies of slavery and settler colonialism, on the one hand, and denialism (of evolution, of climate change), on the other, have been subject to more intense critical scrutiny in the twenty-first century than ever before. The eighteenth- and nineteenth-century thinkers studied here were certainly aware of these fault lines, and in some cases made their analyses of temporality an occasion for critical reflection on colonialism, as I will show. The binary construct of "science vs. religion" is of more recent date, but the production of deep time affords a historical perspective on the tangled history of evolution, colonialism, race, and religion that may be useful for present-day critical engagements. In Chapter 4, I argue that the acceptance of species time in the form of human evolution initially seemed to support scientific racism, and here one of the through lines in this book dovetails with recent scholarly accounts of race and temporality. The performance studies scholar Tavia Nyong'o, for instance, argues that performance art built around the memory of enslavement replaces the historical concept of "change over time" with an emphasis on "time over change." By analogy to deep time, Nyong'o posits a "dark polytemporality" that bridges the "incommensurabilities" of racialized experience, "a deep time that is also a dark time of inhuman concerns."[67] Nyong'o's position highlights some of the ways in which ideology and ideological conflict are integral to the conceptual history of deep time.

Abysses of Time

My last chapter places Darwin's work on human evolution at the end of a conceptual history in which deep time appears as a set of imaginative possibilities not yet tied exclusively to geology. The first chapter engages closely with the Forsters' revisionist history of the species, picking up where this historical introduction leaves off. Using Lehmann and Cuvier as historical bookends—or as neighboring mountain ridges, to extend Koselleck's metaphor of *Sattelzeit,* a "saddle time" spanning the transition from premodern to modern concepts in geology—I have placed the Forsters and Buffon in relation to the history of geology here in order to create a wider context for them in the two chapters that follow. The third chapter, on epic time and oral tradition, departs even further from the history of geology to explore the deep human past increasingly associated with artistic practice, an association that was also of particular interest for Darwin in *The Descent of Man,* the subject of the fourth and final chapter.

Of the authors treated in this study, Buffon is perhaps the most frequently mentioned in specialist histories of geological time, for two reasons. The first is that Buffon experimented with absolute dates for stages in the earth's history, more empirically grounded and therefore less extreme than the conjectural dates in the millions of years posited earlier by more cosmologically inclined Enlightenment thinkers including de Maillet, Denis Diderot, and Immanuel Kant.[68] The second reason is a common but mistaken attribution to Buffon of a phrase that is central to the literary history of deep time, "the [dark] abyss of time" (*le sombre abîme du temps*), which is incorporated into the titles of books on the history of geology by Paolo Rossi (1979/1984), Claude Albritton (1980), and Paul Lyle (2016) and of a work of archaeological theory by Olivier Laurent (2008). Rossi and Laurent explicitly attribute "le sombre abîme du temps" to Buffon but do not cite a location, and neither this phrase nor its variants turn up in an electronic search of the corpus. The "dark" is lacking in Albritton and in Lyle, who focuses on John Playfair's expression "the abyss of time" and its context in Anglophone geological debate.[69] The vertiginous quality of this poetic figure powerfully suggests the dislocation that occurred when students of earth history began to recognize the scalar incommensurability of geology and sacred history.

As useful as it is in a geological context, the dark abyss of time also has a rich literary history, in both French and English, that should not be overlooked. "The dark abyss" (*le sombre abîme*) is one of several epithets attached to time by a contemporary of Buffon, Antoine-Léonard Thomas, in his "Ode to Time," which won the French Academy's poetry prize in 1762. In English, Shakespeare gives us the earliest instance of "the dark backward and abyss of time" (*Tempest,* I.ii), followed by John Dryden, who published an adaptation of *The Tempest* and used the figure of "time's abyss" in a translation of Juvenal as well as (most influentially) in his own tragedy *All for Love* (1677).[70] Variations on this theme occur in the works of a range of eighteenth-century poets. Albritton also notes that "the comparison of time with a fearsome abyss is common in writings from the seventeen hundreds, though this conceit goes back much earlier."[71] In the literary tradition, the dark abyss of time signifies death and the fear of oblivion, and these abyssal associations powerfully inform the metaphor in its new application to the long secular time scale.

The story of this book is an alternative history of deep time—not as a "time revolution," created suddenly in its modern form when Darwin drew out the implications of Lyell's geology for the history of life, but rather as an imaginative framework created gradually through the Enlightenment habit of gazing into the abyss of time. Deep time as an area of uncertainty entails multiple scales and forms of evidence, and even today the figure of deep time is associated as much with the literary science writing of McPhee as with any scientific discovery. For this reason, my third chapter concentrates on literary history, on poetry and the conjectural history inspired by the ballad revival. Blake, Robert Burns, and Herder are among the poets who strive in this context to make the deep past present through poetry. The ballad, according to its many theorists, also does this work as a form. Herder's monumental conjectural history, *Ideas Toward a Philosophy of the History of Humankind* (1784–91), deliberately incorporates human origins into a larger history of the earth, drawing material from both written and oral traditions. For its theorists, and for many of its practitioners since at least the eighteenth century, the ballad instantiates the deep past of the species, as a singing one.

As Blake's fascination with "ancient men" develops in his later work into a sustained engagement with human antiquity—most famously through his allegorical character Albion, the Ancient Man—it becomes easier to see the relationship between poetry and speculations about deep time. But his earlier *Songs of Experience* makes an essential starting point in part because of its direct relationship to the ballad revival. Preliterate human societies were the major concern of conjectural history, the discipline that in turn provided the common ground shared by poets and philosophers. The history of civil society as propounded by Adam Smith and his followers provided a theoretical idiom for theorists and practitioners of the ballad revival in Britain; on the continent, Rousseau and Montesquieu (among others) played a similar role not only for theories of culture but also for natural history, and hence the discourse of species. Even Buffon, as I argue in Chapter 2, depends equally on natural and antiquarian evidence, fossils and conjectural history, to support his dramatic expansion of the time scale in the 1770s.

The first two chapters concern themselves with nonverbal media of the deep past: volcanic soils and practical arts, primitive rocks and primitive customs, hand axes and fossil bones. For this same period, poetry was the verbal medium of deep time: not of geological time, to be sure, but of a deep human past predating the written record, not yet formally understood as prehistory. According to Herder, the self-styled "historian of culture and of humankind," the true history of the species predates the "dead letter." The boundary between geology and human prehistory appears more permeable than ever in our geological moment, and Eric Gidal has made a compelling case for "bibliostratigraphy" as a method for showing how the development of human-induced climate change is inscribed in eighteenth- and nineteenth-century texts.[72] Moreover, geological science today relies more than ever on metaphors of inscription such as stratigraphic "signatures," thereby diminishing the conceptual distance between deep time and the deep human past. The ballad revival, and in particular Blake's complex response to it over the course of his poetic career, demonstrates that poetry—like other deep time media—can serve a speculative reconstruction of the deep past.

The method of Blake, Walter Scott, and other poetic antiquaries is not naturalistic in the same sense as Buffon's, but like him they turn to ancient art as an evidentiary basis for prehistory: putatively ancient oral traditions are their equivalent for the flint arrowheads and similar artifacts used by Buffon and others to argue for a species history much anterior to the written record. Songs and ballads are analogous in several ways to the natural/cultural artifacts elsewhere used to negotiate antiquity. My notion of "deep human time" is perhaps even more tendentious than the proposition that "deep time" in any form was recognized prior to the mid-nineteenth century. It is also, in a way, the keystone for the argument of this book, which is that the modern, post-Darwinian understanding of deep time rests as much on ethnographic disruptions of the Biblical timescale as on any recognition of the historicity of nonhuman nature. Blake and the "balladeers" (to adopt Maureen McLane's capacious term) recognized that poetry, and the arts and sciences more generally, antedated the historical record, and this recognition "crossed over" into the recalibration of geological time in the sense that the expansion of human antiquity provided a powerful impetus to seek for earlier origins within natural history as well. The expansion of the human time scale in this domain of poetic antiquarianism thus offers a paradigm for the expansion of the GTS itself.

If a ballad revivalist such as Joseph Ritson (like Rousseau before him) could argue that writing was a late arrival on the scene of species history, then Cuvier could argue that the human species itself was a late arrival on the scene of earth history. The depth of human antiquity is not merely transferred from "natural" time, once the latter is discovered (this is the premise of Daniel Lord Smail and Andrew Shryock's "time revolution" and, I think, of Foucault's biological turn as well), but rather informs geological time in the first place, creates a favorable condition for the discovery of deep time in its broader sense.[73] In Chapter 4, I argue that Darwin's *The Descent of Man* (1871) preserves this composite character in its attention to ethnographic encounter and to the evidence concerning prehistory afforded by the "comparative method" of making "savages" analogous to prehistoric peoples or even prehuman ancestors.[74] Darwin uses these materials to forge a link between geology and prehistory, thereby placing human evolution, too, in geological time.

Today we have a history of the human species that situates itself deliberately in the deep time of geology, as in the "deep history" of Smail and his colleagues or in the stunning proliferation of theories about the Anthropocene. The conceptual history of deep time reveals it as a story of the earth that reflects the entire history of the human species back to the reader at a glance. Ballard captured the curious composite that enables this "descent into deep time" in *The Drowned World* when he had Strangman, in imitation of Kerans's typical mode, observe that "leaving the sea two hundred million years ago may have been a deep trauma from which we've never recovered."[75] Ballard's novel might be seen as anticipating both the geological idea of deep time popularized by McPhee and Gould in the 1980s and the climate change fiction of the twenty-first century. *The Drowned World*, with its rich texture of allusions to baroque painting, Shakespeare, Keats, and T. S. Eliot, also incorporates a cultural history that predates the scientific revolution. Ballard alludes to a strand of *The Tempest* that is entirely apposite for the fiction of a planet nearly submerged by its oceans: the song beginning "Full fathom five my father lies," Ariel's account of the sea change suffered by a drowned man. Ballard passes over an equally apposite reference from this same scene (I.ii), Prospero's injunction to Miranda to look "in the dark backward and abysm of time" and remember what she can of her life before their shipwreck, before her third birthday. Although it is merely a matter of twelve years, the bottomless depth of these early memories expresses the indefinite power of escalation inherent in time itself. The image calls for an extension of memory into prehistory, an imaginative labor akin to the contemplation of other temporal depths that nearly elude our grasp, including the origin of the earth as it is now understood.

1

Primitive Rocks and Primitive Customs

GEOLOGICAL AND HUMAN TIME IN THE PACIFIC VOYAGE NARRATIVES OF JOHN REINHOLD AND GEORGE FORSTER

IN THE *Journal of Researches* now known as *Voyage of the* Beagle, Charles Darwin observed that the natives of Tierra del Fuego—actually four distinct ethnic groups living at the southern tip of South America—were "savages of the lowest grade." Besides the long legacy of similar remarks about the Yaghan, Selk'nam, and Alakaluf peoples by voyagers, Darwin had the additional motive of disillusionment after seeing the apparently unspectacular reunion between the three Fuegians Captain FitzRoy had taken to England and those they had left behind. In an article on the same voyage written for the *Transactions of the Geological Society*, Darwin described Tierra del Fuego as being "strewed with great fragments of primitive rocks."[1] At this stage the term *primitive* was still reserved for the more demonstrably ancient geological features, but within decades it had crossed the nature/culture divide to describe ostensibly ancient or premodern peoples—not least because of the influence of Darwin's own thought as it informed anthropological studies such as Edward Tylor's *Primitive Culture* (1871). While social scientists from Tylor's time onward adopted *primitive* as the more scientific term, Darwin continues to prefer terms such as *savage* and *barbarian* in *The Descent of Man* (also 1871). The two modes of ethnographic description that we see diverging here both emerged from the vocabulary of the shipboard naturalist in the era of Cook and Bougainville, when Johann (John) Reinhold Forster and his son Georg (George), among others, first began to place their conjectural histories of

Pacific peoples in the context of the emerging geological time scale. As the term *primitive* crossed over from natural to social science, its connotations shifted from "commendatory" to "depreciative," as noted in the *OED*'s most recent definition of this adjective.

Taking note of the many volcanoes in the South Pacific, the Forsters brought the question of how the Pacific islands formed to bear on the question of human origins and migrations. Following George Forster's *Voyage Round the World* in 1777, Reinhold Forster's *Observations Made During a Voyage Round the World* appeared in 1778, the same year as the Comte de Buffon's *Epochs of Nature,* which develops by far the longest detailed time scale offered by a naturalist up to that time. The Forsters drew on and corrected Buffon's earlier work as well as the mineralogical fieldwork of Johann Jakob Ferber and several others in their study of volcanic phenomena throughout the voyage of the *Resolution* from 1772 to 1775. By linking geological and human origins, the Forsters established experimental points of contact and divergence between recorded human time, on the one hand, and deep time with its associated geological terms, including "primitive rocks," on the other. The Forsters were not the only ones to insist on incorporating the history of peoples into the history of nature, but their version of this project made especially vivid use of cultural evidence to explore geological time and vice versa: their narratives deserve a special place in the history of both geological and anthropological time because they document the original proximity of these two seemingly discontinuous scales of time.

The Forsters' successful integration of human and nonhuman time owes much to their historical situation as shipboard naturalists. Their repeated experiences of first contact with Pacific peoples were enacted on a stage already framed by their expertise as naturalists. In a more theoretical vein, Buffon extends human prehistory by expanding the record of geological events implied or suggested by antiquarian scholarship, as Chapter 2 argues, and geological contextualization likewise transforms ethnographic categories in the works of other writers featured in this study, including Darwin. I began with Darwin because this lineage from the Forsters and their contemporaries is particularly salient in his remarks on Tierra del Fuego many years later. Darwin and his contemporaries also inherited the Enlightenment discourse of conjectural history and its comparisons between ancient and indigenous modern populations, formalized by the Victorians as the "comparative method." This chapter introduces an environmental history of the human species that is recast by Buffon, by Herder and Blake, and by Darwin and

Lubbock, as subsequent chapters will show, without departing wholly from the model established in the Forsters' voyage narratives.

Unstable Temporalities

European explorers saw themselves as entering a new order of time in the South Pacific. An anthropological kind of deep time arose from the shock of recognition and the simultaneous reaction that made these voyagers want to suggest that they were encountering only faint traces of themselves, a human past so ancient as to be almost forgotten. This reaction was informed by the much earlier literature of first contact in the Americas and by a new stock of ideas about human antiquity. Eighteenth-century excavations at Pompeii and Herculaneum, for instance, had raised new questions about the once-familiar classical past, fostering the kinds of tactics also brought to bear on Pacific encounters: attention to everyday life, speculation on human prehistory, and especially the contemplation of artifacts as a medium of cultural empathy. This prehistoric turn must be distinguished from the historical turn in the study of nature famously described by Michel Foucault in 1966 as "the breaking up of the great table," the displacement of "static" natural history by biological concepts around 1800. Paolo Rossi has argued that a "dark abyss of time" was opened even earlier when historical methods were applied to fossils and strata, and more recently Martin Rudwick and others have further explored the historical turn in natural science.[2] These are different kinds of arguments, but they all track a movement of chronological organization from human history into the study of nature. The natural history of Pacific populations and environments, however, had the opposite effect of disrupting *human* temporality.

The explorers' perverse insistence on the antiquity of living cultures and artifacts not only flattered the presumption of European superiority but also promoted a prehistoric turn in the understanding of human time itself, challenging the short accepted chronology of the human species without the assistance of a formalized GTS. James Hutton's pioneering account of deep time in 1788 rejected the idea that any extant phenomena could be "primitive, or original." Nearly a century later, Darwin persisted in describing human groups as "savage" and ancestral, in defiance of his own account of evolutionary time. These terms are traces of a rift in time that was triggered by cultural contact and preceded the hard distinction between historical human time and what Johannes Fabian calls "naturalized Time."[3] Time in the Forsters' narratives becomes dilated, I shall argue, because they integrated these two registers, placing environmental and human history in one continuum. Pratik Chakrabarti offers a compelling parallel instance of the "naturalization of antiquity" inspired by excavations in nineteenth-century India. European exploration of the Pacific belongs to a different phase of colonialism, during which the horizontal movement from island to island substitutes for the vertical descent carried out by the more settled colonial regime of the East India Company. Like Chakrabarti's study, however, this chapter also shifts the geographic focus associated with deep time as a Western idea to "trace the discovery of deep time in colonial history."[4]

The chronological distinction between primitive and secondary rocks or mountains was a formative paradigm of early geology, having entered technical English usage in the 1770s by way of a French translation of a German mineralogical treatise. This usage (*OED* #8) clearly derives from the most pervasive general sense of "primitive" (*OED* #1) as "first" or "original." Due to the nature of the so-called primitive masses—unstratified, crystalline, reaching the highest elevations—they necessarily carried associations that were not purely chronological: associations with remote, inaccessible, thinly inhabited places such as the Alps. As Humphry Davy observed in the first English course of public lectures on geology in 1805, the "primitive rocks" are further characterized by their "sublimity"—an aesthetic judgment that also operated powerfully in European accounts of the high peaks on New Zealand's south island and the people inhabiting the shores below them.[5] Both Davy and Georges Cuvier, in his geology lectures given the same year in Paris, attribute the distinction between primitive and secondary formations to the German mining engineer Johann Gottlob Lehmann. Both accept this distinction as a matter of course and pass over the fact that it was Baron d'Holbach who first adopted the smooth "primitive" to render Lehmann's more rugged "uranfänglich" in his 1759 French translation of the latter's 1756 treatise.[6] Reinhold Forster's publications in the years prior to the *Resolution* voyage include his *Introduction to Mineralogy* (1768). Although this work is mostly concerned with the classification of minerals, Forster characteristically chooses a more esoteric term in his brief systematic account of the "original or primogenial" versus the "latter or secondary" mountains.[7] *Primogenial* never entered geological usage, but Forster persists in using it in his *Observations Made During a Voyage Round the World*, where it serves the crucial purpose of distinguishing those few islands that are not of volcanic origin.

By contrast to this early geological usage, the term *primitive* was seldom applied to existing cultures before the advent of Victorian anthropology. The most common use of the term was historical, designating the earliest stage of a process, as in "primitive Christianity," which carries the commendatory associations of "high" and "pure," as noted by the *OED*. The Forsters used a different vocabulary for native peoples, including "savages," "barbarians," and even "Indians," an instructive misnomer that also occurs in Joseph Banks's *Endeavour* Journal and other Pacific voyages. Although they cited conjectural histories of antiquity such as Antoine Court de Gébelin's *Monde Primitif* (*Primeval World*), the Forsters also realized that the equation between ancients and natives did not always hold up under scrutiny. The *OED* cites only three eighteenth-century instances in which "primitive" might have signified "barbaric": a more or less neutral instance of the noun *primitive* used to describe native people in Guinea in 1779; a somewhat ambiguous instance in which Edward Gibbon speculates concerning the "abject" original state of humanity; and an unspecified French source of 1794 which is said to anticipate more derogatory or objectifying sociological usage. If they were looking to French sources, the *OED* editors might also have cited Philibert Commerson, the naturalist on Bougainville's voyage to the Pacific, who insisted that the natives of Tahiti were a "primitive people" and "the genuine sons of their soil" rather than immigrants.[8] Commerson, however, seems to transfer the "commendatory" sense of the word from its ancient context to a living non-European people, using it strictly in a chronological sense.

The Forsters might well have known Commerson's 1769 text—a passage strongly resembling it occurs in Reinhold's *Observations* (*O* 320)—but they were more ambivalent in their view of Pacific peoples, though they did have high praise for the "polished and civilized inhabitants of the Society and Friendly Isles" (9). They were strongly influenced by early geology as well, which complicated their chronology. *Primitive* arguably carries some advantages over *savage,* in the sense that it offers a range of values, from commendatory through neutral to pejorative, in place of the noble and ignoble savage dichotomy. Reinhold Forster's ambivalent view of Pacific peoples derives from this binary, however, and it is further complicated by his commitment to an allochronic concept of "degeneration": the "lowest savages," for Forster, are degenerated, which means they cannot be primitive or original. Forster's innovative fusion of conjectural and natural history offers us a universal cycle of degeneration and civility that extends to animals, plants, and even matters that aren't generated to begin with, such as the southwest coast of the south island of New Zealand, which, he says, "degenerates into rude rocks" toward its southern extremity (117). Unlike Commerson, the Forsters also believe in monogenesis, a single origin for human populations, which creates a constant tension between primitive customs of civility, such as the presentation of green branches universally understood as tokens of peace (*V* I.205), and the "original chaotic state"—or the "annihilating future"—that threatens to swallow them back up (I.106).[9]

Many of the armchair travelers who theorized human origins debated whether cultural difference should be attributed to differing rates of progress or to degeneration. The elder Forster, the first Pacific voyager who was thoroughly familiar with these debates, saw evidence of both progress and degeneration in his broad historical sketch of Pacific peoples, recognizing on the strength of his firsthand knowledge that the evidence was not sufficient to support either hypothesis categorically. Forster argued (and George's narrative confirmed) that empirical evidence of the diverse populations in the Pacific required a careful reexamination of the "stages" of civil society and also proved that existing scholarship had no authority for its hypotheses concerning (in modern terms) racial difference or the human evolutionary process:

> The History of mankind has often been attempted, [but . . .] None of these authors have ever had the opportunity of contemplating mankind in this state [of original simplicity], and its various stages from that of the most wretched savages, removed but in the first degree from absolute animality, to the more polished and civilized inhabitants of the Friendly and Society Isles. Facts are the basis of the whole structure [i.e., Forster's book]. (*O* 9–10)[10]

In Forster's argument, the "most wretched savages," exemplified as in Darwin by the Fuegians, reached that state by degeneration, for which Forster assigned the environmental cause of migration from a warmer into a colder climate. Unlike some anthropological explanations that postdate evolution and genetics, Forster's does not take the most apparently primitive people to be at the earliest evolutionary stage. He argues for environmentally driven transformation of human varieties over long periods of time and compares some living populations to ancient ones, both European and Pacific. Adapting a method he encountered in its historical form as a Fellow of the Society of Antiquaries (from 1766), Forster points out that customs and manners provide the only evidence concerning human antiquity in cultures without written records.[11] This evidence from customs and manners, along with biogeography, informs his conclusion that "the warm tropical climates seem to have been originally the seat of the human race" (342). Their experience in the field gave the Forsters an understanding of climate as environment that is more nuanced than that typically offered by travelers writing in the vein of Montesquieu: their biogeography is more than crust-deep.

Appropriating his experience of first contact as a "great branch of the study of nature" (*O* 145), Reinhold Forster argued that Pacific "arts and sciences" represented a creative response to the natural setting of the islands and had evolved a great distance from the ancient practices of the population that originally migrated to the islands. He also believed, however, that Tahitian religion had degenerated from an ancient Asiatic cult, traces of which survived the process of migration and resettlement (cf. *V* I.228, 281). In other words, Forster vacillates between a long and a short time scale in his history of Pacific peoples. This contradictory temporalization organizes Pacific space by means of "heterochronism"—to borrow a term from Foucault's "Of Other Spaces"—as well as "allochronism," Fabian's term for the separation between "savage" and ethnographic time.[12] In some areas Forster recognizes that change is slow, an insight that favors speculation about remote human origins and provides an alternative to the proto-racial distinctions he makes elsewhere (*O* 172–90). In other cases, such as religion, he attempts to trace Polynesians (for example) to living Malay peoples. Despite their critical awareness of its inadequacy, both Forsters sometimes revert to the chronological habit of other educated voyagers such as Banks and Bougainville by comparing Pacific islanders to Greeks and other ancient peoples.[13] This comparative gesture, though it seems to flatten cultural distinctions, also makes some reckoning with human antiquity unavoidable. One end result is a diachronic turn in the understanding of preliterate societies that precedes the geologically fixed hard dating established by modern archaeology.

Forster's analysis of the different Tahitian "arts and sciences" varies greatly with the subject matter and does not always support his own general premise that knowledge is cumulative: "The more a tribe or nation preserved of the ancient systems, and modified or adapted them to their particular situation [. . .], the more improved, civilized and happy must that tribe or nation be" (*O* 196). While this principle holds in the areas of religion and fine arts, Forster takes a different approach to "mechanical arts" like textile manufacture, emphasizing local adaptation and the use of natural materials. In his meticulous account of bark cloth (*tapa*) manufacture on Tahiti, for example, Forster draws attention to the coconut shell cups used for dye. This is the kind of detail that also surfaces in contemporary theories of the origin of art.[14] In this type of ethnographic account Forster is less apt to refer to contemporary scholarship, whether on ancient art or on the stages of civil society, than to ancient authors such as Pliny and Lucretius, who are frequently cited in the long chapter "On the Human Species" (143–376). Pliny's account of the origins of technology provides the frame for Forster's description of another "mechanical art" of the Pacific, boat-building without the use of iron (281n.). The ambivalence already apparent in this account of "mechanical arts" becomes more pronounced in Forster's openly paternalistic and idealizing narrative of the fine arts, in which he claims that Tahitians' spontaneous "verses [. . .] are always delivered by singing, in the true antient Greek style" (286). This philosophical conjecture relies more heavily on Enlightenment commonplaces and seems to echo early works by Rousseau such as *Discourse on the Sciences and Arts* and *Essay on the Origin of Languages.*

The contradictions in Forster's account support his claim to be testing philosophical conjectures experimentally in the field, but they also reflect the ambivalence inherent in the discourse of primitivism itself. Empirical study of customs and artifacts readily unsettled the orderly sequence of four stages of society partly because the theorists of those stages themselves suspected or recognized that the sequence was not orderly. Nicholas Thomas points out that although both Forsters readily invoke the ancient/primitive analogy to explain violence among the Maori, their response is ultimately ambivalent, mirroring the ambivalence already inherent in stadial histories such as that of John Millar. Thomas notes that Millar made "liberty" characteristic of both early and late stages of civil society and argues that Reinhold Forster shared Millar's anxiety about the slippage from "liberty" to "license" in both stages. Nostalgia for "savage" liberty casts doubt on the celebration of commerce and progress in Forster as in Millar and even in more conservative theorists such as Adam Ferguson. Harriet Guest argues that "Forster was concerned to develop these theories"—Millar's in particular—"and adapt them to his novel experiences." In a similar vein, but emphasizing the Forsters' use of European climate theory, David Bindman points out that "experience made things complex and unstable, challenging the formulae of climate and social life that had dominated views of the 'savage' world."[15] Richard Lansdown expands this broader European frame of reference by tracing the "bipolar vision" of "cultural" and "chronological" primitivism from classical antiquity down to the Enlightenment, emphasizing "the depth of Rousseau's ambivalence" as a major influence on the voyagers. Thus both new empirical evidence and instability in the theories themselves made it far from a "simple matter," as

Thomas rightly suggests, "for Europeans to apply their prior notions of savage or primitive life" to Pacific peoples.[16]

All these interpretations reinforce Reinhold Forster's own distinction between armchair theorizing and ethnographic fieldwork. They also help to contextualize the adaptation of conjectural history to different kinds of knowledge projects, which will be a key theme of later chapters of this book. The Forsters' revision and adaptation of this model for understanding cultural development hinges on observed discrepancies between the stadial paradigm and diverse populations—these discrepancies promoted a keener attention to novel and specific features of each encounter.

Theories of the earth further complicated the Forsters' fieldwork, and vice versa, because they too carried particular investments in the shape of the primitive world. The elder Forster is at his most admiring in his account of Pacific geography, astronomy, and navigation. Their "remarkably strong sagacity" in this domain inspires Forster to credit the islanders with "being the inventors of the whole cyclus of their knowledge" (*O* 320). (This avowal contradicts earlier claims for degeneration and departs from the condescending attitude toward indigenous knowledge that is more typical of both Forsters.) Commerson's "Post-Script" cites the same body of autochthonous knowledge to argue that the Tahitians were "a primitive people" in the literal sense that they did not migrate to the island from elsewhere. In both cases, the semi-obligatory condescension of the voyage narrative gives way before an example that implies a "deep," quasi-evolutionary time scale. Commerson insists that for a naturalist the only tenable view is to regard the Tahitians as an autochthonous or "protoplastic" people, and he ridicules the monogenetic account of human origins as plausible only to a "mythologist."[17] For Forster, taking a naturalist's view means attending to the geology of the islands and its implications for the prehistory of a population. The advanced civilization on Tahiti presupposes the "long existence and fertility" of the island (41), measured in relative terms by its apparently long-dormant volcanoes and abundant "fat" (fertile) volcanic soils. Easter Island, on the other hand, appears to have been devastated recently by a "terrible revolution," as George Forster puts it in the German text (*R* 506; cf. *V* I.320); reestablishing the ancient civilization seemingly implied by the monuments there would have to be a "work of ages" (*O* 198). The Forsters exploit the double sense of "revolution" as a social or natural phenomenon to posit a wide range of dates for the formation and the settlement, respectively, of the various islands. Reinhold in particular insists that lands can become "habitable" and "inhabited" at widely different times (105).

Volcanoes and Revolutions

Both Forsters emphasize the ubiquity of volcanoes and volcanic islands all across the Pacific. Volcanic eruptions, and the formation of volcanic islands in particular, presented a geogony in miniature, as Darwin later noted enthusiastically of the Galápagos. Volcanic rock could not be primitive by definition, but it constituted a suggestive middle ground between stratified and unstratified materials. Plutonism was the theoretical school that exploited the analogy between basalt and granite most fully, arguing for the high internal heat registered near the top of William Buckland's witty Geological Thermometer (Fig. 1.1). The Forsters chose to credit Ferber with the most accurate observations of active volcanoes in Italy instead of Ferber's rival Sir William Hamilton, who appears here in the somewhat cooler region of 65–70 degrees with other Vulcanists. Ferber was the better mineralogist, but the Forsters' philosophical view of volcanoes more closely resembles that of Hamilton, who also happens to be the naturalist credited by the *OED* with introducing *primitive* in its geological sense (#8) in 1778.[18] Like Hamilton, the Forsters take issue with Buffon for positing a molten earth absolutely prior to the molten lava of the present and trivializing the "raw force" of active volcanoes, to borrow Michael Dettelbach's apt phrase (*O* lxxi–lxxii). Although seeking to appear cautious in the geological chapters of his *Observations*, Reinhold Forster devotes much space to the "ravages" of volcanoes and asserts that "volcanoes, no doubt, cause great changes on the surface of our globe" (103–4). These two chapters dramatize the extent of volcanism, finding that only parts of New Zealand and New Caledonia are "undisturbed" by it, or "primogenial" (34, 35n.). Forster further postulates three "classes" of volcanic islands on the basis of both recent and ancient evidence showing that "fires have in all ages raised isles out of the sea" (111). Though based in Europe, Hamilton also drew on evidence from the Pacific, citing the many "implements of stone" (i.e., basalt) collected by the naturalists on Cook's first voyage to argue that volcanic eruptions were geologically formative events not only in antiquity but across the earth and across time.[19]

The conflict between theories of igneous and theories of aqueous formation constitutes one of the best-known chapters in the history of geology, but it is worth revisiting here for two reasons: first, the story is often constructed around a reductive binary that conflates Plutonism with Vulcanism to create a simple antithesis to Neptunism; second, the theoretical spectrum schematically displayed on Buckland's Thermometer offers a range of possibilities for

ART. XXVI.—*Geological Thermometer, shewing the Opinions attributed to various Geologists with respect to the Origin of Rocks.*

Geologists	Scale	Region	Opinions
WHISTON, Theory of the Earth, 1725. BUFFON, Theorie de la Terre. LEIBNITZ, Protogæa, 1768. DESCARTES.	100 95	*Plutonic Region.*	All Rocks affected by heat. The Earth struck off from the Sun by a Comet.
BOUÉ, Essai sur l'Ecosse, 1822.			All Rocks of Chemical origin igneous.
HUTTON, Theory of the Earth, (Ed. Trans. v. i.) PLAYFAIR, Illustrations, 1820. Sir J. HALL, Edin. Trans. vol. vi. 1806. Sir G. MACKENZIE, Travels in Iceland, 1810.	90 85		All the older rocks either fused or softened by heat. Metallic Veins injected from below.
Sir H. DAVY, On Cavities in Rock Crystal, 1822. MACCULLOCH, Various papers in Geol. Trans. from 1814 to 1817. KNIGHT, Theory of the Earth, 1820. BRIESLAC, Journal de Physique, vol. xciii.	80 75		Granitic Rocks igneous: Granitic Veins injected from below. Some Granite and Sienites igneous.
FAUJAS ST FOND, Essais Geologiques. HUMBOLDT, Travels and Memoirs. SPALLANZANI, Sur les Isles Ponces. Sir W. HAMILTON, Memoirs, &c. DOLOMIEU, Voyage aux Isles de Lipare, 1783. SAUSSURE, Voyages dans les Alpes, 1787. W. WATSON, Section of Derbyshire, &c. WHITEHURST, Theory of the Earth, 1786.	70 65 60	*Volcanic Region.*	All Trap Rocks igneous.
CORDIER, Sur les Substances Minerales dites en masse, 1815. VON BUCH, Travels, Memoirs, &c.	55		Augite Rocks igneous.
BUCKLAND, Memoirs. CONYBEARE, Geology of England, 1822. SEDGWICK. HENSLOW.	50 45		Flœtz-trap Rocks igneous. Whin-dikes injected in a fluid state from below. Some Flœtz-traps igneous; others aqueous.
DOLOMIEU, Journal de Phys. vol. xxxvii. 1790. SAUSSURE, Journal de Phys. (an. 2.) 1794. DAUBUISSON, Ib. 1804, Sur Volcans d'Auvergne. DAUBENY, Edin. Philos. Journ. 1821, On the Volcanoes of Auvergne.	40 35		
DAUBUISSON, on the Basalts of Saxony, 1803. DELUC, Treatise on Geology, 1809. KLAPROTH, Beiträge, vol. iii. JAMESON, Edinburgh Philos. Journal, 1819. RICHARDSON, On the Giant's Causeway, Ph. Tr.	30 25	*Neptunian Region.*	Igneous origin of any Trap Rocks questioned. Whin-dikes contemporaneous with the rocks they traverse.
MACKNIGHT, Wernerian Memoirs, 1811. JAMESON, Geognosy, 1808. MURRAY, Comparative View, 1802. MOHS, Memoirs, &c. KIRWAN, Geological Essays, 1795. WALKER, Lectures, 1794. WERNER, Theory of Veins, 1791.	20 15 10		All Rocks (except the Volcanic) deposited from aqueous solution. Metallic Veins poured in from above.
LAMARCK, Hydrogeologie. DEMAILLET, Telliamed.	5		Secondary Rocks secreted by animals and vegetables, and formed out of Water.

FIGURE 1.1. William Buckland, "Geological Thermometer," *Edinburgh Philosophical Journal* 7.14 (October 1822): 376. Courtesy of Peter H. Raven Library, Missouri Botanical Garden, St. Louis, Missouri.

interpreting the relationship between volcanoes and the "primitive" as it informs the Forsters' Pacific natural history. The term *Plutonic*, which first arose in connection with the work of Hutton and James Hall in 1790, was not available to the Forsters, but they would likely have regarded it with the same skepticism that led Buckland retrospectively to place Hutton with Buffon and the seventeenth-century theorists in the "Plutonic" region of his thermometer. The Forster's inferences, based on the widespread volcanism they observed, belong to a more empirically driven approach that originated in the mid-eighteenth century but continued to develop into the nineteenth century. Reviewing this development in 1805, Humphry Davy credited Hamilton, Lazzaro Spallanzani, Ferber, and especially Déodat de Dolomieu with "minute observations" of volcanoes "combined with the most enlarged views."[20] The empirical emphasis of this tradition aligns with the early stratigraphy of Lehmann and Arduino, as outlined in the introduction to this book.

Davy identifies himself with this tradition by echoing the disavowal of theory or "system"—of "enlarged views" in the absence of "minute observations"—that occurs repeatedly in Hamilton and in both Forsters, for example in Reinhold's *Observations*: "I pretend not to defend . . . any theory of the earth" (*O* 24).[21] The Forsters (among others) repeatedly single out Buffon as the primary target of these disavowals, since Buffon dogmatically regarded volcanoes as superficial by contrast to the primitive molten state of the planet as a whole. While their approach lacked the explanatory power and aesthetic appeal that only a system could offer, diligent students of volcanoes in the field, including the Forsters and their preferred authority, Ferber, could cite many impressive local instances of new land produced by volcanoes. In Davy's view, volcanic eruptions were "the most sublime of the phenomena belonging to our globe," and at least in this sense their products could be seen to rival the primitive rocks in importance. Davy dutifully "excludes" lava from "the series of primitive masses" but also notes that volcanoes are typically found in "primitive countries" and argues that their association with "unknown powers" makes them essential for geological theory.[22] Deferring a theoretical solution, Davy dismisses Hutton's Plutonism as "conjecture" but rejects Neptunism even more firmly as "error," a system totally unable to accommodate the geological significance of volcanoes. Though not formative in their own right, volcanoes, for Hutton, were at least the symptom of a far more powerful deep-earth process that he pictured by analogy as "subterranean lava." While this model diminished the importance of volcanoes operating visibly at the earth's surface, it boldly denied the very existence of "primitive rocks" nearly a century before geologists stopped using the term: there was no solid material, Hutton insisted, that had not been subject to alteration by "subterranean lava" at some remote period.[23]

Leaving aside the variety of theoretical frameworks, the Forsters' treatment of volcanoes is consistent with the period's earth science in two respects: they recognize volcanic processes as "secondary" but also accord them a special status as phenomena that may provide evidence of "the original state and formation of the earth."[24] In the context of Pacific exploration, Hutton's skeptical approach merits comparison with Commerson's revisionist treatment of Tahiti's "primitive people": in this secular view, a thoroughgoing naturalist must exclude any sense of "primitive" that derives it from a supernatural act of creation. In his later European travel narrative, *Ansichten vom Niederrhein* (1791), a more cautious George Forster casts doubt on any theoretical connections between volcanoes in the Pacific and the appearance of extinct volcanoes along the Rhine or elsewhere. Just thirty years later, however, Forster's pupil and his companion on this later voyage, Alexander von Humboldt, articulated a more modern geological view by focusing on the essential identity of volcanoes all around the globe based on their "permanent connection to the interior of the earth."[25] Humboldt's theory, based on new evidence from Central and South America, confirms the relevance of the Forsters' work on Pacific volcanoes for the emerging global paradigm of geology and the importance of volcanoes—as formative if not "primitive" in the strict sense—within that paradigm.

For the Forsters, placing volcanoes in relation to the earth system is ultimately less important than placing them in relation to a comprehensive Nature, which includes human populations. They appeal to a wise Providence directing the general improvement of land by way of volcanoes and other secondary processes, an appeal that echoes Hamilton's (and later Davy's) confidence in volcanism as a "creative" power intended for the ultimate "benefit of future generations."[26] The paradigm of improvement extends to all the islands with evidence of both volcanism and agriculture. George Forster takes the active volcano on Malekula as a proof of the island's fertility (*V*II.479) and explains the apparent contrast between the islanders' intelligence and their lack of "conveniences" by conjecturing that they have arrived on the island only recently, since the geology stabilized (II.490). By contrast, Reinhold Forster places great emphasis on the geologically primitive state of New Zealand (the generalization is primarily based on the non-volcanic western coast of the South Island, especially the granite and diorite of Dusky Sound, which the

Resolution explored for two months in 1773). His influential distinction between polished Polynesians and "savage" Maori depends on the geological rate of improvement: "The tropical isles have all the appearance of a long existence and fertility. But the southernmost parts . . . are still unimproved, and in that rude state in which they sprung up from the first chaos" (*O* 41). In the South Island of New Zealand, "Nature has just begun her great work," while the South Atlantic island of Ascension provides evidence of a recent eruption initiating the long, slow alteration of primogenial land for the better (110). Davy offers explicit chemical evidence when claiming that volcanic soils are "by far the most fertile" in his 1805 lectures, while the same providential paradigm informs his view of volcanoes as "compensating" for the loss of arable land by erosion.[27]

Because volcanic eruptions can have revolutionary consequences across the natural and social world, close mineralogical examination of their products is particularly important to the Forsters. George Forster's chapter on the island of Tanna (New Hebrides, now Vanuatu) develops this point most fully. Two unique circumstances of the *Resolution*'s two-week sojourn there account for the unusual length of this chapter, by far the longest in *Voyage Round the World*. First and most important, this landing was the first European contact for this population; second, the island had an active volcano that erupted twice during their stay and became the Forsters' scientific preoccupation. George is especially thorough on both points, and Reinhold incorporates unusually detailed discussion of this volcano, Yasur, in his *Observations*. Both accounts approvingly cite Ferber's *Travels through Italy* for its explanation of the sulfur-rich white clay that they found around volcanic vents on the island.

Although this was their first opportunity to do fieldwork on an active volcano, the Forsters took particular note of volcanic features from the very first landing of the *Resolution* at Madeira. Fourteen months into the voyage (September 1773), we find George making generalizations about extinct volcanoes in the Society Islands, where he notices peaks on four islands that have the typical "appearance of a crater" (*V* I.203). In July 1774, on Malicollo (Malekula)—the *Resolution*'s first landing place in the New Hebrides—he infers a shared volcanic past based strictly on the fertility of the soil, which, he notes, is as "fat" as that of the Society Islands. Empowered in these "conjectures" by the sight of an active volcano on a neighboring island, George speculates boldly on the prehistory of this island's population in light of its high fertility and low population density: "perhaps if we could ever penetrate through the darkness which involves the history of this nation, we might find that they have arrived in the South Sea much later than the natives of the Friendly and Society Islands" (II.490). In fact, the opposite is true, but the Forsters' innovation—clearly illustrating the ethnographic context of geochronology—is to draw the inference about population from the more recent date of volcanic "improvement" in this island group. The *Resolution*'s considerably longer stay at Tanna provided more material, including their analysis of the volcanic soil, to support this conjectural history of the present. George considered the Tannese "better proportioned than . . . the Mallicolese" (II.510)—either less racially marked or less unfamiliar in the wake of first contact on Malekula or both—and his and his father's eagerness to study the volcano led them into ever closer proximity with the people. The Forsters had to content themselves with a close examination of the *solfataras,* or volcanic vents, because the people tolerated this but would not let them approach the mountain itself.

By confirming that the mineral composition of the *solfatara* matches that of South Italian sites, and even by using this Italian name, the Forsters provide a philosophical anchor for their history of this unfamiliar people. Yasur thus enables conjectures about the origins and development of the Tannese, but it also provides a kind of alibi for the Forsters' unusually intensive and sometimes convivial engagement with families and settlements in the interior.[28] Their accidental discovery of these vents happened to coincide with an eruption that confirmed their hunch concerning the volcanic nature of several mounds of white and greenish clay in the forest that smelled of sulfur: they emitted puffs of steam with each loud explosion of the mountain, still out of sight and at least two miles away (*V* II.524). At the time they noted the aluminous taste of the clay, the heat underground, and the proximity of the "fat" red ochre used by the Tannese for face paint. Though they eventually came within sight of a spectacular eruption, a tense encounter with islanders prevented them from approaching the mountain. On a second attempt, they were unexpectedly led back to the beach by a party of Tannese who had seemingly offered to lead them to the crater, and a third attempt with a larger party including Captain Cook was similarly derailed. On both occasions, however, the Forsters took temperature readings and other observations of the *solfatara.*

These investigations prompted wide-ranging speculation on the nature of volcanic islands as well as a rare recognition of indigenous knowledge concerning the site. The investigation was somewhat hampered by a steady rain of ash, which irritated their eyes, but George reflects in the narrative that the volcano "abundantly compensates" the islanders for such inconveniences by fertilizing the soil: "many plants here attain twice the height which they have in other countries" and are "more richly scented" (*V* II.529). He goes on to cite European examples and then the Forsters' own observations in confirmation of this point, adding the New Hebrides, Marquesas, and some of the Friendly Islands (along with the Society Islands) to the catalogue of volcanic islands with "fat" and "fertile soil, in which nature displays the magnificence of the vegetable kingdom" (II.529). As they were lowering their thermometer and digging for samples in the *solfatara*, a friendly group of Tannese asked them earnestly to desist, informing them that "it would take fire, and resemble the volcano, which they called an *assoòr*" (II.530). According to the twenty-first-century editors of this narrative, "a Tannese myth [that] traces the movements of the [still active] volcano from place to place" is still current, causing the same apprehensions today (II.821n37). If the Forsters had credited this belief, they might have been less hesitant to affirm that these vents, as well as the hot springs (II.547), derive their heat from the same lava that is found in the crater (as is really the case). George does, however, conclude that the full range of soils and minerals on the islands "are, to a greater or lesser degree, the work of the volcano" (*R* 804).[29]

The Forsters are mainly concerned with the human population in the New Hebrides, but these volcanic episodes enable them to situate their experience of first contact in the context of more familiar natural phenomena and to associate their fieldwork on the "natural history of man" with the authority of European naturalists such as Ferber and Buffon. Ferber's work conveniently appeared while the Forsters were composing their narratives in 1776 and offered a chemical analysis that explained the phenomena they had seen. Ferber recognized that the color of the *solfatara* near Vesuvius came from sulfur rather than calcium compounds, but the Forsters' extravagant praise suggests that more is at stake—that Ferber, a Swede who was trained by Linnaeus and then traveled extensively before serving as Assessor of Mines, presents a model of rigorous empiricism combined with practical application and service to the state, and hence a role that the Forsters (Reinhold especially) envision for themselves. On the specific issue of the sulfurous clay mounds, Ferber explains their formation and then notes that they might, "by their appearance, be mistaken for lime-stone." Reinhold, evidently impressed by the close correspondence between Ferber's account of Vesuvius and what he had observed on Tanna, relies heavily on Ferber in his chapter on the strata. This "tactless philosopher" also goes out of his way to discredit Hamilton for having mistaken the clay for limestone, just as Ferber anticipated, in his 1771 publication on the *solfatara*.[30] He praises Ferber, by contrast, as "the most intelligent, and most accurate, mineralogical writer of this age" (*O* 33n.), a sentiment perhaps inspired by the utility of Ferber's exhaustive classification for Forster's own project of establishing an overarching volcanic history of Pacific islands. George's narrative echoes this praise, championing Ferber's priority claim and essentially nominating him for the position of naturalist on the next voyage to the New Hebrides: "Their volcanoes, their vegetables, and their inhabitants, would provide sufficient employment for a Ferber, a Solander [another student of Linnaeus], and many of the historians of mankind" (*V* II.561).

Here George echoes Rudolph Erich Raspe, whose English translation of Ferber is cited by both Forsters and who claims in his preface to have recommended Ferber's research program—particularly the search for volcanic "new-raised islands"—to Cook himself as the latter was preparing for his third voyage in 1775–76. The Forsters were likely influenced by Raspe's framing of Ferber's work in the translator's preface and notes, which make explicit the global application of the many local phenomena meticulously catalogued and classified by Ferber. Raspe, for example, presents Ferber's "volcanico-marine fossils" as a geological record of volcanic islands "new raised" in the remote past—a connection left implicit by the author himself. Raspe also argues explicitly that describing the origins of various rock types will produce a system that "is, in respect to the fossils, what the Linnaean Sexual System is to the plants" and will allow naturalists "to judge of their origin and antiquity."[31] Dettelbach has shown how central Linnaean classification was to Reinhold Forster's identity as a naturalist, and Ferber's exhaustive catalogue seems to have provided a model for corresponding sections of Forster's *Observations*. Though proudly "cautious of extremes," Ferber is also moved to question whether the visibly slow evolution of some volcanoes is "consistent with our common Chronology," a doubt that may have inspired the Forsters as well.[32] The casual tone of the question is noteworthy, for it suggests a more gradual approach to the unfolding of geological time than the revolutionary change often projected by narratives of this "discovery."

By juxtaposing Ferber against Buffon, the Forsters effectively suggest that conjectures about geological time should be justified by painstaking classification of the kind that Ferber offers, as opposed to the theoretically based generalizations they saw in Buffon. As often in the *Voyage*, we find George giving a strong literary and rhetorical shape to positions less fully elaborated in the *Observations*. The father contents himself with saying that Buffon's "assertion" concerning volcanoes "seems not to correspond with our experience" (*O* 103); the son impetuously condemns the "many naturalists" who still "blindly reiterate" the "hypothesis" of this French writer (*Schriftsteller*) that volcanoes occur only as surface features and only in the highest mountains (*R* 805).[33] The Forsters' integrated study of geology and culture on Tanna led them—unlike either Ferber or Buffon—to believe that volcanic eruptions can and do constitute revolutions in the natural and human history of a place. Their engagement with both sources helped to establish their authority to make this claim. The verifiable natural history of the island provided a methodological foundation for their innovative ethnographic work, but at the same time the "history of mankind" provided the motive and transformed their natural history and geochronology. The reciprocal effect of these two scales is my central concern in this chapter.

The Forsters recognize the dual social and natural character of revolutions, volcanic and otherwise, in several ways. Concluding his "Remarks on the Changes of Our Globe," Forster père enters somewhat self-consciously "into the land of fables and mythology" to suggest that "the inhabitants themselves have some idea of a great revolution, which happened to their isles" (*O* 112). The story of the god Maui "dragging a great land from west to east through the ocean" indicates that perhaps some of the Society Islands were not raised from the sea by volcanoes but "were only dismembered by the sinking of the intermediate parts." In *Epochs of Nature*, Buffon claims that "the first men were witnesses of the convulsive motions of the earth, which were then frequent and terrible," and that the fear of earthquakes is their enduring legacy (*FA* 381). The Tahitian myth similarly recalls an earlier state of humanity for Forster and may provide evidence of geological upheaval in the prehuman past.[34] The Forsters argued for a "terrible revolution" of much more recent date on Easter Island, where they felt that the state of society directly reflected the impact of a natural catastrophe (*R* 506, *O* 264). Their cyclical vision of social change, perhaps reflecting the volcanic cycles of the terrain, compensates for the anachronism of slow improvement versus catastrophic degeneration. In keeping with the stadial mobility outlined in the advertisement prefixed to Reinhold Forster's *Observations*, the Easter Islanders, like any "degenerated" group, may once again "behold the dawn of civilization" (*O* lxxviii). He extends that privilege to those "sons of liberty" who are "willing to withdraw themselves from the oppressions of growing despotism in Europe" to seek "asylum" in New Holland (Australia), which he apparently envisioned, in good liberal imperialist fashion, as a seat of ancient liberty to rival revolutionary America (24).

The idea of revolutions is a commonplace in early geological thought, but the Forsters greatly expand its semantic field by incorporating their notions of resistance to oppression—viewed by George as the "natural circle of human affairs" (*V* I.200)—and their experience of first contact—with their interpretations of Pacific landforms. Their observations run the gamut from a "great revolution," which gave the continents their present shape (*O* 25), to a vague plan for rejuvenating the "wretched" Fuegians: "a more frequent intercourse with Europeans, or some other unforeseen accident, for instance, the fortuitous invention of iron . . . [or] a discovery of the utility of some vegetable . . . must doubtless, sooner or later, bring on a revolution in their condition" (215). George, in the course of his *Voyage,* continually refines his cyclical model of primitive equality giving way to hierarchy and despotism, early warning signs of which on Tahiti initially prompt him to observe that "a proper sense of the general rights of mankind awakening in them, will bring on a revolution" (*V* I.200). On the Marquesas Islands seven months later, he notes that the Tahitians have more conveniences and greater refinement, but that the "original equality of the social orders is already in greater decline" there, in comparison to the healthier and more tranquil population on the Marquesas (*R* 537).[35] Revisiting Tahiti immediately afterward, the extremely high demand for red feathers picked up elsewhere by the *Resolution* as trade goods (555) seems to confirm this hypothesis of incipient civilizational decline, which in turn casts doubt on the linear progress typically associated with "advancing" stages of society.

Only a generation later, Georges Cuvier helped to establish geology as a science by purging traditional terms such as *primitive* and *revolution* of these ambiguities and locating them on a planetary and prehuman scale. Cuvier famously presented himself as "a new species of antiquarian" who passed over the antiquities of peoples and nations to study "traces of revolutions prior to the existence of every nation."[36] These "revolutions" are catastrophic transformations of the earth's crust that had to intervene, according to Cuvier's reading of the fossil record, in the spaces between sets of strata with totally different fossil fauna. By this logic of succession, "our present societies" necessarily postdate "the last revolution," and the quiescent state of the present earth gives us little idea of the agents of revolution in this strong sense: "Our volcanic eruptions, our erosions, our currents, are pretty feeble agents for such grand effects."[37] Likewise, extant kinds of large animals will not be found to match any genera predating past revolutions. Though readers in 1912 were still eager for Arthur Conan Doyle's speculations that dinosaurs might yet be found, in his novel *The Lost World,* for Cuvier it was already decisive that "Bougainville and Cook found only pigs and dogs on the South Sea islands." Cuvier does resort to the old species of antiquarianism to describe his last revolution, a "sudden displacement of the sea" attested by the legends of "all the nations that can speak . . . that they have been renewed recently, after a great revolution of nature." He even hedges a bit by calling this "the time that human colonies began—or began again."[38]

But the fossil record does not vouch for even this much human antiquity; while Cuvier dismisses absolute dating together with hypotheses about primitive rocks and other conjectures of those who "pile up thousands of centuries with a stroke of the pen," his relative dating of the secondary formations with their fossil record approaches deep time in its expansiveness relative to the "instant" accorded to "man."[39] For the Forsters, however, the presence of "young" civilizations on "young" islands provided viable evidence for conjectures about origins, and at least in this analogical sense volcanic revolutions brought human prehistory into contact with primitive geology. For them, the absence of any quadrupeds besides pigs and dogs meant that cultural evolution had to take different course in the South Pacific, passing over the hunting and herding "stages" of civil society (*V* II.554).

Primitive Systems, or, Stadial Environmental History

For the Forsters, then, human prehistory came into view as a part of geological succession. At Dusky Sound, New Zealand, after four months of dreary sailing in the Antarctic Circle, both Forsters are prepared to be enchanted even by rude rocks and captivated by the one family that seems—Adamically—to inhabit this place. As Nicholas Thomas has noted, William Hodges's painting of this site (*Cascade Cove*, Fig. 1.2) seems to capture the "undomesticated strength" that the artists and naturalists on board the *Resolution* perceived holistically as the natural state of the place, surrounded by forests unchanged since the Flood (as George claims) and teeming with previously unknown genera and even families (which Reinhold celebrates). The Forsters' narratives struggle to map a scale of human values onto an increasingly long and unsettled time scale and onto the complex geographic itinerary of their voyage. They ultimately produce a conjectural history of the present that is animated by their heterodox notions of progress, by a close attention not only to time's arrow (to borrow the terms of Stephen Jay Gould) but also to time's cycle as illustrated by the geological environment. To chart the progress of their voyage, as well as the "progress" of populations and ecosystems that they encountered, they continually revised and amended their criteria and scale of analysis by establishing new correlations—between or among populations and geological features as well as plants and animals in a given climate. In this way, deep time became implicit in human time, and evolutionary time (to use modern terms) could be inferred from phenomena on the "eco-devo" scale of ecology and development.

The following itinerary—from Dusky Sound to Tanna (Vanuatu/New Hebrides) and then to New Caledonia—is intended to illustrate the interplay between these smaller scales and the longer ones considered in the preceding pages. The Tannese, as we have seen, would not allow the Forsters to investigate the volcano Yasur, in spite of their persistent efforts to do so. In returning to this scene, I will be asking what happened when the encounter with this culture, instead of being merely an obstacle, itself became a scientific pursuit for the Forsters in this context. How did their study of the volcanic soil and its vegetable and animal "productions" inform this pursuit and vice versa? This section of the chapter takes up these kinds of questions for all three sites in a broad ecological context, while the last two sections address ethnography more specifically. Aesthetic categories (as in the case of the "wonderful phaenomenon" of the Tannese eruption; *V* II.510) help to a establish a holistic ecological view of each place on this abbreviated itinerary—each of which is also a site of first contact.

Hodges's painting (Fig. 1.2) provides a visual index of the convergence between aesthetic conventions and novel experiences in the field that produced the "savage place" of Dusky Sound. I borrow Coleridge's description of the "romantic chasm" that traverses his response to exotic travel narrative, "Kubla Khan," to indicate the influence of the Gothic and picturesque traditions—besides the ethnographic tradition of "savages"—on the Forsters' construction of New Zealand as the primitive foundation of their social stratigraphy. They draw attention to these influences themselves. The epithet *savage,* so often attached to the paintings of Salvator Rosa in picturesque theory, moved Reinhold to wish for him in Dusky Sound: "Some of these cascades with their neighboring scenery, require the pencil and genius of a Salvator Rosa to do them justice: however the ingenious artist [Hodges], who went with us on this expedition has great merit, in having executed some of these romantic landscapes in a masterly manner" (*O* 51–52). George begins his chapter on Dusky Sound with a similar aesthetic response, declaring it to be "in the style of Rosa" because of its cliffs, waterfalls, and "antediluvian forests which clothed the rock" (*V* I.79). He quickly substantiates this aesthetic response with empirical data: not only is it "historically probable" that these primeval forests are

FIGURE 1.2. William Hodges, *Cascade Cove, Dusky Bay* (1775), oil on canvas. William Hodges (1744–1797), Royal Museums Greenwich, public domain.

"untouched and in their original, wild, first state of nature," but their appearance demonstrates this primacy indisputably. The family depicted here by Hodges thus belongs to a time at least as old as sacred history. The Forsters treat them accordingly as a first family of humans—though, as Thomas points out, it is far from clear in George's narrative, and even less so in the painting, what sort of family this is or whether it is a nuclear family at all. The posture and costume of these figures, the dense "primeval" forest, the foaming cascade, and the snow-covered "primitive" peaks evoke an environmental history of human society that adapts and revises the stadial theory (i.e., the theory, primarily Scottish and French, of civilizational "stages") that Hodges as well as the Forsters brought to bear on the encounter.[40]

With the publication of *The Spirit of the Laws* in 1749, Montesquieu did more than perhaps any other writer before or since to promote environmental history as a way of studying human populations.[41] The Forsters' approach to the study of peoples and climate made biogeography more than crust-deep by incorporating geological questions, as I argued in the previous section.

Incorporating not only rocks but also flora and fauna, along with more immediate intellectual influences such as Linnaeus and Buffon, the Forsters came into their own as environmental historians of the species at Dusky Sound. They rendered this spot on New Zealand's South Island a "savage place" by taking the most comprehensive view of nature, classifying and correlating flora, fauna, and human populations previously unknown to Europeans with climate and geology. George's narrative celebrates the diversity of new plants and animals at Dusky Sound: "among them hardly any that were perfectly similar to the known species, and several not analogous even to the known genera" (*V* I.80). Early in the narrative he expresses anxiety that they would not have the same opportunities that Joseph Banks and Sydney Parkinson had on the *Endeavour* to make discoveries (I.51), but here (as later on Tanna and New Caledonia) he claims a kind of Adamic priority, giving names to things with an enthusiasm inspired by the "patriarchal" family of humans in Dusky Sound. Martin Guntau identifies the Forsters' "holistic concept of nature" as typical for the "bourgeois Enlightenment worldview"; it is also invoked by Alexander von Humboldt as a precedent for his encyclopedic view of global nature.[42] This view of nature generates new metrics by which to gauge the state or "stage" of a society, as Reinhold and George both insist on a correlation between a comparatively smaller number of unknown species and a "higher state of civilization" associated with agriculture (*O* 114). Hence the contrast between New Zealand with its estimated 400 to 500 new plant species and Tahiti, where the scarcity of new species attests to that island's "high cultivation" (*V* I.156).

Thus all the human and natural features of Dusky Sound help to establish one pole of a binary that eventually structures both narratives: the Society Islands are cultured and beautiful, while New Zealand is savage, primeval, and by turns sublime or picturesque. It also happens to be the first significant stage (for the Forsters as naturalists) of the *Resolution*'s itinerary. In this "shaggy country," Nature "was amazed to see herself so correctly copied" in Hodges's "romantic prospects" (*V* I.105). Although the arts and sciences seemed to flower at Dusky Sound during the ship's time at anchor, George anticipates that their harbor will soon "return to its original, chaotic state" (I.106). The younger Forster's predilection for literary topoi such as recognition and reversal is apparent here; his narrative is more skeptical and ambivalent about progress than his father's, both in its style and its philosophical content. In this instance, he displaces onto the landscape an affective response more typically attributed to human subjects of representation, an unusual figure that suggests self-conscious reflection on the inequality of the human encounter.

Nevertheless, the polarity of "savage" higher latitudes versus "cultured" tropics and the framework that follows from it—the developmental spectrum of island societies, the scale of time and values mapped on to the ship's geographic itinerary—often structures George's narrative, while in Reinhold's *Observations* it becomes the burden of the argument.

Reinhold Forster offers a stadial environmental history which, although comprehensive and revisionist, leaves little doubt about the ultimate progress of human and nonhuman nature. His chapters on mineralogy, botany, and zoology present a series of nested progress narratives that begin either with Dusky Sound or with the even more "degenerated" southernmost lands of the western hemisphere. While George worries that their conversion of Dusky Sound from a "wild and desolate spot" to a "living country" might prove to be only temporary (*R* 179), Reinhold assures us that here "nature has just begun her great work" (*O* 41). If the forests of Dusky Sound are left to rot a while longer, the island is merely "hoarding up a precious quantity of the richest mould, for a future generation of men," a matrix which in the meantime has the power to "rescue new animated parts of the creation from their inactive, chaotic state" (43). In their notes to *Observations*, Thomas, Guest, and Dettelbach credit Forster with extending to the plant and animal kingdoms Lord Kames's "polarity" of "torpid" or stagnant versus "roused" or progressive civilizations (416n2), exemplified for Forster in the contrast between New Zealand and Tahiti. Forster not only encompasses nonhuman nature—even the mineral kingdom, if one looks closely—within a dualism of extreme versus moderate climates (which probably owes as much to Montesquieu as to Kames's *Sketches of the History of Man*). He also insists, in colonial fashion, that "the lord of the creation" is a necessary part of this economy; nonhuman nature can only become degenerated or "deformed by being left to itself" (99). Forster glosses his own allegory of the "changes made on the surface of our globe by mankind" as a contrast between "the wilds of New Zeeland, in the Southern isles" and "Taheitee, the happiest isle," where drainage, clearing, and planting have superseded prehuman chaos (100). In this version of the progress narrative, which happens to track with the *Resolution*'s eastward movement in 1773, human agency supplements or competes with geological agency to correct a degeneration that seems at times—as when Forster refers to Tahiti as a "new paradise" (100)—a secular allegory of original sin, suggesting that in its *truly* primitive state, nature already was humanized.[43]

In surveying all the kingdoms of nature, including "the human species," *Observations* amasses the evidence needed to flesh out the savage/civilized binary into a complex progress from one environmental stage to the next. Moving from the mineral into the plant kingdom, Reinhold seems to identify four stages along the continuum from rude to cultured environments:

> We have observed how much the least attractive of these tropical countries, surpasses the ruder scenery of New-Zeeland: how much more discouraging than this, are the extremities of America; and lastly, how dreadful the southern coasts appear, which we discovered [South Georgia]. In the same manner, the plants that inhabit these lands, will be found to differ in number, stature, beauty, and use. (*O* 113)

Because of its retrospective form, *Observations* incorporates all these stages more fully than George's journal. In the chapter on "organic bodies," Reinhold tends to reverse the order of the ship's itinerary in 1773 from Antarctic waters to New Zealand to Tahiti. By starting with Tahiti and the tropical islands that he views as the cradle of humanity (342), Reinhold constructs an "implied history of gradual migratory degeneration" (417n2), as Thomas aptly puts it. His zoological observations further elaborate this structure: "In the same manner the animal world, from being beautiful, rich, enchanting, between the tropics; falls into deformity, poverty, and disgustfulness in the Southern coasts" (128). Reinhold glosses this particularly succinct (and inverted) version of the progress narrative with illustrations ranging from the delicious hogs of the Society Islands to the "degenerated" dogs of New Zealand (130) and finally the "monstrous" seals of Tierra del Fuego (129). This scheme of classification even allows, paradoxically, for the degeneration of *in*animate nature, for Reinhold claims that by contrast to the North Island of New Zealand, which has some fertile ground, the South Island "still degenerates into rude rocks" (117).

Because it follows the itinerary of the voyage more closely, George's narrative shows more clearly how both Forsters gradually revised their itinerary of environmental stages and modified their accounts of places and peoples to accommodate the diversity that they found. At first, the encounter on Dusky Sound conditions his assessment of New Zealand as a whole, figured as the savage home of what he imagines as the first family (Fig. 1.2). Since the *Resolution* next visited the more populous region of Queen Charlotte Sound (at the north end of the island) and anchored there twice more in the course of the voyage, George had ample opportunities to correct his first impression—both by comparing the two populations (*R* 882), much as Reinhold compares the two environments above, and by revising upward his estimate of the islands' biodiversity (*V* II.616). At first, however, the small population at Dusky Sound seemed to both Forsters to correspond at most to the "small band" level of organization familiar both from Buffon and from conjectural history. Reinhold presents his revision in retrospect by situating the "barbarism" of the Maori, considered as "one of the steps" toward "a better state of happiness" (*O* 210), ahead of the "stupid indolence" of the Fuegians (214), whom the *Resolution* actually encountered twenty-one months later. He places the Fuegians, by contrast, in the same situation as "all the nations found by the first European discovery . . . of America," that of "savages, just one degree removed from animality" (205), using the same language that appears in his preface (*O* 9) to signal an expansion and revision of stadial history.[44]

The use of animal products offers another environmental standard for assessing the movement of peoples in both directions along this expanded continuum. On Tahiti, Reinhold finds luxury articles made of "bone, shark's teeth, &c." together with "delicate . . . victuals"; in New Zealand, some "conveniences," together with palatable food; and at Tierra del Fuego only the "bare necessaries of life," "bit[s] of seal skin" for clothing and meat that he (like Darwin after him) considers "putrid" and "disgustful" (*O* 140–41). Thomas, Guest, and Dettelbach again seek to derive this metric from Kames's *Sketches*, observing that this derivation would be "more clearly articulated" if the *Resolution* had encountered a pastoral society in the Pacific (419n11). In fact, the Scottish Enlightenment model could only have been of limited usefulness for a paradigm that was built on natural history rather than political economy and that responded dynamically to the information provided by each new encounter. Environmental conditions were, for Reinhold Forster, "the true cause" both of civilizational progress and of "debasement and degeneration among savages" (207). The Forsters' conjectural history of the ethnographic present thus unfolds in the course of the ship's itinerary—a spatial movement correlated in shifting and unpredictable ways with distance in time.

The eighteenth-century discourse of conjectural history has a prolonged afterlife in several nineteenth-century disciplines, as Frank Palmeri has shown in detail. Due in part to the strong influence of Adam Smith on stadial theory, the economic aspects of conjectural history persist in the political economy of Ricardo and Marx. The ethnographic form of conjectural history practiced by naturalists like the Forsters persisted in the evolutionary theory of Darwin (as I will argue in Chapter 4), and the social science of Auguste Comte and Herbert Spencer occupies a sort of middle ground between economic and naturalistic paradigms in Palmeri's account.[45] Taking into account these ramifications of conjectural history in the nineteenth century also creates a new context for understanding eighteenth-century figures such as the Forsters and establishes continuity in the conceptual history of deep time. The Forsters' localized, naturalistic revision of conjectural history illustrates its usefulness for framing prehistories of indefinite length, a feature that appeals as well to Buffon, Herder, Blake and other writers examined in the present study. In the Forsters' account of degeneration, conjectural history becomes an occasion for skepticism about the idea of civilization as progress. A deeper past implies greater uncertainty about both past and future.

In June 1773, the *Resolution* proceeded to the Society and Friendly Isles before returning to Queen Charlotte Sound and then making a second, much longer circuit to the south and east that took in a number of new island groups including the New Hebrides (July–August 1774). While these were not the *Resolution*'s only experiences of first contact, Reinhold's and particularly George's narratives make much of the insights gleaned from newly encountered peoples in Vanuatu/New Hebrides, especially Tanna, and afterward in New Caledonia. This stage of the voyage revolutionized the Forsters' stadial environmental history, complicating the binary scheme of "savage" and "cultivated" that had been reinforced by repeated visits to New Zealand and Tahiti and by the Forsters' extensive use of earlier voyage narratives. The active volcano on Tanna, as we have already seen, opened a new time dimension for the correlation of social and natural "improvement"; the encounters with human and nonhuman populations during this phase of the voyage transformed the Forsters' understanding of the topos of traveling into the past and focused attention on ongoing processes of adaptation. The environmental histories of Malekula, Tanna, and New Caledonia were less intelligible by analogy to the ancient past and instead made clear how much of this past, including the quantity of time elapsed, remained unknown.

The encounter on Tanna stands out for its novelty—that is, for its abundance of new natural phenomena as well as for the sociability that mediated the Forsters' knowledge of these phenomena and prompted their rich but inconclusive analysis of the "stage" or state of the environment as a whole. The island offered their first opportunity to study an active volcano as well as especially intriguing flora, such as the wild coconut palm—which, as George notes, they encountered nowhere else—and wild nutmeg. Their keen interest in the latter led to a complex engagement with the bird and human populations that used this seed. Their adventure with the nutmeg began the morning after they abandoned their effort to ascend the volcano, Yasur, having

FIGURE 1.3. George Forster, *Tanna Ground Dove* (17 August 1774), watercolor. George Forster (1754–1794), public domain.

again met with strong resistance from the Tannese on their third and final attempt. In the event, the Forsters (Reinhold particularly) were frustrated once again in their efforts to find the source—the nutmeg tree in this case—and it may be that frustration over the volcano spilled over into this encounter. Searching the forest for specimens, George shot a variety of pigeon (Fig. 1.3) that he thought he remembered from the Friendly Islands (actually the Tannese ground dove, now extinct), calling it "the most valuable acquisition" of the day (*V* II.541). After making this beautiful watercolor sketch (now held at the Natural History Museum in London), he and his father dissected the bird and found two fresh nutmegs in its crop, noting the intact mace surrounding the seeds and even their bitter taste. It was presumably due to the money value of the spice that the Forsters doggedly, but unsuccessfully, sought to intimidate the Tannese into showing them the tree. Michael Hoare calls the whole episode a "comedy of errors" because their first informant took them to be requesting information about the dove instead of the tree and was then accused of misleading them deliberately.[46] They had already discovered that the Tannese coconut palm plantations contained improved varieties of the endemic wild species and surely wanted to know if the nutmeg (probably *Myristica inutilis*) was a cultivar as well. Any stadial assessment of this island's ecology as a whole depended on the answers to such questions.

The role of indigenous people as both abettors of and impediments to discovery (only rarely acknowledged by the Forsters elsewhere) renders the savage/cultivated binary moot. On the one hand, the presence of an active volcano signifies ongoing improvement of the land and perhaps a recent "revolution," as I suggested earlier. Similarly, the presence of the wild coconut palms alongside the cultivars suggested an active process of improvement, probably in its early stages, but impossible to date either in absolute or in relative terms. On the other hand, George notes customs as well as physical characteristics suggesting that the main outlines of Tannese culture may predate the migration of this group or groups (three different languages were spoken) to the island (*V* II.554). In the effort to identify a metric for their civilization (*Maaß der Civilisation*), George articulates a comparative ethnography, noting accomplishments (music, respect for property) and lacks (dress, trade) before concluding tentatively that the Tannese have advanced beyond the Maori and are approaching the degree of gentility (*Grade von Sanftmuth*) exhibited by the Tahitians (*R* 812). The long chapter on Tanna presents a mass of observed detail but also acknowledges many contradictions in its account. The Forsters were delighted with the hospitality of the local people, who rewarded them for desisting from their volcanic expedition with prolonged and repeated exchanges of gifts and song. But islanders' resistance to the European knowledge project made it difficult to gauge the state of their (and their island's) "improvement"—and must have struck the Forsters on some level as a mark of sophistication in its own right. Their unwillingness to trade for food accounts in part for the relatively short stay of two weeks that also limited the opportunity for knowledge-gathering; in this case, George seems naively certain that scarcity of food—hence lack of cultivation—must be the cause. Throughout his chapters on Vanuatu/New Hebrides and New Caledonia, George emphasizes how much of their history must remain unknown and recognizes that the lapse of time required for the natural and social processes they observe cannot be measured or even guessed at.

The attempt to correlate climate and culture on New Caledonia met with even greater difficulties. The very large main island of this group, Grand Terre, was the only island on the voyage that showed no evidence of volcanism, or "of any volcanic production," as George notes with some surprise (*V* II.563). Volcanic activity in the remote or recent past provided a rough measure of "improvement" elsewhere, but here the existence of a complex agricultural society in a rugged and seemingly barren environment posed a new kind of challenge. This difficulty, and the apparent lack of any anatomical or linguistic kinship with either the Vanuatuans or the Maori, led him to declare the New Caledonians "a race of men very distinct from all we had hitherto seen" (588) and to increase his emphasis on the uncertainty of Pacific history (II.563, cf. 573, 592).[47] Their small plantations, surrounded by large tracts of desert or "heath," appear to George so unproductive that he ranks the whole agro-ecosystem only "just above" that of Easter Island; every other tropical island boasted "a variety of fruits" (II.571). Instead of volcanic or humus-rich soils, he notices extensive outcrops of what is now known as ultrabasic igneous rock—George, following Ferber, uses the German term *Gestellstein* (*R* 833n.). His fine mineralogical details, such as the prominence of mica and quartz (but not feldspar) in these rocks (837) and the presence of *Hornfels* (hornstone), with its characteristic small "garnet" crystals and prospects of mineral wealth (840), all support his literary construction of the terrain as unfruitful. Once again, Cook and his men are unable to trade for food, and here George bases his inference of scarcity on the soil, which "poorly rewards" the local farmers' labor (*V* II.574). The New Caledonian plantations show evidence of much more laborious cultivation (*Umgraben und Umwühlen*) than any other Pacific Island (*R* 857). Although neither Forster offers an explicit correlation between the land and the people, it stands to reason that this "primogenial" earth—to recall Reinhold's term for land unaltered by any secondary processes—carries a strong association with the deep geological past, a time predating even the volcanoes.

Comparative Ethnography as Conjectural History

George Forster glosses Cook's choice of the name New Caledonia by explaining that, like Scotland, it is a barren land inhabited by a hospitable people (*V* II.579). It is at least a suggestive coincidence that these "New Caledonians"—the Kanak—occasion some of the Forsters' most refined revisions of the stadial theory associated with the Scottish Enlightenment. While retaining the paradigm of stadial progression, they recognize that progression is nonlinear and identify "microstages" of social development, as I will call them, by comparing many different local populations. On Tanna, George had already noted that a truly barbarous "stage" is out of the question where there are no game animals (II.554), and on New Caledonia he imagines that introducing goats on the island might compensate the islanders for having missed the pastoral stage and "forward civilization among them" (II.593). Since the Kanak have no experience of quadrupeds, they assume that the sailors eating salt beef must be cannibals. Their horror at this sight implies a degree of civilization that renders them, in George's view, "farther advanced" than "their more opulent neighbors" on Tanna, who still practice cannibalism (II.592).[48] He adds, however, that they are "not [yet] sufficiently enlightened to remove the unjust contempt shewn to the fair sex." In these analyses, civilization is decoupled from political economy, or "opulence," and the recognized stages of progress become disordered, fragmented, and ambivalent.

The sheer difficulty of ethnographic classification does, at times, prompt broad generalizations in *Voyage Round the World* and more frequently in Reinhold Forster's *Observations*. George reports, for example, that "perhaps each family" on New Caledonia "forms a little kingdom of its own, which is directed by its patriarch, as must be the case in all infant states" (*V* II.593), thereby attempting to subsume his finer distinctions under a purported Pacific universal. The idea of "patriarchal" hospitality occurs frequently in George's first chapter on Tahiti, where Reinhold also argues that "the beginning of their civil society . . . is of the patriarchal kind" (*O* 223). The term in this sense seems to allude to the Biblical patriarchs; at the end of their first long stay on Tahiti (*V* I.199) and again on Tanna (*V* II.544), George also uses the term *patriarchal* in a more ethnographic sense, to refer to a form of social organization that is gradually being superseded. The Forsters' repeated use of this category may provide evidence that a homogenizing idea of "primitive" peoples is beginning to emerge out of the contradictions between the stadial model and observed local differences. This idea of an aboriginal Pacific population in a generally "infant state" remains in tension with the Forsters' more sophisticated idea of highly differentiated populations resulting from migrations and differing environmental histories.[49]

Some criteria for distinguishing populations evolved as metrics for gauging the "level" of civilization on a given island and became increasingly refined in the latter stages of the voyage. These metrics include refinement of "conveniences" and mechanical arts, complexity of diet, music, and sexual mores; and attitudes toward women and toward cannibalism. These last two are not generally related, but George Forster does make this connection in the New Caledonian example given above. He addresses both issues repeatedly in the latter part of this chapter, which sometimes enters the mode of conjectural history. In one of his most moving descriptions of contact on the island, George describes his repeated attempts to observe three women at work, who communicated by signs that they would be strangled if they were caught with him (*V* II.577). "Considering these humiliations and cruel oppressions of the sex," he then reflects, "we have sometimes the greatest reason to admire, that the human race has perpetuated itself" (II.583). In his detailed description of the incident of supposed cannibalism, he achieves something like a reversal of the gaze, contemplating the sailors gnawing on a beef bone from what he takes to be an indigenous perspective. Here too a larger generalization in the mode of conjectural history follows: "men seem to have had recourse to animal food by necessity at first, as the depriving any creature of life is an act of violence" (II.586). Misogyny provokes an unfavorable comparison between the Society Islanders and Tannese, who are not yet as "advanced" on this score (522); because the latter also practice cannibalism out of an instinctive thirst for vengeance—so George conjectures—they "still belong" to the "class" of vengeful peoples exemplified by the Maori (II.533; cf. *R* 773).[50] The binary opposition between this "class" and the more peaceable stage represented by Polynesia is consolidated in another cannibalism episode, when the *Resolution*'s Raiatean passenger, Maheine, is horrified by symptoms of cannibalism during their second visit to New Zealand (*V* I.279). On their third visit, however, two new factors complicated this binary: first, they had now visited the New Hebrides and New Caledonia, where they had experienced a wider range of attitudes toward cannibalism; second, they gathered evidence that crew members from their companion vessel, the *Adventure*, might have been killed (and eaten, as they learned later) at Grass Cove, near Queen Charlotte Sound, in 1773 after the two ships were separated.[51]

Music and chastity also come into play in the complex four-way comparison among Polynesia, Tanna, New Caledonia, and New Zealand that develops in George's account of this third and final sojourn in Queen Charlotte Sound. Returning to the issue of misogyny at the end of his chapter on Tanna, George finally decided that the Tannese were "gradually advanc[ing]" toward the Polynesian standard after all, because they were "less unjust to their women" than the Maori (*V* II.556). The new evidence of large-scale cannibalism among the Maori tended to support that view. At the same time, however, these two populations both had music that sounded relatively complex and pleasing to the European ear. George and the Forsters' assistant naturalist, Anders Sparrman, spent long afternoons exchanging songs with the inhabitants of a Tannese village (Fig. 3.4), who were especially delighted with Sparrman's Swedish songs (II.534). George in turn credits their songs as harmonically richer and rhythmically more complex than anything they had heard in the Society and Friendly Islands. Recounting these events in sequence, rather than presenting a synthetic judgment, George begins at this moment to reconsider his emphasis on the "vengeful" nature of the Tannese and to credit them with "benevolence" and hospitality (II.535). The highly emotive songs of the Maori fall between this complex Tannese music and the "wretched humming of the Taheitian" on the scale of "genius" (II.615). Here George draws on the observations of James Burney, who recorded Polynesian and Maori songs in detail on the *Adventure*, to upend the usual hierarchy of value. This superior "taste for music" among the Maori, he concludes, is one of the "stronger proofs in favour of their heart" (II.616). Even the cannibalistic attack on the *Adventure*'s boat in 1773 can be justified as an expression of the "right of retaliating injuries" consistent with savage liberty (II.610). When it comes to chastity, however, Tanna and New Caledonia—the scenes of first contact—set the highest standard, while the Maori, the Polynesians, and the Europeans themselves all fall short.

As concerned as he is to do justice to the native genius and the savage liberty of the Maori and to refine his naïve initial account of the "first family" at Dusky Sound, George is unsparing with his disgust in narrating the sexual traffic between Maori women and the ship's company. On the positive side, their instinctive sense of justice and their lack of duplicity provide additional "proofs in favour of their heart," besides their musical sensibility. As Thomas has argued, George was already inclined at Dusky Sound to defend Maori characteristics that seemed to him to express an idea of liberty that was both consistent with the values of an advanced commercial society and harder to attain in "civilized" circumstances. Thomas's juxtaposition of "liberty and license" also provides an effective lens for analyzing the final episode at Queen Charlotte Sound, where the Forsters gathered new evidence of both and of the complexity of the dynamic between them. Music and revenge (or resistance to oppression) become aligned with "liberty" in this episode, while sexual traffic becomes associated with the "license" that George had earlier noted as the cause of extreme and violent misogyny at Dusky Sound.[52] The sensibility of the Maori heart, which he holds up as a rebuke to "philosophers in their cabinets," possibly Rousseau in particular (*V* II.616), helps to explain in some measure the Maori leader's "puzzling" refusal to sail with them to England even though "he felt the superiority of our knowledge . . . and mode of living" (II.615). That superiority is, however, very much in question when George slips from a description of the "inconvenience" of domestic life among the Maori into an invective on the "grovelling appetites" of those shipmates (both officers and men) who entered their smoke-filled houses "in order to receive the caresses of the filthy female inhabitants" (II.612). He praises the natives of the New Hebrides and New Caledonia for "having very wisely declined every indecent familiarity with their guests," unlike both the Maori and the Polynesians, and in his New Caledonian narrative he specifically praises the women's skill at eluding the sailors' advances (II.577–78).

The problem of chastity, and the behavior of sailors more generally, provides an opportunity for both Forsters to explore the degeneration of Europeans, a provocative theme briefly mentioned in the introduction to Reinhold's *Observations*. There, Reinhold pointedly sets his "History of Mankind" apart from "systems formed in the closet or at least in the bosom of a nation highly civilized, and therefore in many respects degenerated from its original simplicity" (*O* 9). This is not quite the "gradual migratory degeneration" (417n2) caused by movement away from the tropics. In the penultimate stage of the voyage, at Tierra del Fuego, this scheme becomes a pretext for highly pejorative racist representations apparently more indebted to Buffon's view of the Americas. George Forster claims that here even the sailors are disgusted by the native women (*V* II.631). Much earlier in the voyage, at Dusky Sound, there is the sense that the Maori there have degenerated only a little from the "original simplicity" of the species, and in a different direction from the Europeans. Even though the environment may be seen to "degenerate into rude rocks," the human population at this microstage seems to have retained something of the "original system" that seems threatened or obscured—as George comes to feel on Tahiti—by increasing refinement.

At Dusky Sound, too, the savagery of the sailors is represented in a more positive light. This chapter makes a very artful transition from "savages" to savage landscapes to the rigors of exploration: after an attempt by the Maori patriarch to anoint the captain with "odoriferous" seal oil, Cook and the Forsters take their longboat up into a rocky fjord where they proceed to sleep rough. These conditions, writes George, "soon taught us to overcome the ideas of indelicacy" attached by "civilized nations" to eating with one's fingers, sleeping in a "wigwam," or even to the "droll" antics of common sailors (*V* I.99). Later on he frames the character of the sailors in terms that also resemble Thomas's dynamic of "liberty and license," underscoring the slippage from "civilized" to "savage": "Though they are members of a civilized society, they may in some measure be looked upon as a body of uncivilized men, rough, passionate, revengeful, but likewise brave, sincere, and true to each other" (I.290). In Reinhold's preface, the term *degeneration* seems instead to refer to European refinement or luxury, and George echoes this sense in his first, more idealistic narrative of Queen Charlotte Sound, where he attributes the appearance of prostitution to previous European contact (I.121–22).[53]

The Forsters' innovative use of conjectural history as a form enables highly nuanced, empirically grounded ethnographic comparison but also requires some commitment to human universals and an overarching chronology.[54] Though it shows him at his most prejudiced, George's reflection on the sexual disgust he feels at Tierra del Fuego illustrates the two aspects of this innovation particularly well. Although the Fuegians present "the most loathsome picture of misery and wretchedness," he notes that they nevertheless wear paint and jewelry, and concludes that "ideas of ornament are of more ancient date with mankind, than those of shame and modesty" (*V* II.628). The reflection implies that the Fuegians are in fact more "ancient" and not, as Reinhold more consistently argues, degenerated from a better ancestral condition. The Forsters' complex awareness of environmental factors and cultural differences does not always cancel the demand for linear progression and universals inherited from conjectural history. In his early chapters, George is especially apt to lapse from the study of customs and manners into generalizations about human antiquity, as when he notes that the Maori observe the custom of presenting white flags (or green branches) as tokens of peace. The universal recognition of these symbols, he writes, implies a "general agreement . . . anterior to the universal dispersion of the human species" (I.100). David Bindman argues that in such moments, Forster "speculates that a folk memory of the original state of humanity might remain with the peoples he encounters."[55] In the case of the unfastidious and "keen appetite" that he shares with the sailors at Dusky Sound, or even the impulse toward contact and trust that seemed to be felt on both sides of the beach, these universals extend to Europeans as well.

The stadially oriented comparison of societies also at times implies a universal human trajectory that populations follow (or diverge from) at differing rates. At the end of his first long chapter on Tahiti, George omits his usual detailed overview of customs and manners, deferring to existing accounts, and instead offers a particularly sweeping history of civil society. He specifies the stage of Tahitian politics, in the manner of conjectural history, as incipiently "feudal" with some "patriarchal" qualities remaining (*V* I.199). Although the forms of feudal despotism vary somewhat on Tonga, George argues that Tahiti and Tonga share "primitive ideas" from a common source (I.260)—perhaps the mother country on the Asian mainland that putatively "colonized" both, or perhaps even an original, pre-"dispersion" society. (Reinhold Forster remains committed to this monogenist view of human origins, but George eventually takes a more polygenist view, as we shall see.) Six months later, on the Marquesas Islands, George is more inclined to contrast Tahiti as a scene of luxury against what seems to him the simpler and happier state of the Marquesans, who are prevented by no refinement from "obeying nature's voice" (I.341). Still later on Tanna, he stresses instead the redeeming features of "abundance" as the economic precondition for women's rights. Making an appeal to political economy that recalls Adam Smith, George argues that Tanna (unlike Tahiti) lacks the population and the advanced division of labor required for "refined sentiment in the commerce of the sexes" (II.537). He goes on to note, however, that even "the savage" (*der roheste Wilde*) is "capable of tenderness," which he then illustrates with the local instance of a Tannese man's attachment to his young daughter. Elsewhere the "savagery" of the Tannese is mitigated by "the custom of making friendship by reciprocal exchange of names" (II.519), which the Forsters seem to regard as a Pacific universal. As their field of comparisons expands, they modify the vocabulary of conjectural history, marking local differences as well as introducing new universals, ranging from customs such as this one to entirely new stages—such as agriculture in the absence of a prior hunting or pastoral stage, a pattern they first recognize on Tanna (II.554).

The Philosophical History of Human Nature

Thus far I have argued that the Forsters' innovations in the history of the species depend on their practice of natural history—a science of secondary importance to most other historians of civil society in this period. It is at least a suggestive coincidence that Buffon himself returned to the "natural history of man" in the 1770s and published *Epochs of Nature* (1778), with its seventh "epoch" devoted to the ascendancy of the human species on the planet, at the same time that the Forsters' narratives appeared. For all his commitment to Linnaean natural history, clearly explicated by Dettelbach (*O* lx), Reinhold Forster felt it was a mistake for naturalists to abandon the comprehensive, philosophical view of nature that Buffon offered, as he made clear in his otherwise mixed review of Buffon's *Epochs*. A similar logic motivates George Forster's insistence on the philosophical history of humankind as a priority for the naturalist. In the preface he defends their work as a "philosophical history" of "human nature" (*V* I.5–6) and later insists in a crucial passage that "observations on men and manners . . . ought to be the ultimate purpose of every philosophical traveler" (I.423).[56] This transitional passage introduces the final year of the voyage, which "teem[ed] in new discoveries"—prior to July 1774 the *Resolution* had mostly landed in places where Bougainville, Cook himself, or others had landed before. Though their work in the first two years was still worthwhile because "men and manners" had been neglected by their "unphilosophical" predecessors—as George reflects somewhat ungenerously—the final year was invaluable above all because of the research on human populations enabled by its multiple, sustained experiences of first contact.

Ethnography itself was very much a pre-paradigm science (to use Kuhnian terms) in the late eighteenth century. The ferment of both geological and anthropological ideas before the formal advent of these sciences produced the thought experiments that are the subject of this book, and in the Forsters' case this openness favored the heightened attention to phenomena that their narratives disclose. George Forster's 1786 debate with Kant on the subject of race helps to clarify the stakes of joining ethnography and natural history. In his polemical reply to Kant's anthropology, Forster points to the philosopher's geographically specific misreading of "Indian" in Pacific travel narratives as evidence that the "natural history of man" has not yet reached the stage of an established discipline, also noting that even established disciplines are eventually transformed.[57] The field evidence that Kant marginalizes is precisely what is needed to give the history of humankind a disciplinary foundation. In the final third of the *Voyage*, George increasingly uses the more established branches of natural history as an alibi or pretext for investigating human varieties in the field. The pretext does not always work, however: he notes ruefully that even though they gestured to botanical specimens they had already gathered, he and his father were turned back from the interior of the island by the native people on Malekula (*V* II.485). Initially, the same thing happens on Tanna. Then, in the case of the volcano, a derailed attempt to study the natural world becomes an opportunity for ethnographic work. In this context, botany, zoology, and mineralogy become subordinate to the study of cultures and peripheral to the narrative (though George's notes often refer readers to his father's *Nova Genera Plantarum* for details). Thus the Forsters insist on a naturalistic context for stadial and other forms of conjectural history, mapping a generative area of uncertainty between geological and species time. Their references to and revisions of stadial theory may be compared to Buffon's use of J. S. Bailly, Nicolas-Antoine Boulanger, and other philosophical antiquaries in *Epochs of Nature* (Epoch VII).

The natural history of the species also provides the context for the Forsters' selective and critical use of classical antiquarianism, introduced in the first part of this chapter. I argued there that even such conventional images as that of Tahitians "singing in the true antient Greek style" (*O* 286) necessarily implied some reflection on human antiquity. The experience of contact led to new inflections of this ancient/native topos and to more critical distance, as is evident in George's criticism of the conventionally neoclassical engraving of the landing at 'Eua (Middleburgh), Tonga, that appeared in Cook's account of the voyage (*V* I.232). He does, however, credit William Wales—the ship's astronomer who attacked the *Voyage* in his polemical *Remarks on Mr. Forster's Account*—with an empirically grounded comparison between Tannese warriors and ancient Greeks. Wales, in a journal entry that George translated as a note to his German text (*R* 741n.), exclaims that watching these men with their weapons made him realize that the warlike feats of Homer's heroes are no poetic exaggeration but rather based on accurate observation. George pointedly uses Wales's testimony to verify the accuracy of his own detailed account of these weapons (*R* 740–42; see also *V* II.819n17). Native men and boys repeatedly demonstrated their use and the Forsters collected numerous specimens, including five distinct types of clubs, some of which are still preserved in museum collections in Göttingen and Oxford. Not only does the image of the warrior band support his conjecture that the populations of Malekula as well as Tanna remain at the "small band" level of social organization; it also evokes neoclassical discourse on athletic exercises in ancient Greece, and specifically Winckelmann's rhapsodies on the athletes' nakedness.[58]

The Forsters responded with alacrity to the nakedness of men on Malekula and Tanna as well as New Caledonia, who all wore a similar style of penis wrapper, even though their physical appearance and languages varied widely. The appearance of the wrapped penis, fastened at the top to a rope girding the waist, provoked distinct interpretations from George in each of the three instances, all involving the putatively modern origin of modesty or "decency." Two of them touch on the Greco-Roman deity Priapus and the phallic imagery associated with him: the Tannese men strike him as "a living representation of that terrible divinity" (*V* II.515), whom he then names explicitly in designating every Kanak man as "a wandering Priapus" (*R* 827). Here the classical association most effectively captures the observer's atavistic feeling of sexual intimidation. It also evokes the contemporary work of adventurous antiquaries such as Richard Payne Knight and Pierre Hugues d'Hancarville, who used Priapic worship to posit a prehistoric fertility cult as the original human religion.[59]

The Forsters reacted strongly to physical differences between the unfamiliar bodies of these Melanesians (as they are now called) and the Polynesians, with whom they identified more strongly. The classics (Tacitus, Pliny, Lucretius, many others) provided them with one rich source of responses to human variety—"race" being, as George later decided in his response to Kant, an unscientific term—while natural and conjectural history provided others. The common denominator among these lines of response is time. When he declares the Kanak on New Caledonia to be "a race of men very distinct from all we had hitherto seen" (*V* II.588), George qualifies this claim by showing the inadequacy of climate, diet, and other conventional explanations for understanding their distinctive characteristics. These could only have developed over "a series of many ages" (*geraume Zeit*) (II.592) from a multiplicity of partly unknown causes. In theoretical terms, this passage represents the culmination of George's evolving thesis that "the history of the human species in the South Seas cannot yet be unravelled" (II.563). It is also marked by the only occurrence of the word *Race* (now *Rasse*) in the German text (*R* 861).[60]

He returns to the problem of what we would call evolutionary time in his later systematic treatment of the term *race* in the essay addressed to Kant. According to this essay, the only empirically sound definition of *race* would be "a tribe of unknown origin" (*ein Stamm, dessen Herkunft unbekannt ist*), though he acknowledges that Kant and others have (mis)used the term to make absolute distinctions.[61] Because skin color and other ethnic markers undergo change through reproduction, "races" can never be considered more than "varieties" in the Linnaean sense. The distribution of current populations demonstrates that they ascend and descend the "color scale" (*Farbenleiter*) continually as a result of climate and migration—but, he adds, "we cannot determine the length of time" required for these mutations, which are not recorded by archives or monuments like those of civil history. (Buffon makes a very similar point about the "archives of nature" in *Epochs of Nature.*)[62] In a passage that partakes of both the mathematical and the Biblical sublime, George backdates the ultimate origin of human varieties to a "period veiled in incomprehensibility"—a temporality that leaves open the possibility for human polygenesis, or multiple origins at multiple points in space and time. Although polygenesis has become associated with the scientific racism of the nineteenth century, it is important to note that for George Forster it does not imply species-like differentiation; on the contrary, human species identity is the premise of his revision.[63] The origins of present "races" are unknown simply because we have no record of the process by which they diverged from ancestral "races." In the *Voyage* he is still more inclined to follow his father's commitment to monogenesis—the more conventional premise of only one original "race." For Reinhold, this paradigm also implies an expanded time scale as the only expedient that would allow an environmental explanation of human variety: "if climate can work any material alteration, it must require an immense period of time to produce it" (*O* 182). This recognition leads both Forsters to challenge other naturalists, including Buffon (*R* 699) and Cornelius de Pauw (67), and to reconsider the causes of "racial" difference, which is ultimately relativized by their attention to geological time and environmental factors. Their willingness to consider an expanded time scale as the explanatory context for human variety unsettles the correlation between skin color and "level" of civilization, though without challenging it explicitly.[64]

To mitigate the anachronism of situating the Forsters in relation to "geological time" and "evolutionary time"—categories that did not exist then in anything like their current form—let me offer a comparison with Darwin. Darwin deployed evolutionary theory consciously for the purpose of bridging two dichotomized scales of time, and he relied on language from Charles Lyell's *Principles of Geology* to make the case that historical time, or phenomena at the eco-devo scale, needed to be reconciled with changes occurring at the entirely different scale of geological time. Without this awareness of radical scalar discontinuity, the Forsters saw the problem of temporal confusion differently and found a different solution. Their scale of societies is avowedly at the same time a scale of values and a scale of time, not strictly progressive but rather designed to accommodate natural processes in the explanation of historical processes. In assessing the contrast between Tahiti and Tierra del Fuego, the question is not merely how long it took Tahiti to reach its current stage of complexity, but rather how long it took one ancestral group to attain the complexity from which we now see these two groups diverging in different ways—and what environmental factors influenced the rate of change.

Reinhold Forster explicitly invokes time along with climate as a factor in human variation when he compares the Fuegians with the Maori, who come to occupy a middle position in the scale. The Maori, he writes, are "more immediately descended from such tribes as had more remains of the general principles of education" (*O* 210), whereas the Fuegians are "descended from a degenerated race, and [hence] become the remote offspring of such happy tribes" (206). In his systematic account of the origin and progress of societies (202–38), Reinhold implies that some of these changes happen on a temporal scale compatible with the historical record, by drawing analogies from ancient history. But he also acknowledges that ahistorical factors such as climate and geology strongly influence rates of change and that dating is limited to "distant guess and conjecture" where customs, manners, and spoken language provide all the available evidence (357). George's contempt for the "animality" of the Fuegians, which stands in stark contrast to his relative open-mindedness toward Pacific peoples, leads him to introduce a historical factor that may intensify the naturalized "degeneration" posited by his father.[65] Only exile, being "the miserable out-casts of some neighboring tribe," could have reduced them entirely to the state of "brutes" (*V* II.632). At the same time, George is the one to draw direct inferences about human antiquity from the Fuegians (II.628), so his account clearly demonstrates the temporal confusion resulting from the gradual divergence of two scales of time.

George Forster's disgust reveals an underlying continuity between the Forsters' disorderly conflation of historical and natural time and Darwin's more systematic innovation of evolutionary time. In 1833, a Fuegian man would not eat Darwin's food, which he found disgusting, and Darwin's reciprocal disgust for the Fuegian's touch made him unwilling to eat the morsel of food the man had touched. Darwin's recollection of this incident in his book *The Expression of Emotion in Man and Animals* helps to explain why he feels as much revulsion as wonder when he remembers the Fuegians in *The Descent of Man* (1871), exclaiming, "such were our ancestors!"[66] The disgust that Darwin and the Forsters share in common, though hardly an honorable emotion, registers the affective power of the voyager's experience of encounter. This experience in the context of biological and geological fieldwork motivated the expansion of the time scale—more tentative in the Forsters' case, eventually more radical in Darwin's. Their shared prejudice against "savages" collided with their observed proximity to other human varieties. This perceived cultural distance, which Darwin was still driven to express in temporal form in 1871—though his own theory taught him to assign roughly the same evolutionary "age" to all members of one species—collapsed in the face of the much greater magnitudes that presented themselves to these voyagers in the form of both geological processes and biodiversity. In this way, the experience of human otherness may be seen as informing the alterity later attributed to geological time. This history resurfaces even in a term of art actually intended by the philosopher Quentin Meillassoux to signify the absolute, prehuman past: "ancestral time."[67]

George Forster's response to the people of Malekula (Vanuatu/New Hebrides) presents a contrasting illustration of the ancestor revulsion that prompted voyagers to reexamine the place of human time in natural history. He gives the visual form of history painting to the scene of recognition here that corresponds to Darwin's realization ("such were our ancestors"). The scene imagined as a painting took place on the second day of the encounter, after some mutual suspicion had given way to a festival atmosphere in which large groups of Europeans and Malekulans were engaged in trade and conversation. George uses the picturesque mode here to signal a resolution of the estrangement of the first day, on which shots had been fired and members of the ship's company had cast aspersions on the people's "ugliness." The incident that inspired George's idea of a "superb historical painting" (*R* 692) was a false alarm; the sudden silence that momentarily interrupted the exchange was only due to a man's gesture with his bow, misinterpreted as hostile. The suspense of this moment unites this diverse crowd in a set of attitudes that "afford a proper subject to exercise the talent of painters and poets" (*V* II.486). As a "history painting," it captures the progress of civilization ostensibly embodied in the contrast between the two "nations"—illustrating the way in which painting could serve as a philosophical idiom for the history of civil society, as Guest has argued.[68] By focusing on the universal expression of alarm, however, George also posits the common humanity of both groups here as a corrective to those who had made an "ill-natured comparison between [the Malekulans] and monkies" (II.481). In fact, George takes considerable pains to correct this mistake and discipline his initial revulsion, as scholars have rightly noted.[69] He goes on to call the Malekulans "the most intelligent people we had ever met with in the South Seas" (II.481), which also constitutes a kind of ethnographic priority claim (cf. II.507). In this instance human animality is assimilated into the heroic present of art.

Malekula, unlike Tierra del Fuego, belongs to that large majority of islands that displayed evidence of volcanism, which provided the main nexus of natural and human history in the Forsters' narratives. In the second section of this chapter, I argued that the Forsters recognized volcanoes as involving human populations in a cycle of "revolutions" that extended backward into prehuman time. Although volcanism was a "secondary" process in the language of the period's earth science, and therefore an agent of improvement, many volcanic rocks were also closer in form to the "primitive" rocks than to the stratified secondary rocks. George Forster finds symbolic associations between volcanic landforms and ostensibly primitive features of populations all the way through the voyage, from Madeira, where the *Resolution* and the *Adventure* first anchored after they set sail from England in 1772, to Ascension, the *Resolution*'s second-to-last port of call in May of 1775. The latter island is actually uninhabited, and so devastated that Reinhold Forster is moved to describe it (in allusion to a Roman poem on Aetna) as "nature in ruins" (*O* 110), but George feels that it "might shortly be made fit for the residence of men" (*V* II.670). He may be recalling the Tuamotu Archipelago, where small human populations sustained themselves on circular reefs rising only a few feet above sea level. Reinhold Forster was particularly fascinated by these "low isles" (atolls) and all islands newly "raised out of the sea" (*O* 108), whether by "subterraneous fire" or by coral "animalcules" (in fact both are at work in most atolls). George's emphasis on the "scanty" subsistence eked out by the "shy and jealous" populations on these atolls—and on the obscurity of their history—mirrors the raw state of their environment (*V* I.141). The work of the "animalcules" and the other organisms that gradually accumulate on the reefs promises a kind of "improvement," but at such a slow rate that it seems unlikely to produce adequate ground for civilization in the scope of historical time. Many theories of the earth, from Thomas Burnet's through to John Whitehurst's in 1778, held that land first rose out of the sea in the form of low islands. The population on the atolls, therefore, provided the Forsters with a particularly vivid image of prehistory in real time.

Monsters of a New Museum

Even today, many Cook voyage artifacts remain in natural history museum collections, though modern curatorial practices have done much to counter the naturalization of Pacific antiquity. The Forsters collected extensively in the Pacific. They selected numerous objects for the University of Oxford, and that collection, now at Oxford's Pitt Rivers Museum, remains one of the most notable of all Cook voyage collections.[70] Collecting was a key part of the Forsters' practice as naturalists, though this chapter has focused mainly on their observations in the field and the ways in which their ethnography was shaped by encounters with people in the course of their geological, botanical, and zoological observations. Collections from the colonial era will take on a more central role in the next chapter, which will focus on Buffon, Cuvier's predecessor at the botanical garden in Paris that was and is the home of the French national natural history collections. The British Museum, which housed all British national collections until the separate Natural History Museum was opened in 1881, provided an even more striking array of the spoils of empire, with Cook voyage artifacts and antiquities from around the world cheek-by-jowl with natural history specimens—including, by 1819, a giant fish-lizard discovered by Mary Anning. This fossil skeleton or others like it may have been on Lord Byron's mind when he responded to Cuvier's work on fossils by speculating in verse about future "monsters of a new museum."

By the time Cuvier established the history of life on a sound geological footing, the Forsters' attempt to establish a longer time scale for the history of species, together with their emphasis on volcanoes and volcanic islands, began to seem outdated. The whole secret of deep time—of truly geological time, as Cuvier understood it—was to recognize that the human species was a late arrival on the scene. As far as Cuvier was concerned, there were no human remains to speak of in the fossil record, and it was clear that the advent of the species postdated the last great revolution in nature. Moreover, each successive revolution and the new "creation" that followed it seemed to be smaller and weaker, implying both that proto-humans could not have survived the massive earlier revolutions and that an era of relative geological stability had arrived.[71] Byron was deeply impressed by Cuvier's new time scale, as we know from his drama *Cain*.

In *Don Juan*, Canto IX, however, Byron cast a more skeptical eye on Cuvier and found that he had evaded the real challenge of positioning the advent of the human species within a longer time scale. In his eagerness to make an absolute distinction between the paltry monuments of human antiquity and the majestic "ruins of the great Herculaneum overwhelmed by the ocean"—that is, the fossil record—Cuvier precludes the possibility of biological continuity between humans and earlier species.[72] Byron had no more of a notion of human evolution than the Forsters did, but he understood intuitively, as they did, that there had to be more than "traditions" and "records" (*monumens*)—the conventional evidentiary domain of antiquarianism to which Cuvier himself tried to confine any inquiry into human origins. Thus Byron, in what might be considered a riposte to Cuvier's *Recherches*, imagined a future geological record that included modern Europe, now relativized in relation to "other relics of a former world." He imagines that his culture's remains, too,

> will one day be found
> With other relics of a former world,
> When this world shall be former, underground,
> Thrown topsy-turvy, twisted, crisped, and curled,
> Baked, fried, or burnt, turned inside out, or drowned,
> Like all the worlds before, which have been hurled
> First out of and then back again to chaos,
> The superstratum which will overlay us.

In this context, Byron continues, the fossil skeleton of George IV will appear to inhabitants of "the new creation" as mammoths and "wingèd crocodiles" do to us, thus making the king and his contemporaries the future "monsters of a new museum."[73] As vast as deep time may be, it must be conceptualized in relation to the *longue durée* of the species. The idea of deep time takes shape as the human species is gradually integrated with the geological record, both past and future. Hence my argument in this chapter that deep time, so often constituted in opposition to human historical time, also has roots in the ethnographic and aesthetic experiences of the voyagers, whose "history of mankind" buckled under its efforts to fold in the histories of others.

2

The "Profoundest Depths of Time" in Buffon's *Epochs of Nature*

"That Beautiful Romance"

In their Pacific travels, the Forsters arrived at a deep qualitative understanding of the variability of geological processes and their potential implications for human origins and migrations. Though recognizing that they could not gauge the immense lapses of time implied by these phenomena, they did not radically revise their estimate of human antiquity. Reinhold Forster's resistance to chronological innovation clearly informs his review of *The Epochs of Nature* (1778) by Georges-Louis Leclerc, Comte de Buffon (1707–88), a review published in German just after the family returned to Germany for good in 1779. This review takes issue with Buffon's radical notion that human beings were a late arrival on the scene of antiquity. More like a writer of romances than a naturalist, in Forster's view, Buffon defines the limits of the primeval ocean, produces now-extinct megafauna, and then causes the continents to separate before allowing humans to exist. "At last human beings too become inhabitants of this earth," Forster concludes sarcastically.[1] Zealous to prove that that the written record offers a sufficient account of human antiquity, he disregards the uncertainty concerning human origins that is built into Buffon's account.

The uncertainty is part of the point, however, and Buffon reflects explicitly on the instability produced by situating human origins in a diachronic history of species. *Epochs of Nature* is one of the first geochronologies to insist on a comparatively long prehuman past, and Buffon anticipates some of his contemporaries' objections by establishing a human epoch, a seventh and final age "in which Man assisted the operations of Nature." In the chapter dedicated to this "Age of Man," he locates the first advanced civilization precisely at the date of 7000 BCE—already a bold move. But in discussing the earlier epochs, such as the fifth, in which "the elephants [i.e., mammoths], and other animals of the south, inhabited the northern regions," Buffon raises the possibility of a much earlier human presence, of beings roughly corresponding to what we now term "early hominins."[2] The present chapter examines Buffon's contribution to the history of deep time in *Epochs of Nature*. To Buffon's contemporaries, it was not enough that he affirmed a modern era of European dominance, though in our time he is apt to draw criticism precisely for his anthropocentrism and his colonial mentality. Buffon's racism and colonialism are complicated, though hardly excused, by his argument for a long transitional period between the origin of life and the dominance of the human species.[3] I am concerned here with this transitional period, regarded by many contemporaries as destabilizing the Great Chain of Being even more than the empirically quantified 75,000-year geological time scale or any other controversial innovation of *Epochs of Nature*.

A prehistoric phase follows necessarily from the clearly demarcated prehuman time that sets Buffon's chronology apart from that of the Forsters. My argument in this chapter is that Buffon's geological turn in *Epochs of Nature* is simultaneously a prehistoric turn; the "abyss of time" commonly associated with this work incorporates uncertainties surrounding human origins as well as new fossil evidence of the emergence and apparent extinction of prehuman fauna, the "gigantic animals" mocked by Forster in his review. Whereas the Forsters are content to situate the human species biogeographically, using contemporary field evidence and the written record, Buffon must at times resort to "romantic" conjectures about a multispecies prehistory, as necessitated by his commitment to prehuman time.

I touched on the Forsters' often vexed relationship to Buffon in the last chapter. They disagreed with him strongly on the geohistorical force of volcanoes, critiquing the views put forward in Buffon's much earlier *Theory of the Earth*.[4] They were more content to cite his *Natural History of Man* of the same period, sometimes critically and sometimes less so. Buffon returned to both topics in *Epochs of Nature*, begun while Cook's second voyage was still underway and drawing material from other recent voyages.[5] Despite their differences, Buffon and the Forsters separately recognized that the geological record had to play a role in the "natural history of man," a science that was still very much in its infancy, as George Forster pointed out later. Considering the origin of human races from the vantage of "a point in time entirely veiled in incomprehensibility" in a later work, the younger Forster eventually decouples ethnography from the historical record, dispensing with the certain dates that he believed only this record could provide.[6] Buffon goes even further, insisting that at least a firm minimal date (or terminus ad quem) is needed for natural processes to avoid "contradicting . . . the archives of nature," even though these lack the specificity of the written record (*Z* 37/*EN* 43). Like the Forsters, then, Buffon insists on a naturalistic context for the history of the human species. Unlike George and unlike most of his contemporaries, he is willing to attach hard dates—at least what he regards as the most recent possible dates—to events preceding the written record, events categorically excluded from the natural history of man by Reinhold in his review. Reinhold Forster's conservative critique makes an apt starting point because it draws attention to the transition between geology and prehistory that is one of Buffon's key contributions to the history of deep time.

Widespread criticism of Buffon's speculations aligns neatly with the Enlightenment division between empiricism and theory. As field naturalists, the Forsters identified more strongly with the taxonomic practice of natural history set forth by Carolus Linnaeus, widely understood at the time as opposed to the more philosophical history of nature compiled by Buffon in thirty-six volumes (*Epochs of Nature* was the twenty-fifth volume to appear). For related reasons, Martin Rudwick positions Buffon as a "geotheorist" and therefore a secondary figure in the development of geological time.[7] Forster's review, however, recognizes and questions the strong binary division between empirical and theoretical natural history, despite his criticisms. Buffon's literary style is a major locus of these criticisms. Several early readers of *Epochs of Nature* took him to task for its romantic style, as Forster does, but Forster also recognizes that a certain stylistic flair is necessary for the realization of a natural history with philosophical ambitions. Forster here charges Buffon with overreaching these ambitions but does not question their validity as such.[8] In the last chapter I detailed some of the Forsters' own philosophical ambitions, especially with respect to the "natural history of man," and these help to account for the favorable references to Buffon, in Reinhold's work especially, in spite of its Linnaean commitments.

Forster and other readers would have liked to see a stronger line of demarcation between the romantic or novelistic style and that of the philosophical naturalist, but today the affiliation between these two styles is more apparent. George Forster's *Voyage Round the World* remains central to the German literary canon, partly for stylistic reasons that may reflect an unacknowledged debt to Buffon. Benoît de Baere has pointed out that Buffon's pursuit of a vivid style—the more strongly attached to his authorship by his famous maxim, "the style is the man himself"—is essential to the work of analogy that informs his scientific arguments. From the point of view of literary history, Ian Duncan has argued that the nineteenth-century novel as a form continues the Enlightenment project of the "natural history of man" in an analogical mode for which "Buffon's *Natural History* . . . set the precedent."[9] The present study is centrally concerned with the literary resources that were developed to effect the translation of prehuman time into a spatial register, deep time, so Buffon's style is of positive importance here as well. Joanna Stalnaker observes that although "Buffon's status as a literary writer" remains unquestioned, many of his "imaginative tropes . . . have not yet received the critical attention they deserve." Rudwick is right to point out that Buffon's "epochs" are largely ahistorical—a criticism that has also been applied to Buffon's entire project in the *Natural History*—but his willingness to specify a concrete sequence of changes in the earth's formation as well as its biota nonetheless made him a "pioneer" in the "discovery of time," as Stephen Toulmin and June Goodfield noted in 1965.[10] Similarly, Patrick Wyse Jackson positions Buffon's geochronology "among the most important measures of the earth's antiquity." It seems fair to conclude, with Jan Zalasiewicz and his colleagues, that Buffon's *Epochs* should count as the "first Earth history," in the sense of a geochronology employing some of the units and criteria still in use today, while granting Rudwick's objection that it contributes little to "geohistorical science" as such.[11]

Forster's gently mocking review helps to show that Buffon's contribution is not solely a matter of style, for he identifies Buffon's stylistic overreach with several substantive topics that remain integral to deep time narratives today. For Buffon, deep time is a multispecies affair, punctuated by climate change, extinction, tectonic displacements, and other global-scale changes. The discrepant time scales assumed by modern science were first sketched by Buffon through the history of species, from fossil plants and shells to living organisms, and he argued boldly that the history of other species is indispensable for arriving at a conception of geological time suited to human understanding. In his review, Forster addresses Georg Christoph Lichtenberg, the editor of the journal where it appeared, directly: "No doubt you will already have read Buffon's recently published *Epochs of Nature*, that beautiful novel [*Roman*]." Jacques-Henri Meister, one of the first French reviewers, had called it both a "sublime romance/novel [*roman*]" and a "beautiful poem."[12] These and other critics were especially dismissive of Buffon's conjectures concerning the origin of the earth as an incandescent mass dislodged from the sun by a comet (reviving a century-old hypothesis; see Fig. 1.1), and Rudwick similarly uses Buffon's emphasis on planetary origins to distinguish his theory from geohistory. This model calls for a long period of cooling, which Buffon calibrated experimentally at 35,000 years by using white-hot globes of iron. Forster protests loudly: "This man paints us the glowing earth, he calculates when and how it assumed its spheroidal form, he calculates its refrigeration." Forster objects to the long deferral of the origin of life necessitated by this model, and particularly to the deferral of human origins, signaled by his exclamation, "at last." He also objects to the fairy tale (*Mährchen*) concocted to explain volcanic episodes, megafloods, and other phenomena that are adequately documented (so Forster believes) within recorded ancient history.[13] Forster appears to dismiss out of hand the "gigantic animals" or megafauna that populate Buffon's "fairy tale," though this line of objection would become less tenable after Cuvier's much more extensive work on fossil bones in the next generation.

Gigantic plants and animals are crucial actors in Buffon's experimental prehistory, bridging the distance he seeks to establish between geological time and human history. *Epochs of Nature* is one of the earliest histories of the earth to invoke deep time in the sense in which it is now most commonly understood, as geological in contradistinction to historical time. At the same time, it is close enough to the Enlightenment genre of conjectural history to explain why the expression continues to invite metaphorical uses today, such as "literature through deep time" or "the deep time of the media."[14] It makes sense to reserve "deep time" in the strictest sense for geological discussion, since the recognition of the earth's history, as differing from other histories in scale, depends on it. Temporal "depth," however, comes to geology as a metaphorical legacy along with others, such as (following Stephen Jay Gould) "time's arrow" and "time's cycle," and analogies between geological and other orders of "deep" time remain essential to the histories of species and civilizations. In twenty-first-century parlance, the Anthropocene names an encroachment of human history on the deep time of geology, and to the skeptic this encroachment may well signal category confusion.[15] Buffon himself invokes the paleoanthropological and art historical registers of deep time in constructing his innovative geochronology (a geological time scale, though still far short of the modern GTS in extent). My reading attends to these registers partly because they are resurfacing powerfully at a time when human and natural history seem again to be collapsing into one. Theorists in the Anthropocene, including Daniel

Lord Smail and Andrew Shryock, are likewise turning to human antiquity, or prehistory, to locate a threshold for contemplating human and prehuman scales at once.[16]

The famous opening of *Epochs of Nature* compares "Earth's archives" with the deeds, coins, and "antique inscriptions" of civil history (*Z* 3). This pervasive and productive homology—predating Buffon but rendered by him in an influential new form—enabled the retrieval of nature's revolutions from the "dark abyss of time," as Paolo Rossi has shown.[17] Deploying the nascent visual language of geology, Buffon argues that each epoch has its geological signature. Stratigraphers of the Anthropocene insist on the anthropogenic carbon or radionuclide signature as a legitimate boundary for a geological epoch. The model of inscription carries over from natural history into geology and operates across scales, unlike Bruno Latour's "modern constitution" or Michel Foucault's "discontinuity of living forms," both of which situate the modern/premodern binary entirely within the historical time of human agency.[18] From the geological perspective of deep time, modernity begins with the Cenozoic Era, named after the "young life" that sprang up after the renovating asteroid impact of 66 million years before the present (BP). On the more local scale of epochs, both the Pleistocene and the Holocene mark human time, but human evolution does not provide the main criteria for dating either. Only the Anthropocene will depend centrally on human inscription, but even here modernity is understood through the capacity of differentially resistant media (the atmosphere, polar ice, coral reefs) to manifest traces of this inscription.[19]

Prehistory is implicit in any hard distinction between geological and historical time. Buffon's recognition of these three stages differs from the empirically driven "geohistory" defined by Rudwick, but at the same time it departs from the "eternalism" of other old-earth cosmologies, which did not traditionally allow for prehuman time.[20] By marking the distinctions between prehuman, proto-human, and historical phases, Buffon constructs a paradigm in *Epochs of Nature* that is essential for a conceptual history of deep time. It offers a form of temporalization, like the modern concepts studied by Reinhart Koselleck, since the first two stages are backdated relative to a modernity that can be defined either as the advent of the first civilization or as the advent of global human dominance (Buffon entertains both possibilities).[21] By taking up this challenge of situating human time over against geological time, yet insisting (unlike Koselleck and other theorists of modernity) on the historicity of both, Buffon also manifests the "geological agency" so much at issue in the Anthropocene, an agency that is implicit in geological periodization itself. In addition to material for a history of deep time, *Epochs of Nature* offers lessons for the Anthropocene, which will be a secondary concern of this chapter—partly because the reception history of this work, which I have been tracing from the early reviews, has entered a prominent new phase associated with the Anthropocene.[22] Buffon's geological turn, epitomized in this remarkable work, is equally a prehistoric turn because it preserves the ethnographic content of exploration narratives like the Forsters' while incorporating it into a history of other species and ultimately of the prehuman world. Buffon brings to this task both new ways of conceptualizing fossil evidence and a richly figurative stylistic repertoire.

Assisting the Power of Nature

Recent scholarship in literary history has proposed the name "early Anthropocene" for the early industrial period associated with the late Enlightenment and with Romanticism.[23] *Epochs of Nature* responds to this contextualization particularly well, partly because it defines anthropogenic climate change as the marker of Epoch VII, the "last" epoch, "when the power of man assisted the operations of nature." Fossil plants, and coal in particular, are major actors in Buffon's multispecies romance, coming into their own as drivers of benign climate change, "assisted" by human extraction. The editors of *Epochs of Nature*—the first complete English translation of the work, published in 2018—caution readers against concluding that the Anthropocene was anticipated by Buffon, and scientific consensus has clearly moved on from the early industrial start date proposed by Paul Crutzen and Eugene Stoermer when they introduced the Anthropocene concept.[24] *Epochs of Nature* first appeared in 1778, anticipating James Watt's rotative steam engine (the basis for Crutzen's date) by six years. Rather than portraying Buffon as an Anthropocene thinker, it will be productive to identify what he has in common with Anthropocene theorists as well as their precursors in the nineteenth and twentieth centuries. Like these later geochronologies, Buffon's engages the narrative "challenges of scale" identified by Rob Nixon with environmentally diffused agency or "slow violence," and it merges the consideration of a new geological epoch with a new historical epoch of reflection on the environmental history of the species.[25] In its binary construction, Buffon's chronology also offers a precedent for the new anthropogenic epoch, which responds to new pressures but repeats the gesture of locating history in cosmic time.

To begin with the end, Buffon's Epoch VII offers the concept of a geological period defined by human agency and grounds it empirically in the growing use of fossil fuels. If Buffon noticed a correlation between this trend and changes in air quality—as some of his contemporaries did—it did not shake his belief in progress.[26] While differing on the scope of prehistory, Buffon and Forster both share this faith in "improvement" and derive evidence for it from the correlations they posit between unaltered landscapes and "primitive" peoples. Although they are not named in Buffon's abstract of Epoch V, Buffon's exposition, including his equivocal revisions to the original manuscript text, reveals that human actors initiate the entanglement of "deep" and "human" time at this very early stage. Here, again, is his abstract of all seven (see Fig. 1a above):

> EPOCH FIRST.
> When the Earth and Planets first assumed their proper Form.
> EPOCH SECOND.
> When the fluid matter consolidated, and formed the interior rock of the globe, as well as those great vitrifiable masses which appear on its surface.
> EPOCH THIRD.
> When the waters covered all the Continents.
> EPOCH FOURTH.
> When the waters retired, and Volcano's began to act.
> EPOCH FIFTH.
> When the elephants, and other animals of the south, inhabited the northern regions.
> EPOCH SIXTH.
> When the Continents were separated from each other.
> EPOCH SEVENTH, and last.
> When the power of Man assisted the operations of Nature.

In his preliminary discourse, Buffon deduces these epochs from five axiomatic geophysical facts and five "monuments" (fossil bones and shells) intended to secure his goal of "form[ing] a chain which, from the summit of the time scale [*du sommet de l'échelle du temps*], shall descend all the way to our present."[27] The chain that links these seven revolutions—indebted metaphorically if not substantially to sacred history—is forged of geological evidence. A quasi-geological series of inscriptions enables the retrieval of widely distributed agencies—human, mammoth, volcanic—from the "darkness of time" (*la nuit des temps*; *Z* 4). Buffon's strong antiquarian interest in inscriptions, his half-ironic allusion to the Great Chain of Being, his early use of the term *time scale*, with all its metaphorical complexity, and even his anthropocentrism—all these are formidable contributions to the verbal idiom of deep time.

The abstract specifies only "elephants" (mammoths) and humans, but the diversity of prehuman life, beginning in Epoch III, proves to be a critical factor for subdividing and measuring the epochs of nature. Buffon's abstract belies the extent to which he depends on the fossil record for evidence of the *effects* of epochal changes in the planet's condition. This complex history culminates in the epoch-making activity initially assigned to "elephants" (V) and humans (VII). In Epoch III, we learn that the primordial ocean very soon produced marine organisms, "the first inhabitants of the globe," whose fossil shells Buffon regards as being deposited in situ and marking the highest ocean level. Terrestrial life, especially plants, came into existence at about the same time, roughly 35,000 years BP, on the exposed mountain peaks. But even though he says that "Nature spread the principles of life" from the sea to the exposed land (Z 54), Buffon attempts no explanation of how these "countless" (Z 51) early creatures "were born" (*ont pris naissance*; Z 60/*EN* 98), as he puts it in a typically vague formulation. He turns to the subject of human life first at the end of his discussion of Epoch V, indicating that human beings received "the breath of life" last of all (Z 98), but not long after the mammoth and other megafauna of the mountains of northern Asia, and in the same place. His Biblical language and the implied analogy between geological epochs and the seven days of Creation (not in itself original to Buffon) did little to mitigate this radical departure from Biblical chronology, and we should not be misled by the lower magnitude of Buffon's expansion (from thousands to tens of thousands, rather than the billions of years we now expect) into missing the depth—or rather, the depthlessness—of his "dark abyss of time." Nor was he "moved by religious motives when he lowered his estimate" of the age of the earth in moving from manuscript to print: "for theologians, 75,000 years was as scandalous as ten million," in Jacques Roger's words.[28]

The association between early hominids and megafauna not only backdates humans by dissociating them from modern animals; it is also proto-evolutionary in the sense that a time of human origins is distinguished from a time of human sovereignty, which in Buffon receives an epoch of its own.[29] Just as theorists of the Anthropocene have explored a range of boundaries ranging from the late Pleistocene to two markers associated with Buffon's time—colonial expansion and industrialization—before settling on the mid-twentieth century, so too Buffon entertains possible dates for Epoch VII as far back as 7000 BCE before finally settling on about 1000 BCE (while also noting the perfection of human control in his own present day). He begins his final chapter by outlining a savage state that evidently dates back to the end of Epoch V, making it roughly contemporaneous with the giant fossil bones that Russian and American naturalists had been sending to Buffon in his capacity as director of the Royal Cabinet (now the Museum of Natural History). Epoch VI calls for human migrations postdating (and perhaps even predating) the separation of the continents (*Z* 112, 175) and corresponds in some ways with much later ideas of human prehistory. The proper start date of Epoch VII—added in 1776, seemingly after the work was already complete in six epochs (*EN* xxxv)—clearly postdates early human evolution for Buffon. In a series of thought experiments he entertains origins progressively closer to the present. Epoch VII might begin with the "great increase in population" that succeeded the band-level organization of hunter-gatherer societies—a form of organization that Buffon, in a classic temporalizing move, equates with the form "we still see today among the savages who wish to stay savages," like the Hurons of North America. These "small nations" of "men just emerging from the state of nature" first gave way to population centers in "the northern countries of Asia" perhaps 9,000 years ago: "from this moment the earth became the domain of man. He took possession of it by his work of cultivation" (*Z* 119–20).[30] The temporalizing equation of modern "savages" with prehistoric ones—a strategy inherited by Darwin and his contemporaries as the "comparative method" in anthropology—allows Buffon to amplify distance in time via cultural difference. Unlike more conventional descriptions of progress from the pastoral to the agricultural stage of civilization, Buffon's account incorporates these human revolutions with the deep history of nature.

Geochronology provides a measure of stability for this unsteady cultural genealogy of early humans. The "convulsive motions of the earth" that still terrified the first humans or hominins lingered in memory and slowed their "emerg[ence] from the state of nature" as the planet continued to cool. (In Reinhold Forster's account, by contrast, the memory of a "great revolution" of nature preserved in Tahitian myth (*O* 112) would not fall outside the scope of historical time.) Moreover, the first civilization's "possession" of the earth would have been merely temporary since, as Buffon argues, it would have been totally destroyed by barbarians encroaching from the north as the subarctic regions became too cold to inhabit (*Z* 123). He estimates the intervening Dark Ages at another three millennia, inferring that "it is thus only since around thirty centuries that the power of man has been united with that of Nature, and has extended over the greater part of the earth" (124). 1000 BCE therefore provides another putative start date for Buffon's Age of Man or perhaps of a second, more durable quasi-Anthropocene phase, since "the *entire* face of the earth today [1778] carries the imprint of the power of man."[31] Buffon thus sets human time proper against a deeper (or medial) past in which humans still had to contend with the dynamic processes of a younger earth. His first prehistoric or proto-human period may be correlated with the youngest extinct volcanoes, which precede the durable marks left later by agriculture and those being inscribed in Buffon's present by burgeoning industry. This sequence reflects the changing resistance of the earth as medium or its domestication.

In the absence of any clear (archaeological) concept of prehistory, Buffon shows understandable ambivalence about the period between his two alternative start dates, a time with no historical record. Following the latest antiquarian scholarship, Buffon reclassifies the "lightning stones," long understood as natural phenomena, as flint tools and weapons made by the first independent "savages" (*Z* 119–20). This diachronic narrative is complicated by a synchronic Enlightenment alarm concerning superstition. The fear of earthquakes lingers, and "even at present, men are not entirely emancipated from these superstitious terrors."[32] Though Buffon's account is pejorative, this lingering terror anticipates in some ways the idea of evolutionary memory that is, in my view, embedded in the Anthropocene, which looks back to a stage of human evolution characterized by a more modest ecological footprint. Buffon's insistence that early but temporary high civilizations in northern Asia followed this barbarous stage—despite the lack of written records—might be seen as an early experiment in "deep history," which Smail defines as reaching seamlessly from the archivally recorded into the evolutionary past. (Deep history, too, offers us a kind of middle distance between history and deep time in its purely geological sense.)

Buffon has an incipient sense of fossil and archaeological evidence but no sense (of course) of DNA evidence, to name the three kinds of deep historical inscription on which Smail builds his argument.[33] Buffon's insistence on "monuments" as evidence owes something to the affinities between natural history and antiquarianism. Presenting origins in material form, his foundational monuments—giant fossil bones in northern lands and in both hemispheres as well as shells far from the sea, not always matching living genera—establish the unique geohistorical importance of fossil evidence, and his discussion of stone tools recognizes what would later be regarded as archaeological evidence. Buffon's reliance on an astronomical calculation, "the lunisolar period of 600 years" alluded to by Flavius Josephus (*Z* 121), as his sole evidence of a high prehistoric civilization, looks flimsy next to Smail's provocative use of genetically based paleo-anthropological research. But in the absence of carbon dating and DNA analysis, Buffon's apparently fanciful deduction of prehistoric astronomy from this "lunisolar period" may be the closest thing available to the kind of quantitative deep historical evidence that Smail defends. In their different idioms, both authors recognize that too sharp a divide between premodern and modern humans might derail the project of incorporating the human epoch into the epochs of nature.[34]

Nonetheless, Buffon cannot resist the appeal of a human-dominated epoch. One mark stands out preeminently among the "marks of his power" now exhibited by "the whole face of the earth": climate change. Buffon reasons as follows: "With all beings endowed with progressive movement being themselves nothing so much as little furnaces of heat," it follows that "the local temperature of each terrain" depends on "the number of men and animals" inhabiting it (*Z* 127). After surveying the microclimates of the Paris suburbs to introduce his subject, Buffon concludes decisively that

> many other examples might be given, all concurring to show that man can have an influence on the climate he inhabits, and, in a manner, fix its temperature at any point that may be agreeable to him; and, what is singular, it is more difficult for him to cool than to heat the earth. (*FA* 400–401/*Z* 128)

Buffon does, in fact, extend the scope of this ultimately global argument, which becomes the main focus of his chapter on the seventh and final epoch. His sultry and populous earth seems appropriate for the dawn of the Anthropocene—or the culmination of the "natural history of man," to use the vocabulary of that period. His interest in atmospheric physics and natural processes in general—as much as he overstates human environmental impact—distinguishes Buffon's account from the stadial histories of civil society. This fantasy of civilization as a furnace clearly portrays an agrarian society but also carries evidence of urbanization, the concentration of humans and draft animals that made warming more efficient.

Directly and indirectly, this inscription of progress operates at the expense of "savages" ancient and modern, who impede the project of world domination. Famously, Buffon used the lower population density of non-European places as evidence for a biogeographic and racial hierarchy of populations.[35] A peculiar return to myth that might serve to distinguish Buffon's from other typical Enlightenment projects—a "recursive moment," to use the well-known vocabulary of Adorno and Horkheimer—arises with his notion that resource extraction benefits nature, already hinted at in the chapter heading for the seventh and final Epoch: "When the power of Man assisted [*a secondé*] the operations of Nature." This new myth provides his ideological justification for the "total domination of nature *and* of humans" that Enlightenment aims for, according to Adorno and Horkheimer's critique.[36] Buffon himself protests that colonialism is only an unfortunate byproduct of biodiversity. Whether or not we take social inequality as incidental, the physics of climate change provides the infrastructure: "It is upon the difference of temperature that the stronger or weaker energies of Nature depend. . . . Hence man, by modifying this cause, may in time destroy what injures him, and give birth to everything that is agreeable to his feelings" (*FA* 402). By "multiplying the species of useful animals," Buffon further explains, "man has increased the amount of movement and life on the Earth; he has at the same time ennobled the entire series of organisms and ennobled himself in transforming plant into animal and both into his own substance" (Z 130).

Because domesticated species are "more useful to the earth," climate change becomes an ethical project as well. Buffon dismisses living non-European peoples on this basis: "these half-brute men, these unpolished nations, large or small, only weigh down the globe without comforting the Earth" (Z 125). Critics beginning with Jefferson have shown Buffon's theories of climate and race to be thoroughly complicit with colonialism, though it is also worth noting his explicitly anticolonial sentiments at the close of *Epochs of Nature* (Z 125–26). In Alan Bewell's reading, Buffon's idea of "brutish men" implies that colonial nature "not only needs but wants to be civilized." Hinting at the *longue durée* of this climate change mentality, Bewell further notes that "Americans did not simply reject colonial biogeography, they learned instead to employ it for their own expansionary purposes." Smail and Shryock, however, implicate the whole species by insisting that "scalar leaps in extractive capacity" have been a recurring pattern since the Paleolithic.[37]

Buffon also believes that resource extraction "assists" nature in the more encompassing sense that it enables the progressively fuller inscription of life onto the planet. The earth, for Buffon, originated in molten form and its subsequent history can only unfold as a gradual, inevitable cooling. Living beings strive to slow this cooling and enjoy considerable success—this is one reason why Buffon himself would object to Percy Bysshe Shelley's description of his model as a "sublime but gloomy theory." In the Anthropocene context, the barely superseded idea of the Holocene as just another interglacial episode may seem similarly comforting, if for somewhat different reasons.[38] Buffon's paradigm is topical for us because it offers a clear view of why global warming is so easy. On the premise that humans and animals "may be considered as so many little fires," Buffon goes on to credit agricultural civilization with the ability to "fix" the surface "temperature at any point," arguably a form of geoengineering. Since the earth is cooling internally, it doesn't matter that "it is more difficult for [us] to cool than to heat the earth." For geoengineers today, of course, this is the whole crux of the problem, as indicated by the thoroughly problematic hypothesis that sulfate particles could be injected into the stratosphere to slow global warming.[39] Absent the underlying belief in a cooling trend, the performance of climate change becomes both less impressive and more inevitable than we find it in Buffon:

> Nothing seems more difficult, not to say impossible, than to oppose the progressive cooling of the earth and to raise the temperature of a climate; however, man can do this and has done this. Paris and Quebec are about at the same latitude and at the same elevation on the globe under the same degree of latitude; Paris would thus be as cold as Quebec, if France and all the countries that neighbor it were as bereft of people, as covered with forest, as bathed in waters as are the lands that neighbor Canada. (*Z* 126)

Fossil fuels drive the process of global warming, linked for us with the spike in carbon emissions that began during Buffon's lifetime and for him with their quality as a miraculously efficient and inexhaustible source of heat that might delay the cooling of the earth indefinitely. Because of this affinity, his paean to coal now has an especially uncanny resonance. Justifying his endorsement of deforestation, Buffon celebrates coal mines as

> treasures that Nature seems to have accumulated in advance for the needs to come of great populations. The more that humans multiply, the more forests will diminish. The wood not being able to satisfy their consumption, they [will] need recourse to these immense deposits of combustible matters, of which the use will become all the more necessary as the globe cools further; nevertheless they will never exhaust them, because a single coal seam perhaps contains more combustible matter than all the forests of a vast country. (*Z* 59)

Buffon echoes this theme of superabundance in his rhetorically powerful conclusion to the whole book, which holds out "treasures of [Nature's] inexhaustible fecundity" against time's dark abyss (*Z* 132). Superabundance of resources is a scientific conviction for Buffon. What is now proved—the imminent reality of peak oil and the precipitous decline of other all-too-exhaustible resources—had to be imagined first as a permanent entitlement of fossil plants that might reconcile human and natural history.[40]

Vegetable Earth

Vast beds of "vegetable earth," fossil shells, and "roasted ivory" are among the prehuman monuments that make the rock record legible and bring deep time into view. Buffon's *Epochs of Nature* both consolidates the analogy between "civil and natural history" and strains it nearly to its breaking point by emphasizing the difference in scale. The humanist topos of "ancient inscriptions" from the earth's "archives" provides his opening conceit and authorizes his use of monuments as a major analytic category. But the sharp discontinuity between premodern nature and the modern continents inhabited by humans surfaces quickly in Buffon's revision of a conventional trope associated with travel to "savage" countries: although we may "form an idea of [Nature's] ancient state" from travel in the New World, that "ancient state" itself is very modern indeed (*encore bien moderne*) compared to the prehuman state of the continents and the oceans "prior to the ages of history" (Z 4). To redirect the reader's gaze from human to nonhuman history, Buffon revises the conventional analogy between the "half-savage" ancients of the Golden Age and modern "savages," juxtaposing the ancient world instead with the supposedly "uninhabited" wastes of the New World and highlighting the wild state of the "young" continents themselves.[41] This exaggerated cultural distance between a savage state and modern "humanized" Nature, however, vanishes in the face of terrestrial antiquity. "How many revolutions have taken place beyond the reach of human memory!" Buffon exclaims and proceeds to enumerate rapidly the prehuman "monuments" that allow us "to distinguish four and even five epochs in the greatest depths of time" (Z 10).

In Buffon's thermal-maximum epochs, the "primitive vigor" suggested by the size of fossil bones is confirmed by the record of teeming early plant life deeply inscribed in the coal measures. William Smellie only included a heavily abridged version of *Epochs of Nature* in his nine-volume translation of Buffon's *Natural History* (1785) because he felt that this "theory" was "perhaps too fanciful to receive the general approbation of the cool and deliberate Briton" (*FA* 258). On this basis Smellie omits the majority of Buffon's main text, including his molten earth hypothesis, much of the material on coal and other fossil organisms, the separation of the continents, and more. Still, the two substantive parts of the work to survive his abridgement unscathed are among the most fanciful of all. Smellie includes Buffon's seventh epoch, his romance of inexhaustible coal, in its entirety, along with all his excursus notes (*notes justificatives*), on the grounds that these contain "interesting facts" that are "too important to be omitted." These notes contain traces of the synchronic multispecies narrative that Buffon suppressed in revising the main text shortly before publication, as documented in Roger's critical edition of the work; three of the longest are dedicated to a theory of gigantism encompassing the extinct "elephants" along with giant ammonites and tropical fossil flora associated with the coal measures. As Smellie renders it, "Nature was then in her primitive vigour. The internal heat of the earth bestowed on its productions all the vigour and magnitude of which they were susceptible. The first ages produced giants of every kind" (*FA* 303). In one of the more adventurous passages deleted from his manuscript, Buffon even proposes that "the human species is as ancient as that of the elephants" (*EN* 161n.), thereby correlating both legendary human giants and fossil mammoths with the fossil plants that function as "monuments" of a planet covered with forests like those of modern-day Guyana (*EN* 159n.).

These fossil plants, like the mastodon molars engraved to illustrate *Epochs of Nature*, serve as "monuments" of past events, but unlike fossil bones and shells they have a double function. In their superabundance, they suggest a permanent endowment, an archive conserving the gigantic energy of the young earth. When Buffon discusses the formation of coal in the third epoch, he emphasizes the "fecundity" of the younger planet that produced both plants and animals adapted to much higher temperatures; one excursus note cites numerous analyses of fossil impressions from coal mines that seem to depict fish and plants that "do not belong to living species" (*EN* 78). His revisions show that he did not anticipate controversy over extinction so much as over the issue of human exceptionalism, which his long time scale seemed to threaten. Accordingly, the seventh epoch, written last, returns to the subject of coal at length to show that the superabundance of the third epoch persists as a guarantee that human intervention will permanently transform nonhuman species and render them "useful."[42] Thanks to fossil fuels, the intervention extends as far as climate engineering, for these vast treasuries make it downright easy to heat the earth, as we have seen. Zalasiewicz and his colleagues credit Buffon as the first scientist to recognize in the decay of tropical forests a "modern analogue" for the formation of fossil fuels (*Z* 29).

Buffon's argument depended on the large body of correspondence and other evidence from abroad that came together at the Jardin du Roi, one of the century's great "centers of calculation." As Elizabeth Anderson has shown, Buffon collated material from numerous colonial officials stationed in French Guiana, which was critical for his account of fossil plants.[43]

Other period accounts of the decay and transformation of plant matter recognized the same excess or surplus of energy produced by it but were more apt to conceive of this process in providential terms. Cuvier, writing in 1810, seems to echo Buffon's view when he observes that "nature seems to have put [coal] in reserve for the present age." Commenting on this sentence, Rudwick terms it "a secular version . . . of the view commonly expressed in Britain at this time, that the ancient coal deposits had been stored for eventual human use, by the care of divine providence."[44] John Whitehurst, writing like Buffon in 1778, spoke somewhat more cautiously of fossil plants "not known to exist in any part of the world in a living state."[45] These impressions, however, combined with those of recognizable tropical plants, provide "a certain indication of coal" in the underlying strata. In Whitehurst's idiom of natural theology, these certainties inscribed in the order of the strata correspond more closely to the "book of nature," written for a future reader who would grow more literate as the value of coal increased. The Forsters were both struck by the sheer abundance of decay in the old growth forests on the Pacific islands they visited with Captain Cook, so unlike the managed European forests of that time. Sharing Whitehurst's faith in the divinely authored book of nature, Reinhold Forster expresses anthropocentric confidence in the ultimate meaningfulness of the decay that happens on such a massive scale in an old-growth forest where the trees fall uncut and decompose. Instead of coal formation, he focuses on the nearer-term future of rich fertile soil that the unchecked decay of these forests seems to promise for future European settlers: the islands are merely "hoarding up a precious quantity of the richest mould, for a future generation of men" (*O* 43). George Forster's account of the same "primeval" forest at Dusky Sound (New Zealand) illustrates some of the fault lines between providential natural theology and a more skeptical view, as he succumbs to the gloomy premonition that any improvements to this "savage" place (notwithstanding their documentation of its botanical diversity) will soon "sink back into the original, chaotic state of this country" (*R* 180–81). The mere fact of lushness is ambivalent and betrays an inhuman aspect. In Georges Bataille's more radically naturalistic view, plants "manifest . . . a much more pronounced" excess than do animals, for they are "nothing but growth and reproduction."[46]

From his vantage point in eighteenth-century Paris, Buffon is more attuned to nature as a global economy that creates a secular but progressive balance between past and present epochs. Traveling in place at the Jardin du Roi, Buffon may have come to the specimens and descriptions of exuberant plant life in Guyana with coal in the back of his mind. These hot young forests, he deduced in constructing his system of global natural history, sustained the energy entombed ages ago in the ancient tropical forests now being recovered elsewhere as coal. In this view, the decay of tropical forests produces a truly massive surplus on the longer scale of geological time. The foundations of this deep time economy are laid in Buffon's Epoch III, "when the waters covered our continents." The decay of plant matter approaches its peak only at the end of this epoch, when the waters have retreated sufficiently to allow continental forests to establish themselves. These forests are preceded by another kind of excess, that produced by animals. Although living animals, as Bataille points out, use more energy than plants for locomotion, feeding, and reproduction, Buffon's fossil shells of the "first ages" produce another kind of excess. Their individual bodies accumulate over time to form limestone, a basic building block of Buffon's young earth, and collectively these shells represent entire "species now obliterated" (*Z* 12). Buffon's belief in extinction is unusual for his time; the idea of extinction was rejected by Jean-Baptiste Lamarck a generation later and continued to be challenged in subsequent generations.[47] Nature's "treasury" of carbon guarantees an inexhaustible surplus of energy for Buffon partly because the sheer number of extinct plant species preserved in the coal measures indicate how much time has passed since their decay. His larger account of extinct animals and plants in diverse strata reveals a corresponding surplus of time.

Epoch III covers a long period in the earth's history, about 20,000 years or over a quarter of the total. Extinction renders the lapse of this long period diachronic in both a quantitative and a qualitative sense. Unlike the two epochs preceding life, calculated largely on the basis of Buffon's experiments with globes of white-hot iron, the expanse of prehuman time in the third epoch may be documented through the succession of fossils, and the extinction of earlier species documents the scale of change in the conditions of life over time, in ocean levels and above all in temperature. Buffon's economy of time depends on extinction to demonstrate the inhuman character of these conditions and the qualitative difference between Nature "in its first flush of vigor" (*Z* 15) and its modern humanized condition. Navigating between fossils and the beds that contain them, Buffon also marshals the qualitative experience of stratigraphy in the field, the slow succession of strata, to trace the early history of life from shellfish to fish and plants. In revising his long manuscript (*Cahier* 3) on the accumulation and retreat of the primitive ocean and its denizens, Buffon arrived at a justification for his expanded time scale that was so compelling that he decided to relocate it to Epoch I, where he used it to frame a sustained meditation on the scalar quality of time. This revision shows that the history of nonhuman species is needed from the outset to explain how deep time works, even at a point in the sequence that actually predates the history of life.

Shellfish, "the most ancient species in Nature" (Z 53), are the main protagonists in Epoch III, ushering in the age of plants, vertebrates, and terrestrial life. They make their first appearance in Buffon's preliminary discourse, where they supply four of the ten axiomatic facts and monuments that launch the narrative. He first points to the presence of fossil shells on mountaintops, which fascinated many early modern naturalists, and then enumerates three qualities of shells as monuments, including their ubiquity, their geographic dislocation, and the apparent extinction of some varieties: "these either exist no longer, or are found in southern climes" (Z 9). The probability that at least some of the displaced species are "now lost or destroyed" increases their value as prehuman "monuments" allowing us "to distinguish four and even five epochs in the greatest depths of time" (Z 10). In addition to the ammonites and other marine species, Buffon here specifies the Ohio animal (mastodon) as "up to the present . . . [the] sole example of a lost species among the terrestrial animals" in a long excursus note added to this passage, the second of several on gigantism and extinction (Z 143). In detailing the first phase of extinctions in Epoch III, Buffon invokes climate change as his proof that "no living analogues can be found anywhere" for these species specially adapted to the much higher temperatures that prevailed at this stage of the earth's cooling: "these first species, now annihilated, lived for the ten or fifteen thousand years that followed the times when the waters were established" (Z 53).

Just here, however, Buffon again anticipates the advent of the megafauna in Epoch V and other later developments, displaying the synchronic dimension of his time scale. Here again the mastodon stands in for the broad scope of nature's "primitive vigor." "Nature was then in its first force," according to Zalasiewicz et al.'s more modern rendering (Z 53), but "then" spans the entire long transition between the origin of life and the human era, a transition that is of central importance for the disclosure of deep time. The diachronic narrative of this chapter that proceeds from shells as "the first inhabitants of the globe" (Z 51) to terrestrial life—from clay (or shale) to limestone to coal and slate—intersects a synchronic narrative according to which "the animal population of the sea is not older than that of plants on land" (*Z* 54), originating on the highest mountain peaks, which were never submerged. In this sense, the first five epochs are *all* deep time epochs, the "first ages" set off against the human epoch (VII) as well as the dimly remembered episodes of Epoch VI, some of which approximate prehistory in its modern sense.

Nonetheless, the narrative arc of the chapter depends on a linear succession of shellfish followed by plants and fish and finally by land animals. Stratigraphy (or "geognosy," in the language of the period) is therefore essential to Epoch III, which includes enough fieldwork by Buffon himself and by others to cast doubt on any categorization of *Epochs of Nature* as exclusively theoretical or romantic. The rock described by Buffon as clay (*argile*), best translated as "shale" (consolidated clay or mud), produced by the dissolution of "scoria" or "vitrescible sand" eroded from the primitive igneous matter of the globe, provides the foundation for limestone formed from shells. Limestone overlies the shale "almost everywhere" (*Z* 55), Buffon notes, generalizing from his own experience in the field, as did Whitehurst and other naturalists of this period. Buffon supports this succession with analysis of a fifty-foot shaft dug into the *argile* on his own property at Montbard, in the Côte-d'Or region of Burgundy. Adding the height of the valley wall he arrives at a depth of 130 feet, adducing the early date of this deposit both from the abundant ammonites and other fossil content and from its thickness. In the 1740s and 1750s, Giovanni Arduino and Johann Gottlob Lehmann similarly developed qualitative and relative dating methods based on their observation of strata, as detailed in the introduction to this book. Buffon's fieldwork here and elsewhere is inspired by Lehmann, whose *History of Secondary Rocks* (*Flötzgebürge*; 1756) is cited in a note to this passage and more extensively elsewhere in *Epochs of Nature* (*Z* 156, 147–48; *EN* 310n6). Eventually the shells accumulated in sufficient quantity to form beds of limestone atop the shale, and at roughly the same time the coal seams began to form between the strata of shale as debris from the primitive forests washed out to sea (*Z* 57). Buffon, who built an iron forge in the nearby village of Buffon, from which he took his title, returns to his property for an account of the iron ore transported and lodged in the shale by the currents of the same primitive ocean (*Z* 64). Having deposited the shale, the limestone, and the coal, Buffon proceeds to enumerate the monuments that allow for more precise relative dating within the very long and somewhat amorphous third epoch, compensating in this way for the vagueness in his chronology: shale, limestone, slate, and coal with their fossil content "are the most ancient monuments of living Nature" (*Z* 63), deposited "more or less at the same time"

(Z 65)—that is, the middle third of the planet's history—but nonetheless in a determinate stratigraphic sequence.[48]

The quantitative and qualitative measures intersecting to form this early deep-time chronology correspond in some ways to the mathematical and dynamic sublime (respectively) of Immanuel Kant. Buffon encourages his reader to imagine the sheer quantity of plant matter preserved in a coal seam, much as Kant invokes reason's demand for totality in the face of an empirical infinite as the basis for the mathematical sublime. Just as the imagination "sinks back into itself" in its fruitless effort to produce an adequate presentation of absolute magnitude in Kant, Buffon creates a sense of deep time from the imaginative effort to grasp "twenty or twenty-five thousand years" worth of plant growth and decay: "The single thing that can be difficult to conceive is the immense quantity of plant debris that these coal seams indicate" (Z 58).[49] As with Kant's dynamic sublime, the raw power of elapsed time is contained by the observer's distanced perspective. Taking his own advice to consider the savage state of the Americas as a heuristic for these deep time processes—for the judgment here is ultimately teleological and not aesthetic, in Kantian terms—Buffon points to the Mississippi and the Amazon, utterly clogged with an "enormous quantity of great trees," which they "carry to the sea" (Z 58–59).

To illustrate further the state of a planet covered in trees and other vegetation, "which nothing destroyed except old age" (Z 59), Buffon inserted his paragraph on the Guyanese rainforest as a future "treasury" of coal. Roger shows that his analogy between coal and tropical forests is borrowed from Lehmann and d'Holbach (*EN* 293–94), and this particular example struck Buffon while making a revision—clearly distinguished from the main text of *Cahier* 3 by appearing in Buffon's own hand—that seems to have been prompted by further reflection on the observably slow accumulation of plant debris and other organic strata as a qualitative index of deep time. The conviction that superabundant energy follows from his understanding of geological time, discussed above, may have been triggered by a report he received from Guyana after his second draft was completed.[50] In the second draft, he had already concluded that the "long succession of centuries" allowed for Epoch III seemed "still too short for the succession of effects that all these monuments show to us" (Z 60) and proceeded to study further examples including the erosion and transport of existing rocks and the accumulation of oyster shells to form new rock. The state of the manuscript confirms that the 20,000 years specified in print is more of a nonce figure than otherwise. As with plant debris, the "immense quantities of marine products" across the

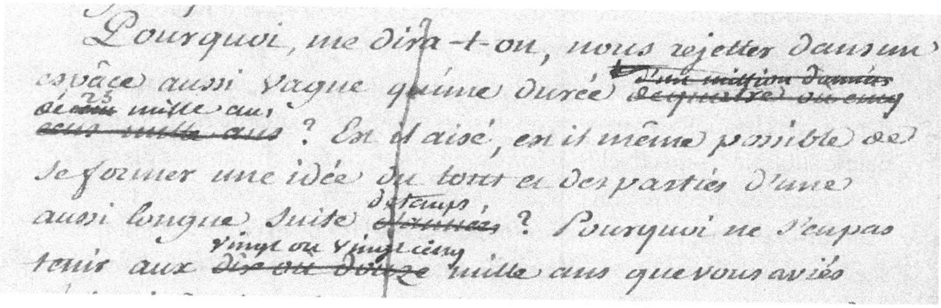
Pourquoi, me dira-t-on, nous rejetter dans un
espace aussi vague qu'une durée ~~un million d'années~~ ~~de quatre ou cinq~~
de 25 mille ans ~~cent mille ans~~? Est il aisé, est il même possible de
se former une idée du tout et des parties d'une
aussi longue suite de temps ~~d'années~~? Pourquoi ne s'en pas
tenir aux vingt ou vingt cinq ~~dix ou douze~~ mille ans que vous aviés

FIGURE 2.1. Georges-Louis Leclerc, Comte de Buffon, "4eme Epoque," fol. 27 (detail). MS 883, *Cahier* 3, © Muséum national d'histoire naturelle, Paris.

geological record must be proportional to the lapse of time. At this point where stratigraphy intersects with the sublime, Buffon inserted a second long addition, building this model of geological dating based on the history of species into a "very philosophical digression" on time, relocated at the next stage of revision to a more prominent place in Epoch I.[51]

In the revision, Buffon elaborates further on the abundance contained in the fossil archive and the inadequacy of any received chronological reckoning as a way of accounting for this magnitude. The life cycle of the individual organism, derived by analogy from a living species, the oyster, provides a critical unit of measure here (*Z* 61). The manuscript (Fig. 2.1) continues at this point with a question: "Why, one will say to me, throw us into a space as vague as four or five hundred thousand years?" (*EN* 39–40n.). The explanation that follows occupies five manuscript pages, including a long passage inserted in the margins during the penultimate stage of revision. Buffon had second thoughts—more than once—about the time scale, and the figure of 20,000 years that we find in the printed text is the end result of a series of at least five alterations. The manuscript shows that "four or five hundred thousand years" is the first figure given in both passages and is successively altered first to one million, then to 100,000, then to 25,000, and finally to 20,000 years in print in the earlier passage.[52] Buffon had second thoughts as well about the content of his explanation, revising his opening question heavily, deleting a long middle section on the arbitrariness of scale, and relocating the result—his rhetorical question about the necessity of a long time scale plus the close of his meditation on the inhuman magnitude of time, illustrated by a new chronicle of sedimentation, in this case of shale and limestone on the Norman coast (*EN* 39–43/*Z* 35–37). In spite of all these changes, the version that appeared in print in Epoch I preserves a central rhetorical feature of the original passage in Epoch III: "Far from having to unnecessarily augment the duration of time, I have perhaps shortened it too greatly" (*Z* 35).

Along with this deliberately overstated caution, the revised and relocated passage also reiterates the importance of prehuman species for marking geological time. We can only begin to grasp the scope of geological time, Buffon argues here—an expanse that for him is at least ten times larger than the period generally accepted in Europe until the mid-nineteenth century, though still short by modern standards—by putting the history of our own species in relation to that of others. The philosophical part of this digression concerns what might be called temporal apperception, or reflection on the cognitive resources available for grasping and marking frames of time beyond human generations (the "life-time" of Hans Blumenberg). This philosophical preamble is followed by a concrete illustration, the final piece inserted in the margins. Here Buffon enjoins his readers to envision a 1,000-foot hill of shale (as at Caen in Normandy) surmounted by a stratum of limestone (*Z* 36). Supposing that the original sediment accumulates at a rate of five inches per year, it will have been 14,000 years before shells began to be transported to this site, and these in turn must accumulate and be turned to limestone before the sea level drops and the hill can assume its final form. By framing this process as a prolonged repetition of miniscule increments, Buffon anticipates a major trope employed by later geologists, including James Hutton and Charles Darwin, and epitomized in John Playfair's famous response to a view of horizontal strata gradually deposited on the broken and upturned edge of a series of much older strata at Siccar Point in southeast Scotland: "the mind seemed to grow giddy by looking so far back into the abyss of time."[53]

Deep time in its geological sense is the product of generations of writing and fieldwork by many geologists and the naturalists who came before them. I have chosen Buffon as a subject here because his argument discloses deep time as a disjuncture between "our all-too-short existence" and the succession of ages (*siècles*) "that were needed to produce all of the animals and shells with which the Earth is replete" (*Z* 36). Hutton and Lamarck are among the early geologists who offer even greater expanses of time, but their steady-state models of the earth rule out extinction, or even clearly demarcated epochs, as way of structuring these expanses. Paleontologists such as Cuvier and William Smith, more widely recognized by historians of geology (and rightfully so), go far beyond Buffon in their empirical recording of fossil species but for a variety of reasons reject absolute dating. Buffon's characteristic alternation between empirical precision and philosophical speculation, combined with his literary approach, produce a narrative of prehuman time that is neither modern nor premodern. Inspired by freethinking naturalists such as Baron d'Holbach and Nicolas-Antoine Boulanger, Buffon takes full advantage of the imaginative freedom made possible by the absence of a consensus or critical mass of data on geochronology.

In the most daring part of the manuscript passage under discussion here, deleted in revision, Buffon defends his proposed chronology as already "abridged" from a scale forty times larger that would have been philosophically more adequate (*EN* 40n.). Characteristically, he altered "forty" to "four" before deciding to reject the paragraph entirely. Even a casual reading of the manuscript shows that almost every specified span of time is struck out and then replaced, quite often three or four times, as shown in Fig. 2.1. In his critical edition, Roger tabulates these changes to document a series of five time scales—one short and three long, followed by a final short time scale—that Buffon entertained at successive stages of composition (lxv). Scaling up or down seemingly at will, Buffon here reveals that his final figure of 75,000 years for the age of the earth is a compromise, a figure low enough to be commensurate with "the limited power of our intelligence," but probably far short of the earth's true age. Far from being exaggerated, this "little" scale is simply necessary to preserve the clarity of the idea. The rhetorical question that follows—"why does the human mind seem to lose itself in the expanse of duration much more readily than in that of space or number?"—is inspired by Boulanger's assertively secular account of time in the unpublished *Anecdotes de la Nature* (1753), as Roger shows (*EN* 285). Buffon reformulates this question in his final printed text in such a way as to redirect attention from the arbitrariness of scale to the observable histories, especially those of other species, that make deep time divisible and intelligible (*Z* 36). This deleted passage clearly shows the abyssal character of time by insisting on the arbitrariness of scale. Buffon suggests that the evidence then available readily accommodates any conceivable expansion of the time scale; the epochs of nature are distinct enough to require absolute dates, but the scope of change implied by each epoch is so vast that geological time is effectively bottomless.

Mammoths and Other Giants

In composing *Epochs of Nature*, Buffon turned his attention from the superheated environment that nurtured the first plants and shellfish, now extinct, to the regions that are coldest in modern times, such as Siberia and upper

Canada. The chapter initially conceived as Epoch IV turned to these regions for fossil evidence of the megafauna, especially mammoths, and for contextual evidence of the global cooling that began at the poles, as called for by his geophysical model. At this stage, Buffon remained unsure about where to locate volcanoes, though he was convinced that their main period of activity lay in the distant past. As Roger notes, Buffon located this intense period of volcanism originally in Epoch II and subsequently in Epoch III. Roger found no manuscript evidence for the stage at which Buffon decided to insert this epoch between the epoch of fossil plants and shells and that of megafauna, as suggested by the eventual retitling of the latter epoch as Epoch V. In 2000, three years after Roger's death, the museum acquired a new gathering (*cahier*) of manuscript pages, also titled Epoch IV.[54] This new portion of the manuscript documents Buffon's later creation of a separate epoch for the volcanoes, and the printed text follows it quite closely. Buffon reminds us that the most elevated land "became covered with great trees" and that the "universal sea was filled everywhere with fish and shellfish" (Z 70), and other references to the super-heated world of Epoch III preserve a record of the original context of this development. The formation of "fuel" for the first terrestrial volcanoes follows a pattern similar to the formation of coal reserves: plant debris mingled with mineral substances on the ocean floor caused volcanoes to form once the sea retreated. Buffon may have expanded this section in part because the volcanoes would serve to slow the cooling (*refroidissement*) of the globe that followed the drop in sea levels. This mechanism helps to explain how the megafauna were able to thrive in the most rapidly cooling (subarctic) regions and effectively extends the "first age of a living nature" (14), the age of giants already discussed in Buffon's preliminary discourse, and a key synchronic dimension of his geochronology as a whole.

The mammoth is arguably the main actor of Buffon's *Epochs of Nature*, even though the work covers the whole history of life as Buffon understood it. Buffon himself does not use "mammoth" to describe the megafauna associated with find spots in Siberia and North America but rather "unknown animal," "hippopotamus," and especially "elephant," which sometimes serves as a metonymy for all three. "Elephant" in the specific sense of "Siberian mammoth" is used in conjunction with the substance that Buffon refers to as both "*ivoire fossile*" and "*ivoire cuit*," with the latter term being credited explicitly to the ivory merchants he consulted in Paris. Smellie offers the poetic if eccentric translation "roasted ivory"; Claudine Cohen's modern rendering, "baked ivory," more accurately captures the probable association with fired earthenware, and "cured ivory" might have better etymological support.

Precisely as an eccentric translation, "roasted ivory" captures the uncertain status of the mammoth as a mediator. Defining the "agency" of objects as one of the areas of uncertainty that disclose networks in action, Bruno Latour uses fossils to illustrate the way in which distance in time can highlight the object's role as mediator: "Even the humblest and most ancient stone tools from the Olduvai Gorge in Tanzania have been turned by paleontologists into the very mediators that triggered the evolution of 'modern man.'" Cohen argues that the uncertainty surrounding fossil ivory inspired *Epochs of Nature* itself: "it is not too much to claim that the question of the Siberian mammoth and the 'unknown Ohio animal' is the keystone—maybe the key—of this masterwork."[55]

Buffon calls "elephants" into service just a few pages into his preliminary discourse to support his main hypothesis of global cooling, but in the process he also gives them their own rich history. Their remains first appear as the third and fourth of five "monuments" that "prove" Buffon's account of the earth's history (*EN* 10). Because they are found in the northern parts of both the Old and the New World, he argues, the fossil tusks and bones of giant elephants and other megafauna support his thesis of the separation of the continents. The wealth of available evidence enables Buffon to construct a major portion of his narrative, his account of Epoch V ("when the elephants, and other animals of the south, inhabited the northern regions"). Their history unfolds in the course of the chapter-length narrative devoted to this epoch and in parts of the following epoch as well, but Buffon provides a surprising amount of detail on these animals even at the outset, when he is simply outlining his history of the earth. The established trade in "roasted" ivory from Siberia and Tartary, together with recent discoveries in North America that had been widely debated in London and communicated to Buffon, along with specimens, in the late 1760s, increased both the visibility of the mammoth and its contemporaries and the uncertainties surrounding them. He harnesses this visibility to illustrate the magnitude of the geological transformations that would have needed to occur for large land animals to flourish.

Buffon monumentalizes their remains in a more literal sense, too, in the form of engravings that dramatize their scale while at the same time linking the categories of natural and civil history, fulfilling a promise made in the first sentence of this work. The idea of fossils as "the antiquities of the earth" was not new, but Buffon escalated the analogy to posit geological "revolutions" of incomparably larger magnitude than civil ones. He accentuates this unconformity in his initial reckoning of the "monuments" of the six "first ages" of nature:

> [W]e have six . . . spans of time of which the limits, though undetermined, are no less real, because these epochs are not like those of civil history, marked by fixed points, or limited by centuries or other segments of time that we can count and measure exactly; nevertheless, we can make comparisons among them, and evaluate their relative duration, and relate other monuments and facts to each of these time intervals, which will reveal their contemporary dates, and maybe also other intermediate and subsequent epochs. (*Z* 15)

The "elephants" and other megafauna receive their own epoch partly because their remains seem to support the idea of a major episode of residual heat following the decline in primitive ocean levels. This long independent history is the main point of a very long excursus note on elephants that follows the initial discussion of the third and fourth "monuments," their fossil remains. In all, four of thirty-six essay-like *notes justificatives* that accompany the more philosophical narrative (or "romance," to its critics)—and form the basis of the first English translation by Smellie—concern these megafauna. Unlike most of the other volumes of Buffon's *Histoire naturelle*, which are richly illustrated, *Epochs of Nature* has only five engravings, and all five were made to accompany this one note (*Z* 140). The first of these (Fig. 2.2), like three of the others, shows the pointed molars of the "unknown animal" (mastodon) to demonstrate the scale of life in earlier epochs and support the controversial idea that these animals and others had become extinct as global climate changed. These fossil teeth and the engravings function as monuments or mediators in both the material space of print and the conceptual space of natural history, where they enact the declension from a gigantic past into modernity.

The long note (n9) quotes extensively from correspondence Buffon received from around the world in his position as director of the Jardin du Roi.[56] On the authority of these observations and specimens from the Ohio River, Siberia, and Quebec (then a part of New France), Buffon argued that megafauna "species lived, fed, and multiplied there, just as they live, feed, and multiply today in the countries of the south" (*Z* 12). By insisting on an independent history for these species, Buffon is not only substantiating his new geological time scale but also rebutting historical and catastrophist explanations of this seeming biogeographic anomaly. Prevailing theories held that the animals had been driven there by humans or by a mega-flood in the Indian Ocean, but according to Buffon the extent of the remains precluded the idea that only "some individuals" had migrated. On the contrary, the ancient elephants lived in the North "in a state of nature and in complete liberty" (12). These elephants and

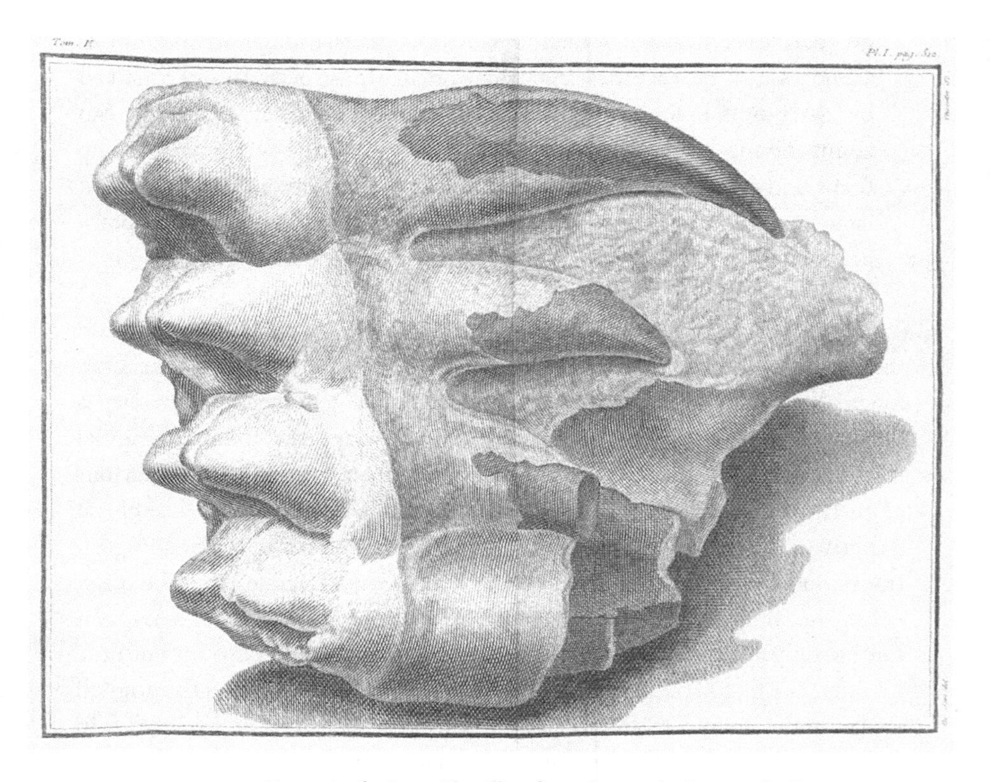

FIGURE 2.2. Engraving by Juste Chevillet after a drawing by Jacques de Sève. Georges-Louis Leclerc, Comte de Buffon, *Epochs of Nature* (1778), Plate 1. Courtesy of Special Collections, Ellis Library, University of Missouri.

kindred species established themselves and flourished over hundreds of generations in the locations where their remains were found, before migrating southward (and, in at least one case, becoming extinct). Hence "there has, perhaps, already been more ivory found in the north than all of the elephants of India living today could provide," and a great deal more remains to be discovered (11). Moreover, a substantial length of time was required for this fossil ivory to "roast." The ivory merchants, Buffon reports, distinguish fossilized tusks by the quality of their matrix and the length of time they remained buried in it, saying that in most cases "one can make beautiful artifacts from them" (136). This lapse of time gives him room to maneuver in his narrative of these animals' subsequent fate, including his account of their attempt to pass the Isthmus of Panama during the sixth epoch. In spite of the size difference and the lapse of time, the merchants still recognize fossil ivory as identical in structure with modern ivory: apart from size, these elephants were elephants, so their habitats must have derived their warmth from the interior of a younger, more vigorous globe (14).

Since the earth could only have cooled very gradually (as verified by Buffon's experiments on iron), these super-heated conditions persisted for long enough to infuse the very first living creatures, followed by a whole succession of giant flora and fauna, and perhaps even—this is Buffon's most radical suggestion—a giant race of early humans. The conundrum of the "elephants" prompts his first analysis of these conditions, and they make such apt protagonists because they lend support to a larger narrative arc extending from the gigantic fossil shells of the third epoch all the way to the first humans of the sixth. The animals were larger during the "long tract of time" it took for the earth to cool (*FA* 286) because "Nature was then in her primitive vigour" (303)—but as we saw in the case of fossil coal and shells, this language marks a synchronic dimension of Buffon's time scale, in the absence of which human origins might seem all too belated. In his gloss on the "monuments," Buffon concedes that the epochs of nature, unlike those of civil history, are relative and not clearly demarcated (*Z* 15), but the cool temperatures of recorded history seem to have arrived only recently. In the main body of the narrative, versions of the same argument—residual internal heat producing gigantism—occur as early as the third epoch and as late as the sixth, which contains a third long excursus note (n30) developing the issues raised in the initial account of "elephants." This narrative strategy lends a certain coherence to the long prehuman past that Buffon proposes. The multispecies "archive of nature" displaces the human record (and with it sacred history), while also establishing geocentric and ecocentric criteria for locating human origins among those of other species.

Buffon conscientiously takes up the discourse of gigantism elaborated by scientifically minded conjectural historians from Lucretius to Vico. He gains a critical purchase on it by introducing a much longer time scale, including the notion of extinction, and by distinguishing more sharply between descriptions of fossil bones of humans or hominids and those of other species. Pursuing an Enlightenment program inspired by his contemporary Jean-Étienne Guettard (*EN* 291n16), Buffon singles out certain descriptions from the so-called *Gigantologie espagnole*, arguing that some of the bones described in this text must be of the same species as the "unknown animal" from the Ohio River (*Z* 154). Although sacred history and "superstition" are displaced by this approach, they are not evacuated entirely. On a psychological level, the nostalgia for nature's "primitive vigor" and its gigantic forms suggests desire fueled by a trauma associated with "convulsions of nature." Here again Buffon took inspiration from the anticlerical *philosophe* Boulanger, whose writings bridged natural history and antiquarianism. In his published work, *Antiquity Unveiled* (1766), Boulanger reinterprets the Biblical Deluge in naturalistic terms, explaining human history as the effect of an original, traumatic flood. In *Anecdotes of Nature*, which Buffon read in manuscript, Boulanger reinterprets natural history in the same skeptical light (*EN* lxxvi–lxxvii).

Skepticism vis-à-vis natural theology does not necessarily extend to other kinds of quasi-legendary material. The selective use of premodern philosophers by Buffon, Boulanger, and their contemporaries illustrates the importance of ancient and early modern ideas for the evolution of deep time as a concept. Playfair's later rejection of premodern theories and histories of the earth as "mental derangement" stands in contrast to the more eclectic practice of the Forsters, Buffon, and others at the earlier stage of this conceptual history that is my focus in this study.[57] Playfair's and Lyell's consolidation of the uniformitarian idea of deep time was carried out in the more familiar Baconian spirit of the nineteenth century. The earlier efforts of Buffon and his contemporaries are critical because they mobilize a broad range of humanistic traditions as well as empirical data to open the imaginative space of deep time.

Though more circumspect than Boulanger, Buffon does not rule out the possibility of human giants, and the manuscript of *Epochs of Nature* reveals that he entertained the possibility of human beings contemporary with giant northern elephants but then suppressed it in the printed text. Buffon concludes his discussion of the fossil evidence of giant tropical animals and plants in the North with an abrupt turn to the human species, asking if the earth could have been populated with "men" at the same time that it was populated with these plants and animals, before the separation of the continents (*EN* 159n.). In the printed text Buffon strategically placed this question in the mouth of an imaginary interlocutor (one of his favorite rhetorical figures) and answered in the negative, citing a separate act of creation and firmly deferring the discussion of human origins to its proper epoch (*Z* 97). In the manuscript, however, both the rhetorical tactics and the long pious interpolation are absent; instead, the narrator raises this question in his own voice and answers very differently, citing the uniformity of the laws of nature and the worldwide distribution of human populations as premises for this conclusion:

> Without wishing to affirm it we therefore presume, according to our hypothesis, that the human species is as ancient as the elephant; that, being

> able to sustain the same degree of heat and perhaps an even higher degree, man will have penetrated the southern lands before the animals; but that his first long sojourn was in the northern lands. (*EN* 161n.)

The manuscript proceeds to introduce Buffon's thesis that this first civilization was established on the high plateaux of central Asia, though even at this stage he planned to defer the details and the "proofs" to the discussion of his last epoch (162n.). By deleting this material and revising the quoted passage, Buffon does much to create the impression that God deferred man's arrival till the earth was "worthy of his dominion" (*Z* 98).

It is striking that even this rhetorical pose of anthropocentrism was not enough to satisfy Buffon's contemporaries, including Forster and, more impressively, the Sorbonne, which condemned parts of the seventh epoch (*EN* 306n1). The deletion of manuscript material alone provides adequate evidence for questioning the "sincerity" of the published version, as Roger pointed out in 1962 to caution those who would "accuse Buffon of anthropocentrism" (*EN* 302n29). And even though he consolidated his thinking on human origins into the seventh epoch, written last, significant material referring to human or semihuman giants was allowed to stand in the sixth epoch and the excursus notes—further suggesting human proximity to other species. In the printed text Buffon retracted his claim that "monuments and even traditions" attest that "the human species has followed the same course and dates from the same period as other species," rendering it purely a matter of "analogy" (*EN* 159n.). Several of these monuments and traditions are reported, all the same, in the material on giants that was not deleted. As Buffon puts it in one of the deleted passages, all these indications give us strong reason to presume that gigantic humans, too, existed in "these primitive times close to the origins of animated nature and on the same terrain as the first elephants and other animals" (*EN* 162n.). In light of theological objections to the published work, it is thoroughly ironic that he tried to be *more* orthodox by *post*dating the species in his revisions.

The most salient episode to have escaped these revisions is the migration of the giant species, including humans, prior to the separation of the continents. The primary evidence for such a migration, once again, is the presence of fossil remains of the same megafauna in both northern Eurasia and North America. As Roger points out, this is essentially the same body of evidence as that put forward by Alfred Wegener in the twentieth century in support of the supercontinent Gondwanaland (*EN* 275n56). In the sixth epoch Buffon proposes that "a few giants, like the elephants, passed from Asia into America" via the same overland route, finding there the "liberty [and] tranquility" they required to "propagate their gigantic race" at a time when lowland South America was still entirely covered by the primeval ocean (*Z* 112). Buffon points to the Patagonian giants, whose existence had been corroborated as recently as 1767 by John Byron on HMS *Dolphin,* as the last vestiges of this race. The Asiatic parent stem died out much earlier with Goliath, according to the excursus note accompanying this passage (*EN* 163–64). Buffon cites scripture infrequently, but in this instance it bears out his general claim in the manuscript that written tradition conserved the memory of giants; in conjunction with the Patagonians, he uses the stories of Goliath, the sons of Anak, and others to posit "constant and successive races of giants" (*EN* 167). The original account of human origins, partially overwritten by the seventh epoch, is thus a synchronic one, in which "the origins of animated nature" describes an indefinitely long process that leaves traces, such as gigantism, in most species. In *Epochs of Nature,* gigantism provides a secular vocabulary for the proximity of species as well as their collective proximity to origins. The corresponding vocabulary of domestication, though it has a Biblical warrant in Genesis, now appears to connote the belatedness of modern civilization in relation to the deep gigantic past.

The memories recorded in ancient texts pass through several stages between the epochs marked by fossil "monuments" and their later preservation in writing. These stages may be traced more indistinctly in oral tradition and in a kind of race memory, or what might today be termed evolutionary memory. Evolutionary memory will take on an increasingly central role in the second half of this book, which articulates the category of "species time" as a Romantic and post-Romantic contribution to the development of deep time as a concept. Oral tradition and species memory are crucial for Buffon because they substantiate the apparent agreement between the monuments and the earliest recorded traditions, such as the stories of giants in the Hebrew Bible. Memory in this extended sense provides evidence of the prehistoric phase needed to bridge geological and historical time in Buffon's multispecies history. He added a substantial excursus note after the rest of *Epochs of Nature* had passed through the proof stage to report on an Arctic travel narrative delivered to him by the Russian count Andrey Petrovich Shuvalov in autumn 1777. This narrative (compiled by Sergey Domashnev from firsthand accounts by the navigator Aphanassei Otcheredin) includes material from oral tradition among the Chukchi people on the Asian side of the Bering Strait as well as observations on the current state of their trade with people on the American side. Buffon concludes that since the Chukchi have carried on this commerce "from times immemorial" (*Z* 175), it could be regarded as a survival of the traffic occurring during the separation of the continents. More general human memory of "races of giants," having been preserved at least until the time of the earliest writings, confirms the hypothesis of gigantism in similar fashion. The high antiquity and gigantic size attributed to the first humans by tradition corroborate the idea that they "received" great size and strength "from nature" during the same extended "primitive" moment in which the first elephants acquired their "prodigious" tusks (*EN* 162n.). Buffon also takes the testimony of ancient geographers such as Strabo concerning the uninhabitability of the "torrid zone" as a kind of species memory of the fifth or even earlier epochs (*Z* 87; *EN* 300n1).

This type of geological memory "preserved by tradition" operates most powerfully at the beginning of the seventh epoch, introduced earlier in my discussion of human belatedness. This bleak narrative of human origins declares that "the first men" were "witnesses of convulsive motions of the Earth, then still recent and very frequent" (*Z* 119). Reminiscent of the state of humanity in Plato's *Protagoras* before the intervention of Prometheus, these early humans or giants were "naked in spirit and body, exposed to the curses of all the elements, victims of the fury of ferocious animals, of which they could not avoid being the prey; [and] all equally penetrated by a common feeling of baleful terror." Small wonder, then, that even the first nations remained "profoundly affected by the calamities of their first condition"; "still having before their eyes the ravages" of flooding, earthquakes, and volcanoes on a giant scale, "these men . . . conserved a durable and almost eternal memory of these misfortunes of the world" (*EN* 120). The secular convictions underlying this grim origin story must have been readily apparent to the Sorbonne, and Roger again cites the influence of Boulanger (*EN* 306n1), who traces all superstition back to the originary catastrophe of the Flood or its analogues. The narrative arc of "giant race[s]" enjoying "liberty and tranquility" before declining into a cooler modern world, the onset of which is loosely marked by a "convulsion" or series of catastrophes, bears some general resemblance to an idea that Boulanger shared in common with other, more orthodox antiquaries: that laws, religion, and other institutions originated from, and commemorated, a traumatic dispersal of humanity associated with the Biblical Deluge.[58]

Boulanger's influence is not as alien to the general trajectory of Buffon's thought about human origins as Roger suggests; on the contrary, Boulanger's skepticism allowed Buffon to preserve a deep ambivalence about human hegemony at the beginning of the very chapter that concludes by celebrating the domestication of nature. The idea of species memory, in particular, goes against the grain of some of Buffon's late theodical revisions. Using the same word, *témoin*, to designate human witnessing of larger planetary processes, Buffon's manuscript expresses a kind of relief that human beings came too late to witness the planetary upheaval of the fourth epoch, "when the volcanoes began to act." Even this relatively late draft of Epoch IV was revised in print to recast these "petrifying and terrible scenes" of volcanism as "heralding the birth of intelligent and sensible Nature."[59] The opening of the seventh epoch suggests that "the first men" did remember, on some level, the great eruptions of the fourth epoch, or at least their aftermath. Throughout the work, then, a synchronic, multispecies narrative that decenters human hegemony infiltrates and complicates the stock Enlightenment narrative of nature subdued.

This Enlightenment narrative prevails at the end of the final epoch, but the triumph of domestication is nonetheless instructive as an aversive reaction against the half-unwilling discovery of human animality. When he abuses "barbarians" of various eras (prehistoric, Germanic, Native American) as "animals with human faces" (*Z* 125), Buffon is, in part, exorcising a kind of newly discovered terror inherited by the entire species from its deep, turbulent past. The same goes for his extravagant claim that the twenty or so species entirely domesticated by humans are "more useful to the earth than all the others," a claim that on some level contradicts the work's preoccupation with ancient megafauna.[60] The earlier, more nuanced account of domestication, though still teleological, depicts its history as embedded in the collective history of species and dependent on a long prehistory of sedentism (*EN* 161n.). I argued in the second section of this chapter that the late arrival of humanity in *Epochs of Nature* creates the distance that Buffon needs for his categorical distinction between geological and historical time. The synchronic multispecies narrative of human origins examined here complicates this linear model, which not only caused controversy among Buffon's readers but also posed an imaginative challenge to Buffon himself. He addressed this challenge by populating the abyss of time to a greater depth than the written record allowed.

The Romance of the Dark Abyss of Time

Forster's rebuttal, in his review of *Epochs*, depends on his reading of ancient geographers, who provide him with adequate evidence for every kind of geological upheaval within the confines of the written record. In leveling this criticism, Forster capitalizes on the ambivalence displayed in Buffon's handling of memory and tradition. It is unclear, in the end, how much geological upheaval the "first men" witnessed or how far back their origin may be traced. Buffon calculates that after 60,000 years the planet reached the state of "repose" necessary to "its most noble productions," the land animals (*Z* 116). But he acknowledges that "this repose was not absolute," since the separation of the continents, and other "great changes," were still to come—some potentially of much more recent date. Thus Forster does not dispute the thesis of geological epochs but rather the chronological scope of the evidence. Challenging Buffon on his own ground, he marshals evidence from ancient geographers, including Strabo, to support the notion that a megaflood originating in the Black Sea basin produced the Mediterranean Sea, and offers the sum of recorded (or at least remembered) "great changes" of this kind as adequate to the task of explaining the earth's present landforms.[61] Although he does not refer to it explicitly, Forster's Pacific voyage may have conditioned his defense of a young earth model, since he had learned to identify atolls and recent volcanic islands in the course of the voyage. More fundamentally, though, he uses his evidence to insist that the shaping of the continents should not be made the stuff of "*Mährchen*" (fairy tales), for "it is history!"[62]

In the twenty-first century, deep time has been claimed as the province of epic both "high" and "low," and Buffon's *Epochs* in particular has been declared "an epic, in scope if not in length" by Zalasiewicz and his colleagues (*Z* xxxiii).[63] Although Buffon's tendency to confuse history and fable made Forster indignant, the latter's view of this work as a beautiful romance (*wunderschöner Roman*) is not unmixed with admiration, as I argued earlier. At least two reviewers writing in French (Meister and Guettard) use the same ambivalent word *roman* (romance/novel) to signal both the literary quality of Buffon's prose and their skepticism about his rigor. The critical distinction between epic and romance offers a way of understanding the reception history, not only of this work but of stories about deep time more broadly. Forster's more local distinction between fairy tale or fable and history makes it especially clear that he is categorizing Buffon's deep time narrative as a romance based on its improbability or perceived lack of truth value. Once a consensus is reached concerning the true age of the earth, the transactions of the rock record take on a more epic character—more like the heroic *Iliad* than the romantic *Odyssey*—as we see in John McPhee's description of the geological events that created the western U.S. basin and range topography as "violence on an epic scale."[64] Even the romantic figure of the "abyss of time" is invoked in modern scientific contexts as simply a vivid expression of the true mastery of time attained by geology. This consensus is of recent date, and correspondingly romance runs deeper than epic in the strata of deep time.

This chapter concludes with an assessment of romance as a structural element of *Epochs* and its reception history in order to locate points of intersection between the history of deep time and literary history more broadly. As an author known for his reflections on style, Buffon would have been attuned to the critical distinction between epic and romance, which dates back to the Hellenistic period, and to the related debate in his own time over the new prose fiction form known in English as the novel. The stakes of this debate might have seemed higher in the Anglophone context, since in French the same word (*roman*) continues to be used for the works of authors from Jean de Meun to Balzac and beyond. In *The History of Tom Jones* (1748), Henry Fielding inserted a critical preface at the beginning of each book to make sure his readers knew that his novel was indeed a "history," based on "the vast authentic Domesday-Book of Nature," and no mere "romance." At the end of the century William Godwin took note of the rising popularity of historical fiction, stating provocatively that "the writer of romance is to be considered as the writer of real history." I believe Buffon would have sympathized with this view, since he willingly employs vivid literary tropes in the service of a true history of the earth.[65] Many of the key tropes in this narrative partake of the marvelous character of romance, including its expansive scale, its multispecies orientation, and its concern with past extinction as well as the distant future. In spite of the epic turn in the reception of deep time, elements of romance survive in popular science writing and in the vocabulary shared by deep time narratives since Buffon: the "epoch," the rock or fossil record, the book of nature, the primitive past, and the abyss of time.

For critics of scientific romance, any exploration of the deep past was suspect because the subject is so bound up with myth and tradition, and because it diverted energy from fieldwork to writing. Cuvier, who opted for a close reading of the fossil record instead of remote conjectures on the *terrains primitifs*, probably had Buffon in mind when he heaped scorn on "naturalists [who] rely a lot on the thousands of centuries that they pile up with a stroke of the pen."[66] One of Buffon's fiercest contemporary critics was Jean-Étienne Guettard, whose mineralogical fieldwork is referenced in *Epochs* (*Z* 156). In a scathing letter that was probably never sent, Guettard exclaims, "More Buffonades, my dear Count, how much longer will you play Cyrano de Bergerac?" Guettard punningly links *Epochs* to the Robinsonades inspired by

Robinson Crusoe as well as to the scientific romances written by Cyrano de Bergerac (1619–55). Citing Buffon's "hypothetical ideas" about the formation of planets and their long, slow extinction, Guettard charges him with getting "carried away into romance" and predicts caustically that his *romans* will "ravish the antechambers" and pass through ladies' toilets to be "devoured" by the "chambermaid" and to "amuse" the "lackey."[67] Playfair, who is now remembered in part for his use of the phrase "abyss of time," rejected all previous theories of the earth even more sharply as "derangement," but his logic is anticipated by readers who took Buffon to task for repackaging obsolete late seventeenth-century theories. Guettard's criticism also emphasizes social distinction, suggesting a gradual increase in the prestige of popular science writing as a possible reason for the more balanced later reception of Buffon and early deep time narratives, in addition to the much wider acceptance of deep time as a reality.[68]

The Anthropocene, in large part a phenomenon of popular science, presents a new motive for engagement with the history of deep time. Zalasiewicz et al. clearly state the Anthropocene project as a motive for their new translation of *Epochs*: "the ideas of Buffon take on a new significance as stratigraphers reconsider the idea of 'epochs,'" all the more because in their view "natural history and cosmology have reunited in planetary systems science" (Z xiv). Partly for this reason, they suggest, even scientific readers of today are less apt to dismiss the work for being accessible or popular, or for its literary qualities. Historical distance also makes it apparent that obsolete science can turn out to be "to an almost uncanny extent correct," as Elizabeth Kolbert puts it in *The Sixth Extinction*.[69] Kolbert limits Buffon to a cameo appearance, but Cuvier's theory of extinction via geological "revolution" (to which the quotation refers) structures her entire narrative; the role of Cuvier (also a "poet") resembles that of a romance hero whose quest for scientific truth can finally be validated. As noted above, Buffon plays a somewhat similar role for Cohen in *The Fate of the Mammoth*. Kolbert's extinction narrative integrates these historical models with current research on extinction as well as human origins, raising the stakes of the extinction threat while also preserving some of the uncertainty surrounding the history of our species in relation to others. Recognizing that Buffon's theory of extinction allowed the human species to appear on the stage of geohistory, she exaggerates his caution considerably to produce a romance of discovery: "with great trepidation, Buffon allowed that this last species—'the largest of them all'—seemed to have disappeared. It was, he proposed, the only land animal ever to have done so."

There is an implied contrast between Buffon's hunch and the confidence that Cuvier was able to build as he added species after species to the rolls of extinction after Buffon's *incognitum* (the mastodon or "Ohio animal").[70]

Cuvier is undoubtedly the more significant figure in the history of extinction theories, but Kolbert's curious treatment of Buffon is worth noting because she allows him to enter as an actor on the scene that he first sketched, in which extinct "elephants" are the major protagonists. The upended mammoth skeleton on the cover of Kolbert's "unnatural history" continues this multispecies narrative. Kolbert also chooses Big Bone Lick on the Ohio as the setting for her own chapter on the mastodon, the site from which Buffon obtained (via intermediaries) much of his evidence. She inherits the topoi of sympathetic identification and interspecies recognition—also suggested by Buffon's narrative of a joint human-elephant migration from Asia to North America—from the eighteenth-century response to the sheer quantity of remains at this site. Cuvier evokes some of the same pathos by describing the fossil record as "the ruins of the great Herculaneum overwhelmed by the ocean"; James Macpherson's lament (in the guise of "Ossian") over a vanished Celtic warrior race may be an equally pertinent example from Buffon's own time.[71] Macpherson surreptitiously places himself on the scene by adopting the persona of the fourth-century bard, somewhat as the modern scientist actually enters the "lost world" of the dinosaurs in real time in Arthur Conan Doyle's extinction narrative *The Lost World* (1912). Buffon's romance of extinction is less sentimental in the sense that the human actors present on the scene of megafauna extinction really are peripheral to it, not because they are "savages" but because they are shadowy figures, proto-human in the uncertain sense that obtained before the advent of prehistory as a science.

Buffon's style is a large subject, and my goal here is to identify a limited subset of figures affiliated with the romance tradition, figures that fall between literary history and the history of science. Having established radical distance between geological and historical time, Buffon saw the need for prehistory as a logical consequence. Since he lacked the strong archaeological evidence or the systematic dating needed to support a prehistoric period, his contemporaries cannot be faulted for seeing Buffon's "giants" and some of his other figures as nothing but "huge cloudy symbols of a high romance." Buffon adapted his argument to the historical circumstances of his science by using literary resources, including romance; his masterful use of classical and modern rhetoric, in *Epochs* and elsewhere, has been studied extensively by Jeff Loveland and Benoît de Baere. "Throughout the *Histoire naturelle* Buffon used striking images, stylish writing, and sublime views of nature to attract a broad readership," Loveland observes, adding that this strategy extended to Buffon's treatment of "God creating the world," also the title of an allegorical engraving that accompanied Buffon's *History and Theory of the Earth* in his first volume (1749).[72]

Buffon's use of the sublime merits special attention here. Nicolas Boileau's 1674 translation of the first-century (CE) treatise "On Sublimity" is widely credited with sparking the European recovery and revival of this source of grandeur and mark of genius in rhetoric and poetry. The influence of Boileau's translation is apparent in Bernard Le Bovier de Fontenelle's *Conversations on the Plurality of Worlds* (1686), a key text in the scientific romance tradition, and the sublime is also essential to literary texts written to popularize Newtonian science in the eighteenth century, including works by James Thomson (1700–48), among many others. The sublime touched almost every area of eighteenth-century thought including the philosophy of Kant, whom I have already linked to Buffon's use of magnitude. All parties agreed that the Hebrew Bible was an important, if not the ultimate, source of the sublime in writing. "On Sublimity" is one of the earliest pieces of literary criticism to cite an example from the Hebrew Bible, including verses from Genesis 1 later glossed by Buffon in *Epochs*.[73] Buffon's use of creation imagery from Genesis throughout his career may be opportunistic, as Loveland suggests. His extended gloss on Genesis 1:1–4 in the preliminary discourse that begins this work, nonetheless, offers compelling exegesis as well as scriptural authority for his extended time scale.

Buffon goes far beyond the conventional interpretation of the "days" of creation as extended periods to derive a long-term sequence of events from textual evidence including imagery and diction, specifically the Genesis compiler's use of verbs in the imperfect. Anticipating the objection that his chronology "contradicts" that of Scripture, Buffon's ingenious gloss highlights the uncertain form and duration of the "abyss" that exists between the origin and the separation of light and darkness, when "the earth was without form, and void" (KJV Genesis 1:2). "The earth was [*étoit*], darkness covered [*couvroient*], and the spirit of God was [*étoit*]," Buffon intones: "these expressions, using the past tense [*imparfait*] of the verb—do they not indicate that it was during a long interval that the Earth was formless and that darkness covered the face of the abyss?" (*Z* 16/*EN* 19). I include the French here because English does not capture in the indefinite duration associated with the French imperfect as well as the Hebrew verb forms, including the crucial instance of "to create" (בָּרָא) that immediately precedes them.[74] "And God said" (*Or Dieu dit*; Genesis 1:3) signals another epoch to Buffon, the end of one process and the beginning of a new one, as do the subsequent verses, and in this manner his sequence of major epochs becomes a more faithful reading of the text than a more literal interpretation would be. The first epoch of raw matter that preceded a cosmos with celestial bodies, taken from the initial emphasis on the chaos and darkness of the abyss (תְּהוֹם), informs the abyss of time that is so central both to the text of *Epochs* and to its reception. The sublime retrospect, a survey of the previous epochs included at the end of several chapters, also owes something to God's recurring review of his creation in the first verses of Genesis. Buffon stands out as a pioneer of geochronology because his time scale is secular, but Pascal Richet still overstates the case when he suggests that "Buffon's epochs did not owe anything to Moses."[75]

Buffon was orthodox enough to accord a special status to the sublime of the Bible, but *Epochs* also embraces material from other cultural traditions. Notable examples include religious reverence for mountains in West Africa, the oral tradition of the Chukchi regarding intercontinental trade, and the "memory of the submergence of Ceylon [Taprobane]" (*Z* 109), an indigenous tradition from Sri Lanka on which Buffon places considerable emphasis as "the most ancient tradition" and thus the strongest available corroboration of another hypothesis concerning the separation of the continents (which Buffon extends here to the Indian Ocean). This folk memory of the sinking of a larger land mass, thus reported in 1611 by a Dutch navigator, was available to Buffon from the *Encyclopédie* as well as from Boulanger, Buffon's main source here according to Roger (*EN* 303–4n20). Here again Forster might object that Buffon is spinning a romance of geological time that could be verified in the written record—in this case, ancient accounts of Taprobane as a lost continent akin to Atlantis—except that Forster himself puts some faith in folk memories of geological revolutions in his own travel narrative. Forster's discussion of the "Great Year" of the ancients in his review, though based in part on classical references, partakes to a considerable extent of the ethnographic attitude toward myth that sometimes authorizes its application to natural knowledge. Another German Lutheran pastor, Johann Gottfried Herder, relies more strongly on tradition, especially Hindu tradition, in forging connections between prehistory and geology (as detailed in my third chapter). Even Charles Lyell, writing in 1830, examined ancient Hindu cosmology with some care before deciding that the true "history of the globe was to them a sealed book" because of the lack of evidentiary support for the generous time scale of this and other ancient cosmologies.[76] Earlier theorists of deep time such as Buffon and Herder were more willing to consider the role of oral tradition as a shaping force of both scriptural and folkloric accounts of the earth's antiquity.

The dark abyss of time is not invoked directly in the Bible or in any tradition available to Buffon. This contemporary trope was widely available, however, in the poetry, natural philosophy, and other literature of his time. The synthesis of these contemporary sources with a longer tradition in the *Epochs* is one of his lasting achievements. "Dark abyss" (*sombre abîme*) is one of the epithets used by Antoine-Léonard Thomas in his "Ode to Time," awarded the poetry prize of the French Academy in 1762. Buffon may have known it, but the example is instructive primarily because it is so conventional. Thomas addresses Time in the first three and last three stanzas of this superficially irregular but deeply orderly ode (in eighteen six-line "tail rhyme" stanzas).[77] The "dark abyss" of time becomes suddenly less fearsome when Thomas turns to address the Deity, who is impervious to the "ocean of ages" (l. 19) and dictates its "law" (l. 16), but it remains intimidating to the individual soul of the speaker struggling to erect a "barrier" between himself and time's "vast expanse" (*vaste étendue*; l. 28). Surveying the daunting ruins of time, the speaker discovers his mastery over time in "active thought" (l. 41) and returns to the theme of "eternal night" in which Eternity will drown Time when it sees fit—precisely the moment of Time's "death" that inspired the image of the dark abyss earlier in the poem. Thomas is remembered as well for his eulogies (*éloges*) of academicians and others, and the figure of time's devouring abyss defeated by the soul of genius seems appropriate to that genre. The poem includes a footnote indicating that Thomas is following "the opinion commonly received among philosophers," that time is "dependent on the existence of created beings" (l. 7n.). The "dark abyss" also occurs in English verse in a context of scientific popularization—as an epithet for the vastness of space in James Thomson's "To the Memory of Sir Isaac Newton" and for the earth's interior in his *Autumn*.[78] This last usage is more consistent with Buffon's rendering of the deep past, but both the French and the English analogues betray an association with conquest and heroic discovery, a quest romance motif that shines through Buffon's figures for deep time as well. Touting the explanatory power of his thesis of the separation of the continents in Epoch VI, he claims that it "confirms" his chronology at that point "where one begins to fall into such a depth of time that the light of the imagination [*génie*] appears to be extinguished" (Z 14).

The "darkness of time" (*nuit des temps*) that Buffon sets out to "penetrate" (Z 4) ultimately has less in common with the *éternelle nuit* of the poets than with the vocabulary of his contemporaries in natural philosophy, including

Boulanger and Nicolas Desmarest. The poetic image is of great service to naturalists seeking to express the qualitative impact of duration made evident by close study of the strata. To reiterate the main argument of this book, the history of deep time before modern geology and its radiometric dates depends on qualitative measures in literary form, such as this one. Where Thomas and others used this imagery to promote a metaphysical view of the *end* of time, Boulanger was one naturalist of the mid-eighteenth century who reconfigured the same semantic field around geological origins, influencing both Buffon and Desmarest. Buffon's virtuosic gloss on the first verses of Genesis borrows so much from Boulanger that it led to a charge of plagiarism in 1779, discussed in detail by Roger and other scholars. Boulanger's discussion of Genesis is much longer and instructive here for the explicit connections that it draws between the Biblical or "Oriental" sublime and the true extent of geological time. Boulanger is ultimately concerned in *Anecdotes of Nature* to backdate the history of the earth by showing how grossly inadequate strict Biblical chronology is to the geological evidence. He is therefore much more concerned than Buffon to refute works of natural theology and is at the same time less optimistic about establishing a definite sequence of epochs or even gaining general acceptance for a long time scale. These and other underlying differences, according to Roger, make Buffon's obvious "borrowings" inconsequential (*EN* lxxvii). Nonetheless, he borrows extensively at several points in *Epochs* that have been central to my analysis: his meditation on the scale of time, his link between the coal measures and tropical forests, and his final account of the condition of the "first men."

Boulanger's reading of Genesis, like Buffon's, points out the implied lapse of time between each of the first four verses, which he considers indispensable "vestiges" of recurring geological revolutions that can be glimpsed as we "search through the shadows of antiquity" (*cette ténébreuse antiquité*). Boulanger likewise focuses on the statement that "the earth was without form and void" but ultimately carries his search for images of chaos and devastation much further, though the Psalms as well as Jeremiah and other prophets. The Hebrew Bible is full of images that are "rendered all the more sublime by the style of Oriental genius" but are "figurative expressions," nonetheless, of matters of which one can only learn the details from nature.[79] Boulanger's point is that these examples may be multiplied indefinitely, but it is hard to imagine that the forty-second psalm did not occur to him and to Buffon at some point in their investigations. Here "deep calls out to deep" (*abyssus abyssum invocat* in the Vulgate), echoing the depth of the Psalmist's despair. Boulanger does ultimately despair of plumbing the "impenetrable abyss of past ages" and contents himself with a time-lapse condensation of all the changes of the globe as they would have unfolded if compressed into 6,000 years—a counterfactual narrative that anticipates Lyell—to show the absurdity of the "vulgar" and literal Biblical chronology. Desmarest, a more meticulous field naturalist than either Boulanger or Buffon, internalized the value of Boulanger's imagery and deployed it in his last work, *Géographie physique* (1803), which includes a long entry titled (after Boulanger) "Anecdotes of Nature." In publishing this work many years after it was first drafted, Desmarest reflects on the continued resistance to the idea of deep time: "The depth of the abyss of time into which our spirit is obliged to plunge from here appears so immense, so little adapted to our way of thinking, that it is not surprising that most people are little disposed to believe in these revolutions, however verified by numerous monuments."[80]

As noted in the introduction to this book, "the dark abyss of time" is sometimes misattributed to Buffon, and it is easy to see why. The above sketch is meant to show the wider semantic field occupied by this expression, a set of conventions well established by Buffon's time and used widely by writers in French and English. Buffon used a number of analogous expressions to help recast the topos of the abyss in secular and naturalistic terms while still profiting from its association with the sublime. Rossi and Laurent Olivier, who attribute "the dark abyss of time" to Buffon without citing a location in his works, may have been misled by Roger's use of quotation marks around this phrase in the introduction to his redoubtable edition of *Epochs*. Buffon's contemporaries, he claims, "could not imagine the 'the dark abyss' of such a prodigious antiquity, an abyss in which man no longer signified, an abyss even more inconceivable than that of infinite space, the eternal silence of which terrified Pascal" (*EN* lxvii). I suspect that these internal quotation marks merely indicate that this was already a conventional expression. The resulting link between a thinker, a key phrase, and a concept amounts to a priority claim: just as Gould (and others following him) assert a priority claim on behalf of McPhee's "deep time," others have asserted an analogous priority claim on behalf of Buffon and the "dark abyss of time." Richet credits Buffon as a pioneer in mapping "the newly discovered immensity of time" but uses "the dark abyss of time" in the first page of his own account.[81] In my introduction I surveyed the importance of this phrase as a recurring topos in modern scholarship, a new form of the convention that testifies to its enduring power.

Conclusion

Rossi groups Buffon with a small radical coterie, including Kant, Denis Diderot, and Thomas Wright, who were willing to contemplate a time scale in the millions of years as early as the mid-eighteenth century.[82] Kant, Diderot, and Wright, however, based their much higher estimates on astronomy, and Buffon's figure of 75,000 years for the age of the earth pales even more in comparison to the cosmological "deep time" at issue in the debate between Mark McGurl and Wai-Chee Dimock over the scale of literary representation. McGurl initiated this debate in 2012 by critiquing Dimock's notion of "literature across deep time" on the basis that literary history is vanishingly small by contrast to geological time. He proposes to "reclaim the term *deep time*" in its "original geological meaning" by attending to literary texts that meet this "representational challenge" by thematizing deep time in this prehuman sense.[83] McGurl chooses the science fiction of H. P. Lovecraft for this purpose. His virtuosic reading of the stories shows that Lovecraft's images of uncanny embodiment can serve as a metonymy for the otherness of prehuman (and posthuman) time. There is more at stake in this juxtaposition of genre fiction ("low epic," as Dimock diplomatically calls it in her response, published in the same journal) against the more canonical examples originally offered by Dimock in her book, such as Ralph Waldo Emerson's remediation of the Vedas and other engagements spanning many centuries or even millennia of literary history. In her reply, Dimock offers examples of the sublime in poetry to bridge the distance between these two competing understandings of deep time, while also questioning the ethical imperative that seems to inform McGurl's vision of deep time, in its modern scientific sense, as a "representational challenge" for writers.[84]

The sublime in its Kantian guise is anthropocentric, but it is an ancient trope with a long, ambivalent history and played a vital role in making the representation of geological time possible in the first place. In concluding this chapter, I have argued that the sublime as well as romance topoi are part of the fabric of geochronology, offering examples from Buffon and his contemporaries. What I have called the "prehistoric turn" in Buffon establishes an area of uncertainty between historical and geological time. However ancient they may be, Lovecraft's alien races too are prehistoric and as such can bear comparison with both the proto-human "giants" and the lost high civilization of Buffon. They serve to extend the scale of cosmic time by backdating the advent of intelligent life, which in turn backdates the origin of the cosmos in a secular system.

Granting a literal geological (or astronomical) sense to deep time should not preclude recognition of its metaphorical structure. Buffon's time scale is foundational not in a geological sense but rather because it enabled a differentiation of human and prehuman times and hence reflection on their differences. Buffon reflects on the stages needed to lead from the one to the other, which I have described as a multispecies prehistory. While this transitional idea of prehistory is historically specific to Buffon's secularizing age, the use of ancient texts and oral tradition as potential source material in areas of uncertainty, including human origins and global-scale geophysical changes, or "revolutions," positions Buffon and his contemporaries in something like Dimock's literary deep time, even as they stand on the brink of the abyss of "natural" time that now looms ever larger. I have suggested that Buffon's engagement with nature's archives produces geological time as a surplus, analogous to surplus energy. The fantasy of a legible earth history, which the Anthropocene returns to us in a new key, constitutes another kind of abyss, a history of nature in which we see the history of our own species reflected. The felt experience of deep time, however, is hardly limited to the specific historical actors who inhabit the Enlightenment, or scientific modernity, or the Anthropocene. Herodotus had the discontinuity of human and natural history very clearly in mind when relating his epiphany concerning the Nile Delta that begins his account of Egypt: "the Delta is alluvial land and has only recently (if I may so put it) appeared above water." In contemplating how much sediment the Nile might have transported during "the vast stretch of time which has passed before [he] was born," Herodotus arrived readily at the figure of tens of thousands of years that Buffon struggled so hard to justify and that his contemporaries decried as romance.[85]

3

William Blake, the Ballad Revival, and the Deep Past of Poetry

> Hear the voice of the Bard!
> Who Present, Past, & Future sees
> Whose ears have heard,
> The Holy Word,
> That walk'd among the ancient trees.
>
> —WILLIAM BLAKE, "INTRODUCTION," *SONGS OF EXPERIENCE* (1794)

Deep Human Time

With the possible exception of "ancient," Blake's vocabulary in the opening stanza of *Songs of Experience* seems entirely foreign to the modern secular idiom of deep time. We seem to be firmly anchored in sacred history, in which temporal duration vanishes in the presence of the word. The poem's eponymous voice, however, soon unravels into two distinct strands, notwithstanding the speaker's appeal to the Hebrew tradition of the prophet who speaks in God's voice. To make matters worse, Earth replies to the anguished appeal of this divided voice in "Earth's Answer," the next poem in the volume, by undermining the "Introduction" completely. She exposes the "starry floor" of heaven as "starry Jealousy" and impugns the Holy Word as spoken by a "selfish Father" of "ancient men" (E 18).[1] Readers of *Songs of Innocence and Experience* know that the descent into Experience is precipitous. In spite of their deceptively simple diction and stanza form, this pair of opening poems introduces a host of unresolved ambiguities, plunging readers into the "endless maze" of doubt from which "The Voice of the Ancient Bard" actually promises to deliver us elsewhere in the collection. Students of the combined *Songs*, advertised by Blake as "Shewing the Two Contrary States of the Human Soul," may have recourse to the structural logic of the collection, juxtaposing (for example) this "Introduction" against the "Introduction" to *Songs of Innocence* (1789), a clear first-person account of inspiration and aesthetic unity. Amidst the confusion generated locally by these two poems, however, three facts stand out that make their subject less foreign to the matter of deep time than it may have appeared. Blake's repetition of "ancient" draws attention to the distinction between sacred and secular time that informs the conflict between the Bard (or Word) and Earth. The very form of a dialogue between Bard and Earth evokes the situation of deep time at the juncture of geological time and human prehistory. Finally, the stanza form of both poems is derived from the ballad, a type of song or poem—as Blake well knew—that was associated since the early eighteenth century with the origin and early history of the human species.

Later in his career, Blake created an allegorical figure for human antiquity, the "Ancient Man" Albion, who became the protagonist of major epics that engage deeply with questions of time and scale. *Songs of Innocence and Experience* shows more clearly the role of the ballad revival in prompting Blake's concern with these questions. The ballad revival depended on conjectural history and on the "comparative method" (of inferring ancient from primitive states), both key methodological features of the natural history and ethnography of the Forsters and Buffon, as I argued (respectively) in the first two chapters of this book. My last chapter will trace this comparative method into the Victorian period, when it was formally recognized and practiced by social scientists including John Lubbock, who particularly influenced *The Descent of Man*. Oral tradition, still entertained as a potential source in these later accounts of prehistory, will be a central focus in the present chapter on the ballad revival, a literary movement dedicated to the recovery of oral traditions. The theorists of the ballad revival brought conjectural history to bear on poetry persuasively enough to inspire literary practitioners including James Macpherson and Robert Burns as well as Blake, whose work is central for this chapter.

The work of Johann Gottfried Herder (1744–1803) is equally central, both for this chapter and for my larger argument, because he synthesized the ethnographic tenets of the ballad revival with the calibration of geological time. In his consideration of geological time and its relation to ethnography,

Herder engaged closely with the Forsters' voyage narratives and with Buffon's *Natural History,* including *Epochs of Nature,* the major texts examined in the first half of this book. He was also a key contributor to the ballad revival who published the first major collection of folk songs in Germany (1780) and went on to write one of the most important conjectural histories, *Ideas toward a Philosophy of the History of Humanity* (1784–91). In the latter work, Herder recognized, like Buffon, that both natural and antiquarian evidence was needed to support the expansion of the time scale. A poet himself and a scholar of the Hebrew Bible, Herder preserved a space for the verbal media of the deep past alongside nonverbal media such as ancient artifacts and fossil bones. In his complex response to the ballad revival, Blake developed another version of the argument that poetry can serve a speculative reconstruction of the deep past. Reading Blake with Herder restores a connection between poetic and naturalistic histories of the human species that remains relevant for the evolutionary thinkers treated in the next chapter, on Darwin and Lubbock.

This chapter will read poems by Blake and others alongside traditional ballads and theories of the ballad, with frequent reference to the conjectural histories and theories of language upon which ballad editors drew. Herder's conjectural history is a special case because he augmented it with a consideration of geological time. Although Thomas Percy, Walter Scott, and other editors of ballad collections saw themselves as voyaging into the primitive past, they did not generally look to the earth's history for a scale by which to measure the antiquity of poetry and song. Apart from Herder's later work, geological time as such is not in play in the ballad revival. Rather, this chapter makes the case for deep *human* time independent of, and prior to, any geologically inflected models. In the last chapter, I showed that Buffon double-dips in establishing the antiquity of the human species, backdating it from both geological and antiquarian evidence (such as prehistoric knowledge of astronomy). The English naturalist John Whitehurst gives even more prominence to prehistoric arts and sciences in his account of the deep past, and Blake draws exclusively on this body of evidence in his reckoning. Blake, however, further cultivates the awareness of temporality as such, of temporal scale as a product of reflection. Thus in Blake we have deep time without extension, the "moment of pulsation" that enfolds past, present, and future into an expanded heroic present. Strictly speaking, Blake's "moment of pulsation" may be the antithesis of deep time. In *Milton* (1804–11), Blake uses this memorable phrase to describe the time in which "the poet's work is done," but in *Songs of Experience* he is more concerned with the disjunction between bardic time and natural time caused by religious morality. Earth concludes her "Answer," therefore, by demanding a restoration of "free Love" as the birthright of the "ancient men" or human species.[2]

The poems of Blake and other contemporaries influenced by the ballad revival, including Burns and Macpherson, set out to make the deep past present. The ballad as a form also does this work. For its theorists and for many of its practitioners since at least the eighteenth century, the ballad instantiates the deep past of a human species defined by its capability for song. For a twentieth-century critic such as Albert Friedman, the "inorganic refrain" that accompanies many ballads (also referred to in the literature as a "nonsemantic burden") constitutes the survival that proves this case.[3] In Blake's "Introduction" to *Songs of Innocence*, it is the nonverbal music of the pipe that constitutes this survival: "Drop thy pipe, thy happy pipe," urges the child-muse of this poem, "sing thy songs of happy cheer!" The first two stanzas of this "Introduction" instantiate the primitive emergence of song-speech out of sound, or at least thematize a performance of that transition. In Burns's "Epistle to Davie, a Brother Poet" (1786) we see a similar transition: "On braes when we please then, / We'll sit and *sowth* [hum] a tune; / Syne *rhyme* till't, well time till't, / And sing't when we hae done." One common source shared by Blake and Burns on this point is Joseph Ritson, whose work Blake illustrated and whom Burns acknowledged as the source of "all [his] critic-craft."[4] Ritson stands out among a long line of poets and theorists in the eighteenth and nineteenth centuries who applied the method of conjectural history to the origin of the arts in order to insist on the primacy of song and the musicality of song. Burns and Blake, in turn, are notably preoccupied with song, and although they themselves probably did not read the more extensive philosophical treatments of the role of poetry and song in the origin of language, such as Vico's *New Science* and Rousseau's *Essay on the Origin of Languages*, Ritson's extensive citations made it clear enough that this was a philosophical issue. The peculiar balladic turn given to this argument by both theorists and practitioners of the ballad revival also paved the way for later developments such as Percy Bysshe Shelley's history of poetry as a "history of our species," and perhaps even for Bruce Chatwin's *The Songlines* (1987).

The poets and theorists who made the ballad revival saw the ballad form as an avenue into the early history of the human species. Although it is anachronistic, the metaphor of "deep time" makes sense here because their conjectures about the origin of song led them to dig beneath the seemingly firm boundary set by the written record. According to Ritson, "all writers agree that Song is the most ancient species of poetry. Its origin is even thought to be coeval with mankind: to sing and dance seeming almost as natural to men as the use of speech and walking. Hence we find the dance and the song wherever we find society; in the least polished, or most savage nations."[5] The frontispiece (Fig. 3.1) preceding this declaration in *Select Collection of English Song* (1783) features an exceptionally rustic shepherd pair; though this is not one of the eight plates that Blake engraved for this book in his capacity as a commercial engraver, the association with pastoral was not lost on him—as we can see (for instance) from the child-muse's first direction to the piper of *Innocence* (1789): "Pipe a song about a lamb!" In looking to pastoral for traces of poetic origin, Ritson, Blake, and others actually skip over the first, barbarous stage of civil society mandated by the standard four-stages theory, the hunter-gatherer stage that notionally precedes husbandry. In her study of the ballad revival, Maureen McLane has noted this "resistance to the unilinear march of stadial history" while assuming the claim for the ballad's originality as the burden of her own argument. Poetry, McLane argues, "is that imaginative genre or medium which extends back into what Beattie, Scott, and after him Shelley all called 'a rude age': the time before print, indeed before writing, before history as we know it."[6]

Introducing his *Collection of Old Ballads* (1723–27), a work recognized by Scott a century later as the "first" ballad collection, Ambrose Philips stated that his primary motive was "to shew their antiquity" and recover the artistic freedom of "our poets of old."[7] Later editors followed Philips in establishing the antiquity of ballads as a special, almost absolute kind of antiquity and in using ballads to pursue history by other means. Burns echoed these theorists by noting archly that "Homer is allowed to be the oldest ballad singer on record" (*BLS* 86n.), though in fact Philips and Scott, among others, gesture toward a much deeper unrecorded past of song. The more idealistic defenders of the ballad's antiquity embraced the doctrine of survivals to celebrate the "true eloquence" of modern-day "savages" among the folk, as Herder put it. Others, including Ritson, Wordsworth, and Scott, were made skeptical by what they saw as evidence of degeneration, making the survival of "ancient customs" dangerously close to atavism. Joseph Addison, one of the first champions of the ballad tradition, introduced this ambivalence by admiring traditional ballads such as *Chevy Chase* as specimens of "nature in her simplicity and nakedness." The polysemous association with "nature" stuck to the ballads, linking the conjectural history of early arts and ballads in particular and the doctrine of survivals ultimately to the species question as focalized by what another ballad revivalist, John Aikin, called "the native species of poetry of this

FIGURE 3.1. *Frontispiece*. Unsigned engraving. Joseph Ritson, *A Select Collection of English Song* (1783), vol. 1. Courtesy of HathiTrust.

country." Writing in the 1960s like Friedman, Roger Lancelyn Green explained what Friedman called "the ballad's otherness" in quasi-evolutionary terms: "The invention of ballads to celebrate great deeds has no beginning; doubtless the cave-man made up a ballad after an outstanding adventure with a sabre-tooth tiger or a neighboring tribe."[8]

Though it is seldom made explicit, there is an underlying connection between the ballad's orality and the deep human past that pervades ballad scholarship from the eighteenth century to the present and deeply informs original poetry by Blake, Burns, and others influenced by the ballad revival. In *The Ballad Revival* (1961), Friedman defined the ballad's otherness in terms of its "subliterary" qualities, preferring a kind of sociological aesthetics to the archaeological idiom exemplified here by Green. When Friedman declares that "proofs of [Blake's] ballad culture are everywhere," he refers to the influence of stall ballads, jingles, and street cries on Blake's *Songs of Innocence and Experience* and some of his other lyrics, including those in *An Island in the Moon*.[9] However, "Introduction" and "Earth's Answer," among other poems, show Blake as equally attentive to the ballad's association with prehistory. In this respect, Blake's mythopoeic practice is consistent with Friedman's own view of contemporary ephemera as survivals from the deep strata of the human past, a view inherited directly from nineteenth-century anthropologists such as Lubbock and more indirectly from some of the same Enlightenment histories that influenced Blake. After examining Blake's "ballad culture" in greater detail in the next section, this chapter elaborates the six topoi associated with the ballad tradition that have been outlined so far: the "rude age" of early humanity, conjectural history, the precedent of Homer, the origins of art, survivals, and human species identity. These topoi (explored in the following three sections) create an interface between ballad literature and human prehistory, or species time. The chapter then returns to Blake to show how these topoi from the ballad tradition informed Blake's reflections on time, scale, and human antiquity in his later mythopoeic epics.

Blake's Ballad Culture

Blake's "ballad culture" encompasses not only the echoes of popular song and the vernacular meters noted by Friedman but also his concern with oral composition, his poetic investigation of earlier stages of society, and even his technique of illustration. Blake composed the earliest poems for *Songs of Innocence*, his first full-length illuminated book, soon after completing the engravings for Ritson's *Select Collection*. Although many of the most famous ballad collections (including those of Percy and Scott) are not illustrated, Ritson's decision to produce an illustrated collection has an important precedent in Philips's *Collection of Old Ballads*, which is profusely illustrated and offers editorial comments in justification of the pictures (*COB* III.v–vi).[10] Some of Philips's illustrations emulate the woodcuts printed with broadside ballads in the seventeenth century, and the history of ballad illustration matters for Blake's *Songs* perhaps almost as much as the medieval technique of manuscript illumination most commonly cited as his model.

The two poems from *Songs of Experience* with which we began also evoke the early modern broadside ballads as print artifacts. By dividing one of the standard four-beat lines in each stanza of these two poems into a rhyming pair of short (dimeter) lines, Blake follows a common printing convention that may be traced at least as far as the Jacobean period. In many broadside ballads, internal rhyme within a standard tetrameter line is accentuated by inserting a line break after the first rhyming word, just as Blake does in the middle of what would otherwise be the third line of the quatrain here and in every stanza of both "Introduction" and "Earth's Answer":

> Earth rais'd up her head,
> From the darkness dread & drear.
> Her light fled:
> Stony dread!
> And her locks cover'd with grey despair. ("Earth's Answer" 1–5/E 18)

Blake may have been inspired by one of Ritson's ballads or by a broadside circulating in his own time, but the precise source for this formal technique is less important than Blake's adoption into his own printing method of a convention for highlighting sound and hence musicality.[11] The repetition of "dread" effectively extends the A-rhyme across the first four lines, while the isolation of "stony dread" accentuates the clash between this exclamation and the grammar of the sentence. (The repetition also retains something of the nonliving quality of Earth from the draft version of line 3, "her orbs dead.") Reading these two poems against the "Introduction" to *Songs of Innocence* (E 7), we might say that these complexities demonstrate the textual corruption that arises in the transmission from music to song to writing, a corruption that the *Innocence* poem barely acknowledges.

Along with this hint of a formal resemblance to printed ballads, these poems touch on some of the traditional subject matter of ballads. By casting the pair as a dialogue between a male and a female speaker, Blake evokes familiar ballads of courtship such as "The Nut-Brown Maid" and "Child Waters." A search of the English Broadside Ballad Archive (EBBA) confirms that "earth" occurs rarely in ballads but typically occurs in a religious context when it does so. As Blake's editors point out, his Earth may allude to Biblical usage in which "Earth" stands for a sinful humanity (e.g., Jeremiah 22:29).[12] If so, Earth's putatively sinful refusal of the Holy Word also parallels her rejection of the possessive monogamy upheld by the courtship ballads. Conjuring up a kind of paradox that is characteristic for the *Songs* as a whole, the call and response structure of these two poems, together with their use of a metrical form indebted to common meter, makes them sound a bit like hymns, whereas Earth's lament for the loss of "free Love" (25/E 19) and her idiom of plowing and sowing (18–20) evoke a more playful secular ballad tradition less of courtship than of country matters. The notebook page containing the original draft includes another poem in this idiom, "Thou hast a lap full of seed." "Introduction" and "Earth's Answer" engage with the ballad tradition by deforming it, perhaps with a view toward disrupting the polished regularity imposed on the form by Percy and other collectors. The sheer difficulty of these two poems, in which speakers are not clearly distinguished, verbs lack clearly defined subjects ("Introduction" 6–11), and pronouns lack antecedents ("Introduction" 8/"Earth's Answer" 7), artfully recalls the bits of doggerel nonsense that are so pervasive in older ballad texts, though valiantly explained away as corruptions by most eighteenth- and nineteenth-century commentators.

In "Earth's Answer," the earth is entangled in time. The "Introduction" seems to offer the "lapsed soul" (6/E 18) a chance to right itself and "fallen fallen light renew" (10). The very next line, the first to be spoken in the Bard's own voice, laments Earth's fallenness by repeating this structure, exclaiming, "O Earth O Earth return!" Earth answers that it cannot extricate itself so readily from its material condition. It replaces the Edenic "ancient trees" of the first poem with the "ancient men" thrown into the world at the dawn of history, insisting on a secular time scale. The bondage placed on "free Love" under theological pretensions seems to operate as well through agriculture, condemned to function in the dark as the earth is seemingly consigned to fallenness by the first poem: "Does the sower? / Sow by night? / Or the plowman in darkness plow?" (18–20/E 19). This stanza was a late addition inspired by the adjacent notebook poem "Thou hast a lap full of seed," which underscores the sterility of "seed" in the context of civil society. All the drafts on the adjacent pages (N113–N109) address sexual freedom and repression, either through the theme of husbandry (cf. "O lapwing") or through the more Biblical idiom of sin and prohibition (the draft "in a mirtle shade" as well as "London," which ultimately became one of the best-known *Songs*).[13] The introduction of agriculture—whether seen as a step toward civilization or a step toward inequality—marks a crucial stage in the history of society, and "Earth's Answer" glances at conjectural history, with its long time scale, in raising these questions. Earth's petrified condition, its "stony dread," secures its entanglement in time—if not literally in geological time, then in the deep secular time that the previous poem had aimed to transcend.[14]

These two poems are also roughly contemporary with *The Marriage of Heaven and Hell* (1790–93), which adopts a more philosophical idiom in its analysis of time and eternity. It includes the "Proverbs of Hell," one of which—"Eternity is in love with the productions of time" (E 35)—might serve as a gloss on the dialogue between the Bard/Word and Earth in *Experience*. The Bard calls Earth into the linear time of sacred history, but ineluctably Earth "turns away" (E 16) because natural time is cyclical, dependent on the planet's motions.[15] *Marriage* shares a spirit of "self-satire and revolutionary naturalism" with *Songs of Experience*, as E. D. Hirsch aptly put it, but the former poem explores a visionary dimension of temporality that complicates the binary seemingly at issue between the Bard and the earth. Blake famously declares in *Marriage* that "if the doors of perception were cleansed everything would appear to man as it is, infinite" (E 39), fusing immanence and transcendence into a temporality that might be distinguished as "stop time" from the linear and cyclical models on offer in *Experience*.

Blake's reflections on temporality in both these early works also bear on his own artistic practice and its idiosyncratic emulation of "ancient" art. Noting another sense in which Earth cannot "return," Roger Lüdeke argues that Blake foregrounds the contradiction between the repeatable sign or Word and the singular line or inscription in the Bard's call and throughout the first two *Experience* poems. The result is that the "prophetic subject" is "thrown back" on the difference between language and perception as divine vision collapses into the two-dimensional space of inscription. In Lüdeke's reading, Blake's *Marriage* again confronts the linearity of Lockean perception, which is supposed to unfold in time, by reformulating this "paradox of singular repetition." The accumulation of sense impressions is stopped (*stillgestellt*) in the simultaneity (*Gleichzeitigkeit*) of all perception.[16] In other words, the "infinite which was hid" (E 39) cancels time's arrow. Lüdeke draws attention to the "Printing House in Hell" and other episodes in *Marriage* that reflect explicitly on Blake's method of printing, and these might be said to disrupt cyclical time as well. For as Lüdeke argues, Blake's artisanal practice of "printing from autographs," in which no two letter forms are exactly alike, forestalls the cycle of mechanical reproduction, producing instead a materiality that is nonreproducible and hence constitutes a biopolitical form of dissent. Lüdeke's discussion of Blake's etched line as a "bodily signature," an aesthetically mediated experience of "bare life," is relevant for my purposes inasmuch as it shows how deeply Blake's "preindustrial mode of production" inflects the subject matter of his poetry.[17]

Blake inserts himself as a character in the ballad of early man not only through his method of relief etching but also through the designs that accompany the words in all of his illuminated books. Ambrose Philips, whose work in the 1720s established the paradigm for illustrated ballad collections such as Ritson's and Blake's, is hardly "early" in the same sense, but "early man" may serve as a fair approximation for a set of visual motifs that is common to Philips's *A Collection of Old Ballads*, Ritson's *Select Collection of English Song*, and *Songs of Innocence and Experience*. While the oldest printed ballads are accompanied by woodcuts, increasing their appeal to a semi-literate readership, their visual/verbal character was marginalized by connoisseurs and collectors such as Percy and Scott, who sought to trace in ballads a more exalted oral tradition dating back to medieval minstrelsy. Ritson, who is notable for his polemics against this position, chose to include illustrations (albeit modern ones) in his collection, and his publisher Joseph Johnson hired Blake to do some of these engravings. Philips's collection predates this dispute over the relative merits of broadsheets vs. oral tradition; his extensive use of early modern broadside ballads as source texts reflects both the practice of earlier collectors (such as Samuel Pepys, who started his famous collection of broadside ballads in the 1680s) and the assumptions of earlier critics (such as Joseph Addison, whose pioneering defense of popular ballads deeply influenced Philips).

The visual motifs linking these three collections form an amalgam of pastoral, history, and archaeology. Philips begins his third volume (1725) with "A Song of the strange Lives of two young Princes in *England*, who became two Shepherds on *Salisbury Plain*." The text refers to the reign of King Stephen, but Philips in his lengthy headnote retells a somewhat similar story from the reign of Stephen's successor, Henry II, before admitting that any link between this story and the ballad is "only conjecture" (*COB* III.4) and that the story might either be entirely "romantick" or "history . . . so very much altered" as to be unrecognizable. The engraving facing the title and headnote takes full advantage of these ambiguities, displaying a shepherd and a shepherdess in classical and eighteenth-century dress (respectively) seated on fragments of columns while their sheep graze amidst the trilithons of Stonehenge (Fig. 3.2).

FIGURE 3.2. Unsigned engraving. Ambrose Philips, ed., *A Collection of Old Ballads* (1723–25), vol. 3, p. 1. Courtesy of the University of Missouri, MU Libraries.

Megalithic monuments begin to occur in Blake's illuminated poetry somewhat later, but the *Songs* are full of shepherds and pastoral scenes that, like the one in Philips, refuse historical specificity while evoking a generalized—and possibly very high—antiquity.[18] The association between herding and earlier stages of society is more pronounced in Ritson, who observes that "in the earliest ages of mankind, the chief employment of all ranks was the care of their flocks and herds," giving rise to pastoral and love songs (*HE* iv). Ritson's frontispiece (Fig. 3.1), not engraved by Blake, shows a decidedly archaic male figure with a wooden flute and a female auditor, underscoring the association between pastoral and early humanity. The similarly archaizing design that accompanies Blake's "Laughing Song" (*Innocence*) is based directly on a design by Thomas Stothard that Blake engraved for Ritson's *Collection* ("Drinking Song").

Blake engaged deeply with images of antiquity, from Stonehenge through medieval monuments, during his apprenticeship to the antiquarian engraver James Basire. The more or less sentimental tradition exemplified here through Philips and Ritson is important for Blake, but there is good reason to attribute some of his "ballad culture" to the empirical study of artifacts also associated with antiquarianism. The title page to *Songs of Experience*, for example, seems to respond directly to a famous early modern broadside ballad, "The Children in the Wood," rather than to the eighteenth-century mediations of Philips and Ritson. This plate shows a man and a woman mourning on either side of the deathbed of their aged parents. In the broadside printing of "The Children in the Wood" preserved in the Crawford Collection, the second of the small in-text illustrations presents a similar scene: the mourning children on either side of the bed are much younger (as are the parents), but the composition is similar, and details such as the drapery along the side of the bed and the paneled wall behind evoke the old woodcut.[19] Soon after his apprenticeship, Blake began producing paintings and drawings of Jane Shore, the mistress of Edward IV, who has a special importance for the early ballad revival. Philips's *Collection* takes its epigraph from Nicholas Rowe's *Tragedy of Jane Shore* and includes two of the old ballads referred to in this epigraph (*COB* I.145–58). He also argues that "historical ballads" are the best source material available, for children and historians alike, on royal concubines and other figures from unofficial history (I.vii). Philips makes a similar pedagogical argument for the illustrations (III.v). Blake's rationale for including designs in his poetry was on a more ambitious scale, but the history of ballad illustration remains a significant contextual element for understanding his verbal-visual compositions, not least because it helps to explain what he means by insisting on the antiquity of both his method and his subject matter.

In his original definition of Blake's ballad culture, Friedman rightly draws attention to ephemera, to street cries and jingles Blake is likely to have heard, and to the cheap print vehicle of broadside ballads (still alive and well in Blake's time and beyond). Friedman points to Blake's early manuscript satire "An Island in the Moon" as one of the best sources of evidence on this point. This burlesque narrative, a sort of intellectual satire featuring a salon loosely based on Blake and his circle, includes many interpolated verses. Some of these, including the first "Nurse's Song" and "Holy Thursday," are first drafts of poems ultimately included in Blake's own *Songs of Innocence*. Others, such as "This frog he would a wooing ride," are very much in the public domain. Friedman notes that this ballad, known as "Froggy Went A-Courting" in its popular American versions, was first cited in print in 1549 and exists in many early modern versions. Still others, such as "Fa ra so bo ro" (E 450), are entirely nonsemantic, evoking the ballad refrain and nonsense poetry as well as a mode of prelinguistic utterance that humans share with nonhuman animals. Friedman argues that Blake's metrical practice owes much more to street ballads and nursery rhymes than that of his literary contemporaries, and I have emphasized formal aspects of his revisionist ballads in my discussion partly because I believe that Blake recognized—paradoxically—some of the most ancient features of poetry in the "temporal musical scansion" of these ephemeral works.[20]

The Ballad of Early Man

"Introduction," accompanied by a nude figure floating on a cloud, and "Earth's Answer," accompanied by a serpent, together hint comprehensively at two or more competing origin myths. The text of "Earth's Answer," with its references to "ancient men" and the introduction of agriculture, situates itself in the era of early humanity explored by writers of conjectural history. The "Introduction" begins by enjoining us to "hear the voice of the Bard," already familiar both from "The Voice of the Ancient Bard," included with *Innocence* in most copies, and from iconic ancient bards such as Homer, the touchstone for many theoretical and historical accounts of the ancient ballad. The Bard's proximity to the Word in this poem also carries implications for the origins of art—sacred origins in this case—while promoting the survival of art's original functions. These too are matters of concern for many of the ballad revivalists. The question of human species identity, one of the largest questions pursued by conjectural historians both inside and outside the context of the ballad revival, is placed respectively in sacred history and in a kind of secular deep time by the dialogue between the Bard and Earth that these poems construct. In this way, some broad affinities may be seen between these two representative poems from Blake's early period and the ballad literature—both verse and prose, both popular and literary—that takes "early man" as both the author and the subject of ballads.

Human antiquity is a pervasive concern of ballad literature, but there is little consensus either on the antiquity of ballads themselves or on their capacity for mediating ancient events or ancient practices of composition. It makes sense to classify Blake as a ballad skeptic like Ritson, who fiercely rebutted Percy's arguments for the medieval origins of English ballads and relegated ballads as "meer narrative compositions" to a subcategory of song.[21] Like Ritson, Blake embraced song as a more capacious medium for investigating the antiquity of vernacular forms, maintaining a connection between oral tradition and the poetic curriculum inherited from classical antiquity. This connection is embodied in pastoral figures like the "young shepherd" who purports to be the author of "Laughing Song" (E 715) and the pipers depicted in the frontispieces to both *Songs of Innocence* and *Select Collection of English Song*. Because Blake and Ritson both look beyond the ballad in their pursuit of ancient song, they are able to incorporate and transform neoclassical conventions and antiquarian methods more fully than the authors of literary or sentimental revival ballads such as Robert Southey, Wordsworth, Coleridge, or William Hayley, whose *Ballads* Blake illustrated in 1805. Although Ritson turns to the larger category of song and Blake eventually turns to epic in his quest to produce "poems of the highest antiquity" (E 542), they inherit their commitment to verse as a medium for preliterate human antiquity directly from ballad collectors and editors such as Percy and James Macpherson.[22]

Writers in the main line of the ballad tradition from Percy to Scott focus their energies more exclusively on the ballad form and its roots in oral tradition. In their accounts of preliterate authorship as well as the traditional subject matter of the ballads—including supernatural elements and untraced historical events—they borrow freely from the discourse of the primitive that circulated between voyage narrative and conjectural history from the early modern period, as detailed in Chapter 1. Percy's "Essay on the Ancient Minstrels in England" credits medieval minstrels with the original authorship of the manuscript ballads in his collection, referring to the "rude admiration" excited by "heroic ballads" in those "rude times." He makes these minstrels, in turn, "the genuine successors of the ancient Bards"—not Greek but Anglo-Saxon, reflecting the "rude, uncivilized state" of the first Angles and Jutes who colonized England.[23] McLane argues that these claims for the ballad's antiquity make the discourse of ballads and minstrels available as a general "way of talking about the cultural and historical situations of poetry." To illustrate this point, McLane positions James Beattie's *The Minstrel* (1771)—a long narrative poem written in Spenserian stanzas rather than ballad form—between Percy's *Reliques* (1765) and Scott's *Minstrelsy of the Scottish Border* (1802):

> If Percy sang the praises of medieval English minstrels, Beattie gives us . . . a Scottish minstrel "born in a rude age": the "rude age" signifying not a recognizable, date-able historical period—say, the twelfth-through-fourteenth centuries—but rather the first primitive stage of conjectural history.[24]

McLane shows how the Scottish Enlightenment historiography of Adam Smith and his followers influenced both Beattie and Scott in their responses to Percy's model of the minstrel ballad. She argues that this synthesis ultimately flattered the ambitions of the Romantic imagination: "Large zones of what Shelley later called 'the history of our species' are inaccessible to the historian, the minstrel protests, and here the imagination . . . must do its work."[25]

It seems fair to say that the ballad tradition secured "poetry's claim on history" for the Romantic poets, but what McLane calls the "story of Man" also provided expression to a deep-rooted ambivalence about primitivism and a critical awareness of multiple, conflicting temporalities. Percy's repeated and apologetic use of the word *rude* is one mark of this ambivalence, as is his unwillingness to settle entirely on one temporal location for his origin story. The interplay between late antiquity, the Middle Ages, and the early modern period does seem to promote a vague, undatable origin, but at the same time it establishes a productive area of uncertainty that clarifies the importance of scale for competing origin narratives. Percy's contradictory impulse to polish and modernize the manuscript verse in order to compensate for its supposed "degeneration" from a high oral culture—an impulse condemned by many later editors, from Ritson to Child—lingers on in Beattie's progress-driven "story of Man" and in Scott's deliberately elegant editorial interventions a generation later. As McLane puts it, "the emerging ballad chronotope put pressure on stadial theory, since what the ethnographic turn in balladeering suggested was the possibility of a heterochronous history, several 'times' of cultural development co-existing, albeit under the teleological sign of progress, [with] the British Union and Empire."[26] This formulation captures something of the temporal uncertainty surrounding ballads but overstates the consensus among this group of writers, who disagreed widely on the nature of progress, the connotations of primitive culture, and the scale of human time.

Several poets in the course of the ballad revival brought a high degree of self-consciousness to the chronological implications of their engagement with vernacular culture. Burns's poetry is especially noteworthy for its ironic distance from questions of antiquity and origins. He engages in friendly mock-rivalry with the antiquary Francis Grose in "On Captain Grose's Peregrinations through Scotland, Collecting the Antiquities of That Kingdom" (1790). Burns met Grose while the latter was collecting materials for his *Antiquities of Scotland* and supplied Grose with traditional stories as well as an original poem, the mock-heroic narrative "Tam O'Shanter." In the manuscript poem on Grose's "peregrinations," Burns implicitly compares Grose's collecting activity and his zealous pursuit of the most arcane antiquities with Burns's own collecting of traditional songs and ballads for James Johnson's *Scots Musical Museum* and other venues. Burns broadly exaggerates the antiquity of Grose's artifacts, and hence the scope of his antiquarian inquiry, by situating them at the origin of sacred history. His eclectic collection, according to the poem, extends from medieval arms and armor to utensils in use "afore the Flood" (*BLS* 36), and Burns's catalogue culminates in a virtually uncontestable priority claim: "Of Eve's first fire he has a cinder" (37). As old as the traditional Ayrshire ballads may be, their antiquity pales next to Grose's relic of the first woman's ostensible first experiment in domestic labor—but the comparison plays on the ballad's association with early human history.[27] Burns mocks and yet reinforces the deep secular time implied by the antiquarian pursuit of primitive culture, precisely by locating Biblical time anthropologically. Although he was inspired by Burns's self-representation as a modern primitive, Wordsworth framed his anthropological pursuits with less ironic distance, claiming for example that authentic ballads are characterized by "nakedness and simplicity" (as Addison had already stated in 1712) and that he had achieved the same effect with his own "Goody Blake and Harry Gill," which he calls "the rudest of this collection" (*Lyrical Ballads*).[28]

Scott's more complicated attitude toward the temporality and authenticity of ancient ballads comes through in the thinly veiled "improvements" added to his version of the traditional ballad "Tam Lin" (Child #39) in the second edition of *Minstrelsy of the Scottish Border* (1830). Burns's satirical idiom and Wordsworth's philosophical one acknowledge in different ways the possibility that ballads and other domestic antiquities may hold clues to an original human nature. By adding a lengthy headnote on the history of a superstition and revising the ballad text, Scott used "Tam Lin" to a different end, problematizing the "ancient" status of the ballad and marking its depiction of human nature as archaic and culturally relative. From his point of view, the "nakedness and simplicity" of this supernatural tale of a changeling knight released from bondage to the Fairy Queen hold more cultural embarrassment than they do legitimate survivals of a deep human past. Scott alludes ironically to the history of debates over authenticity that characterized the ballad revival when he acknowledges Ritson's criticisms of his earlier version, but then cites a highly suspicious manuscript source as the authority for some emendations in the new version:

> The Editor has been enabled to add several verses of beauty and interest to this edition of Tamlane, in consequence of a copy, obtained from a gentleman residing near Langholm, which is said to be very ancient, though the diction is somewhat of a modern cast. The manners of the Fairies are detailed at considerable length, and in poetry of no common merit.[29]

The passage in question comprises six stanzas describing the fairies' approach near the climax of the tale, deviating noticeably from the Scots dialect and not included in any other recorded version of the ballad. As early as 1839, William Stenhouse defended the authenticity of Burns's version of "Tam Lin" by pointing out that "many of the stanzas in Sir W. Scott's version, . . . if not by himself, are evidently the work of a modern hand," and they appear nowhere in the nine traditional versions authenticated by Child.[30]

One verse in particular describes the music of the approaching fairies in pastoral terms. "Their oaten pipes blew wondrous shrill," declares the narrator, and when he adds that "solemn sounds, or sober thoughts, / The Fairies cannot bear," it becomes especially clear that he is editorializing in a manner that is not typical for the ballad as a whole (*MSB* 332). As it happens, this interpolation is consistent in tone and content with the long headnote Scott prefixed to his version of the ballad, weighing in at thirty-nine pages and titled "Introduction to the Tale of Tamlane: On the Fairies of Popular Superstition."

The reception history of "Tam Lin" aptly illustrates the lack of consensus concerning the primitive authorship and subject matter of popular ballads. The first editor to print a portion of the story, David Herd, prided himself on having "recovered" many "fragments that appear of some antiquity" and included it among those never printed before but did not specify a source or period of origin.[31] For Scott, it appears, the ballad's antiquity is a useful pretext for illustrating a quaint "article of the popular creed" (*MSB* 288). The length of his prefatory essay is warranted partly by its position as the first substantive item in *Minstrelsy,* Part II, "Romantic Ballads," as distinguished from the "Historical Ballads" collected in Part I. This positioning and the title of the essay itself sufficiently indicate Scott's allegiance to progress and to empire, as diagnosed by McLane. These ideological commitments are much less clear, however, in other editions, which construct the ballad's antiquity in diverse ways. In the late nineteenth century, Child offers the strongest priority claim, tracing "the principal feature in the story, the retransformation of Tam Lin," to a "Greek popular tradition older than Homer." Stenhouse, who provided commentary for the second edition of the *Scots Musical Museum,* contents himself with noting that "this romantic ballad or tale . . . is of unquestionable antiquity." The ballad was one of many contributed by Burns to the fifth volume of this work, published in 1797, just after his death. James Johnson's preface eulogizes Burns, noting that he not only contributed original songs (such as "Comin thro' the rye") but was inspired "to collect and write out accurate Copies of many others," including "Tam Lin" (#411), "in their original simplicity."[32]

In each case the ballad is an ancient artifact that helps to secure an account of human development, much as geological specimens came to support the architecture of geological time. The key differences are differences of scale, which here depend on whether the appeal is to relics of pagan belief (Scott), to archaic Greece (Child), or to living tradition (Johnson/Stenhouse). Similarly, while the assumption that ballads have primitive origins is widely shared, the connotations of the primitive depend on each author's philosophical and political views. This variety—the heterochronicity of ballad literature as opposed to the chronotope, defined by Bakhtin as a "concrete whole"—creates the area of uncertainty that favors speculation about the scope of prehistory.[33]

Scott had no more reason than Ritson to suppose that 40,000-year-old bone flutes would one day be found in Germany, but it is instructive to compare the "oaten pipes" of Scott's fairies with the early wind instrument in Ritson's frontispiece (Fig. 3.1).[34] Both allude to one of the most pervasive tropes of pastoral, the shepherd's use of materials ready to hand for music-making, but Scott uses the trope allegorically to signal the developmental stage of a society that believes in fairies, while Ritson uses it in a more literal sense. The frontispiece of a rustic flute player presents a conjectural image of "the practice of mankind in the infancy of creation" (*HE* iii) in much the same way that Paleolithic stone tools or musical instruments serve today to provide access to a "deep history" far beyond the reach of the written record. There is no direct evidence that Ritson gave instructions to the artist (Stothard), but the image is consistent both with pastoral lore and with Ritson's account of prehistory, as when he asserts that "song . . . preceded the use of letters" in ancient Greece (*HE* iv).[35]

Ritson relies more heavily than Percy or Scott on the conjectural mode for this prehistory. When it comes to English song, he roundly rejects evidence from oral tradition, insisting on archival sources to support all claims for the antiquity of songs and ballads. In his eagerness to refute the exalted claims made for the antiquity of oral tradition by Percy, Macpherson, and others, Ritson comes to the surprising conclusion that "not a single [song] with the smallest degree of poetical merit" is "discoverable" prior to the reign of Queen Elizabeth (*HE* lvi). Concerning ballads in particular, he adds that "not more than three are so old as the sixteenth century" (lviii). Ritson studiously rejects any source that he cannot authenticate himself, even as he probes more deeply into the manuscript archive in the more antiquarian anthologies that follow the *Select Collection*—itself confessedly a compilation of previously published songs—such as *Ancient Songs* (1790) and *Scotish Song* (1794). In the latter work, while granting a few additional centuries in age to the Scottish ballads, he still declares that "tradition . . . is a species of alchemy which converts gold to lead." Ritson places all his trust in textual sources for songs and ballads associated with British literary culture, mocking those who "pretend" that the "poems of Ossian have been preserved immaculate [in popular memory] for more than a thousand years."[36] This modernism jars against Ritson's placid certainty that "song is the most ancient species of poetry" (*HE* i) and that "in the earliest ages of mankind, the chief employment of all ranks was the care of their flocks and herds" (*HE* iv). His aggressive skepticism regarding oral tradition makes "ancient" British songs and ballads more recent but thereby creates a *deeper* past for the origins of song as a human behavior. His frontispiece testifies eloquently to this idea of a premodern origin for song, predating by an indefinite epoch the advent of national history and literature. Scott, by contrast, accepts Percy's premise that a lost high culture of medieval minstrelsy secures direct continuity between the deep epic past of song and popular ballads, however barbarous these may now appear.

Ritson's conjectural history of song depends more explicitly on comparisons between ancient peoples and modern "savages," a methodological foundation he shares with Herder, the German inventor of the term "folk song" (*Volkslied*). Examining this shared ethnographic approach, later dubbed the "comparative method," will help to clarify how a more radical primitivism underlies the pugnacious anti-primitivism of Ritson's attacks on the antiquity and authenticity of songs and minstrels.[37] On the one hand, he is totally unwilling to accept his colleagues' redactions of "ancient" ballads; on the other, he eagerly accepts as authentic whatever he can glean concerning preliterate cultures from travel narratives. His use of travel narrative is reminiscent of Rousseau's, and he relies heavily on French sources, including Joseph-François Lafitau and Louis Jouard de La Nauze. Ritson's understanding of "the practice of mankind in the infancy of creation" is directly inspired by a sixteenth-century report of Spanish discoveries in the New World, and he endorses early modern accounts of first contact along with ancient accounts of Arcadian shepherds as records of "the human mind in [a] state of nature and simplicity" (*HE* iii–v). Ritson assures us that "the practice of the native Americans is much the same" (iiin6) and even includes "The Death-Song of a Cherokee Indian" from a late eighteenth-century manuscript, though he acknowledges that this "translation" may prove to be inauthentic.[38]

Herder extends the ancient/primitive comparison to living European peoples, defending his endorsement of Macpherson's Ossian poems as living Scottish tradition by comparing them both to ancient Scandinavian epic and to "another nation which still lives and sings and acts on earth today," the "five Indian nations of North America."[39] Like Ritson, Herder draws on Lafitau's and other North American ethnographies to establish the primitive genres of song. Unlike Ritson or any of the English ballad revivalists, however, Herder is optimistic about the vigor of "barbarous" survivals in Europe, using the term with none of the ambivalence that we find even in authors (such as Percy and Ritson) who disagree on most other points. In "Extracts from a Correspondence on Ossian and the Songs of Ancient Peoples" (1769), Herder defines "barbarous" positively as "more alive" and "more freely acting," granting that his correspondent may "mock" his "enthusiasm for these savages almost as Voltaire scoffed at Rousseau for wanting to go on all fours" but defending this primitivism all the same: "But alas too for the philosopher of mankind and culture who thinks that his scene is the only one, and misjudges the primal scene to be the worst and most primitive!"[40] While Herder grants to the living oral traditions of Europe the "lyrical, living, dance-like quality" that he sees as essential to primitive song, Ritson restricts the field of modern comparisons to the New World: "We are, therefore, to look for the simplicity of the remotest periods among the savage tribes of America, at present; or at least before they were civilised—perhaps corrupted—by their commerce with Europeans" (*HE* ii). Ritson's primitivism is therefore more radical in the sense that he sees European popular culture as corrupted by the same civilizing force that defines European civilization as a whole.

Unlike the British ballad collectors, Herder made his exploration of *Volkslieder* into a springboard for a sustained intellectual career in conjectural history and ethnography, producing a body of writings that placed him at the forefront of Enlightenment anthropology.[41] Even in this early work, he betrays a self-consciousness about the appearance of Rousseauvian idealism created by his distinction between the "primal" and the "primitive." This very optimism, the notion that "barbarous" European and non-European peoples preserved more of the "primal" energy of early humanity, expanded the field of evidence available to Herder for mapping the early history of the species in a series of ambitious works, culminating in *Ideas toward a Philosophy of the History of Humanity*. The British ballad collectors, by contrast, all held to some form of degeneration theory, in common with a more skeptical strain in conjectural history.[42] Even Ritson's dissent, his skepticism about the antiquity of ballads, simply proceeded from a stronger form of degeneration theory than the Percy-Scott paradigm of medieval minstrelsy unraveling into print culture: he argued that popular culture, especially in urban settings (disregarded by other collectors), had degenerated so far that its primal origins—which, by the same gesture, he set farther back in time—could no longer be traced. Herder, convinced early on that folk song was a genuine barbarous survival, expanded the repertoire of survivals to include legends, myths, and fairy tales. His conviction that the songs of all "unpolished nations" should be committed to print directly "from the mouths of the people" may have prevented him from collecting more ballads—Scott for one found the German portion of Herder's *Volkslieder* of 1778–79 disappointingly slender—but it informed the pioneering critique of colonialism for which he is still remembered and further demonstrates that the Enlightenment pursuit of a deep human past was an ideologically diverse field.[43]

On the Margins of History

Herder comes the closest to producing full-blown species time out of the ballad tradition by taking seriously the ethnographic premise that oral tradition preserves elements of human prehistory. Herder's initial engagement with vernacular culture was directly inspired by Macpherson's Ossian poems and by Percy's *Reliques*, on which he mounted an aesthetic campaign to renovate German national literature. These concerns soon led him beyond the domain of literary culture strictly so called. Unlike the British ballad theorists, Herder did not merely borrow from conjectural history to create a model for the development of literary art from folk poetry. In outlining the plan for his anthology, titled *Volkslieder*, he recognized that songs, tales, and legends are themselves "a great subject for the conjectural historian [*Geschichtsschreiber der Menschheit*]."[44] By this point Herder was already a contributor to this field—the history of humanity, now more generally known as conjectural history—and he published his monumental *Ideas* in parts between 1784 and 1791. The 1777 essay quoted here, synthesized by Herder from a series of prefaces originally written for *Volkslieder*, marks an important step along the way. He argues that even though modern Europeans (as opposed to the ancients) have known of a greater variety of cultures ever since ethnographers began to expand the map of humankind, deep knowledge of these peoples will come only through their songs, *das Archiv des Volks*.[45] Herder declares that this "archive" contains all preliterate peoples' sciences, religion, history, and more—what J. R. Forster, writing up his Pacific field notes at roughly the same time, called "the whole cyclus of their knowledge" (*O* 320).

Using the ballad revival as a springboard, Herder departs from media history to posit oral tradition as a source of unmediated access to the deep past of the human species. Defending the evidentiary richness of folk song in the same passage, he compares its objectivity to that of natural history: "as natural history describes plants and animals, so do peoples represent themselves" in their songs.[46] In practice, he learned that the ballad collector necessarily trades in artifacts with a long history of mediation, which today is just as interesting to scholars as the traditional content. In his mature thinking about human variety, however, Herder maintained an emphasis on firsthand testimony from non-European peoples as the most reliable source both on their cultures and—in a real sense, among cultures with no written record—on their prehistory. While conjectural historians such as Rousseau and Ferguson and ballad theorists working in this vein such as Ritson habitually relied on travel narrative to integrate non-European peoples into their stadial models, Herder stands out as a skeptical consumer of travel narrative who eventually insisted on a less hierarchical vision of human variety as the foundation of conjectural history. Thus he holds up the ideal of song as true autoethnography and condemns European travel narratives as "one-sided," rejecting the accompanying illustrations as "caricatures" of the non-Western people they claim to represent. McLane argues that the "transnational vernacular cultural imaginary" created by the British ballad revival marks it as a colonial project, but the reciprocal influence of the ballad revival on the history of the species, as we see it in Herder, complicates this equation.[47] By valorizing song as a cultural archive, Herder also establishes an attitude toward European popular culture that differs meaningfully from Percy's and Scott's ambivalence vis-à-vis these "rude" traditions and also differs from Ritson's hardline degeneration theory.

In *Ideas,* Herder turns from songs and ballads to other forms of evidence concerning the primal state of the human species. As suggested by the analogy between ballad autoethnography and descriptions of flora and fauna, natural history proper is an important source for him, and the first book of *Ideas* is devoted to the history of the earth. Here and throughout the work, Herder engages closely with Buffon's *Epochs of Nature* as well as the Forsters' Pacific travel narratives, so his conjunction of conjectural and natural history marks an important locus of continuity between the present chapter and my earlier discussions of those authors. As a historian of humankind seeking a foundation in natural history, Herder freely acknowledges his debt to Buffon, though he is also critical (*I* 209). The expansive space of conjectural history gives Herder room to situate the origins of both song and species in relation to geological time—which was not a factor in the regular ballad histories.

Folklore comes into play in the *Ideas* as a site of contact between geology and prehistory. Concluding a sustained argument that identifies "the mountains of central Asia" as the scene of human origins, Herder turns to ancient Hindu tradition as a proximate source of evidence from natural history to corroborate this prehistoric geography, which strongly resembles Buffon's. He retells an Indian tale (*Märchen*) in which Prassarama (an avatar of Vishnu) creates the first habitable land by striking a bargain with the god of the sea. Prassarama proposes to shoot an arrow as far as he can from the top of a mountain into the sea, and the god promises to make the sea retreat an equal distance from the mountain—thus creating the Malabar coast (*I* 400). Herder cites a reading of the myth by the naturalist Pierre Sonnerat, who explained that it was an origin story about the coastal plain, indicating that its elevation above sea level postdates the formation of the mountains but also predates—the implication is clear—the creation of the human species. Herder argues that such traditions, though inadmissible as history, are better than conjecture (397–98) and therefore deserve "unprejudiced" consideration (402). Ultimately, he determines that these traditions must yield in historical validity to the written record, the Hebrew Bible. He is careful to note, however, that it may be a long time before Europeans gain access to the true text of the Vedas, adding that the first part of these scriptures is considered lost by the Indians themselves (399). Although non-Western traditions cannot be firmly dated or authenticated (by him), Herder nonetheless uses them to create a space for symbolic representations that may reveal more than the Bible about chronology and human origins.

In a synthesis that predates the putatively binaristic conflict between religion and science by more than half a century, Herder privileges the Bible as the foundation of his philosophy of the history of humanity but effectively gives equal authority to natural history over the prehuman past. The uncertain antiquity of oral tradition—which he presents enticingly as a kind of ore potentially fraught with golden survivals of the most ancient traditions (*ein Goldkorn historischer Ursage*; *I* 399)—offers a bridge between the two. Herder's emphasis on natural history is corrective to some degree; he insists at the outset that human beings are "inhabitants of the earth," not purely spiritual beings like angels (23). His history of the earth in Book 1 depends on the classification of primitive rocks or *Urgebirge* (42) put in place by early stratigraphers such as Johann Gottlob Lehmann and Giovanni Arduino in the 1750s. Rhetorically hedging his commitment to Buffon as may have behooved a Lutheran pastor, Herder nonetheless accepts the basic framework of *Epochs of Nature*, including the comparatively young age of the human species that had offended the theologians of the Sorbonne as well as another German Lutheran pastor, Reinhold Forster. As in Buffon, however, the species here is ancient enough to have experienced the most recent revolutions of the earth directly and still harbors an evolutionary memory of those upheavals (31).

Buffon also seems to have been the decisive influence on Herder's choice of a location for human origins, the mountains of central Asia, but Herder draws inferences for human prehistory that go beyond Buffon's account, identifying the *Urgebirge* not only as "the first chosen dwelling place of humankind" (*I* 43) but also as an enduring repository of "wild men" who serve to repopulate the plains and "preserve the species" (43–44) after floods and other catastrophes. "In this way," he concludes, "Nature used the mountain chains . . . to produce the rough but firm outline [*Grundriss*] of all human history" (44–45). Herder supports this geophysical determinism with numerous examples, observing that the laws of geological formation remain largely unknown (49, 54), and then concludes his account of the earth's history with a poetic apostrophe that effectively merges geology—that is, the mineral and fossil content of the secondary strata—with prehistory. Addressing "Man" as an "ephemeron," he writes, "you do not tread the ground of the earth, but rather the roof of your house, created only as the result of many floods" (58).

Setting himself apart from other conjectural historians and from scholars of oral tradition, Herder relies on fossil evidence to situate the emergence of culture against a backdrop of explicitly prehuman time. He points to the absence of fossil content, established by the geognosts as a defining quality of primitive rock, to posit an earth before life and argues from the earliest fossil record for a marine origin for life when it emerges (*I* 381–82). Granting to geology an honorific more often reserved for the Bible, he cites the authority of the "oldest book" and its "pages of clay, slate, marble, limestone, and sandstone" for his statement that fossilized human remains—if ever found—could only be classified with the youngest mammalian fossils (382). If these statements are more unequivocal than Buffon's, this is because Herder is even more dedicated to the principle of anthropocentrism—not an ideology, in his view, but rather the only divinely warranted philosophy of the history of the human as an earthly creature elevated by its immortal soul (386, cf. 409). Even the crown of creation needs an earthly point of origin, however; "the construction of the earth speaks with likelihood" of the mountains of central Asia as this point of origin (*wie schon der Bau der Erde mutmaßlich saget*; 386). Earth's answer to this question of origins strikingly anticipates the premise, if not the content, of Blake's poem "Earth's Answer."[48]

The location of these mountains in central Asia remains necessarily vague, and Herder exploits its vagueness by using a range of "Oriental" traditions, including the Hebrew Bible, to negotiate the boundaries shared by geology, prehistory, and history. Departing from the stadial theory associated with the Scottish Enlightenment, Herder repeatedly positions Europe as a youthful, barbarian continent populated only after an inundation that is recent enough to have appeared in the historical record.[49] Just as Herder situates the dawn of human prehistory relative to geology, he here situates the written record relative to the larger expanse of prehistory. Thus, although he prides himself on his own exegetical skill as a reader of Genesis, he attributes true mastery of temporal scale to the Brahmins, whose civilization is presumably much older: "The Brahmin computes enormous sums from memory: the units of time from the smallest to the largest astronomical period are present to him and despite the lack of any European techniques, he errs seldom" (*I* 395). The Brahmin's fluency in the idiom of cosmic time is of a piece with the geological origin story of Prassarama's arrow. In his experimental mapping of the territory between geology and prehistory, Herder goes beyond the analogy between modern "savages" and human ancestors posited by the comparative method of nineteenth- and twentieth-century anthropology. He looks to oral tradition for traces of direct continuity—however fragmentary—between non-Western peoples and their ancient ancestors, understood as prehistoric if not datable. Chatwin's account of Aboriginal culture in *The Songlines*, briefly examined with other more recent literature in the afterword to this book, follows in this tradition. Like many nineteenth- and early twentieth-century geologists and archaeologists, however, Herder is conservative when it comes to hard dates, venturing only "millennia, without doubt" as a measure of the "large spaces of time" corresponding to the first day of Creation (405–6).

Herder arrives at an elegant solution to the problem of geological time in a chapter dedicated to the age of the earth and the location of Eden. Only history can be quantified, he argues; qualitative measures are wholly adequate to prehuman time. Fables like the one of Prassarama's arrow remain merely suggestive, and in this respect—according to Herder—they are essentially similar to Buffon's "imaginary" period of 75,000 years in *Epochs of Nature* (*I* 412). There is room in geological time, and even in human prehistory, for vast expanses of imagination, but the reason is that prior to history there can be "no chronology worthy of the name" (412). But because Herder reads "the oldest written record," Genesis, as the nationally specific account of a tribe of herdsmen who postdated the most ancient Eastern peoples (422), he still needs natural history and its adjunct, fable, to establish the mountains of central Asia definitively as the ancestral home of the species (423).

Herder turns from natural history in general to the human species in Part 2 of the *Ideas* (Books VI–X), but derives the first principles of his anthropology from natural history as well. Modestly proposing to follow the textbooks of natural history (*die Lehrbücher der Naturgeschichte*)—of which Buffon's *Histoire naturelle* is the archetype—Herder begins with human variety because these authors make it the foundation of their "natural history of man" (*I* 209). His rapid survey of human varieties around the globe allows him to develop the prehistoric time scale established in the preceding books, a realm abandoned for strict chronology in the second half of *Ideas* (Parts 3–4), which is a massive and much more detailed survey of all the ancient and early modern national histories. Repeatedly citing both George and Reinhold Forster—he even calls Reinhold the "Ulysses" of the South Pacific (239)—Herder calls for an "anthropological map of the world" to represent the environmental causes of ethnic and cultural variety (250). He develops and strongly defends both the principle of monogenesis and the thesis of migratory degeneration put forward by Reinhold Forster. "The New Zealand cannibal and Fenelon, the miserable Pesserai and Newton, are creatures of one and the same kind," he declares, all free and capable of reason (147).[50]

Mobility, the signature trait for Herder of our bipedal and artisanal species, has greatly altered those peoples who have wandered far from the scene of origin (though the locale is temperate here rather than tropical, as in Forster). The people Herder calls Eskimos, for example, now face a struggle to preserve (*erhalten*) their humanity (*I* 213) that essentially recapitulates the history of the species. The Arctic environment as a whole is "early" for Herder. He creates a prehistoric geography in which the Arctic Ocean presents a survival of the very first scene of life, "still [serving] even now as the womb of life" at the "boundary of living creation" (210). Although he objects to the term "race" on principle as suitable only for lower animals (150), his depictions of Eskimos and other peoples in "extreme" climates are deeply ethnocentric, and on one occasion he favorably cites Johann Friedrich Blumenbach, one of the founders of "scientific" racism (294). His treatment of Africa, however, accentuates European ignorance of this continent and even places an alternative account of human origins in the mouth of an imaginary African: "I, the black man, am the original man [*Urmensch*]!" (228). In the South Pacific, he happily cedes the authority to Forster's *Observations*, calling this work "a foundation stone of the history of humanity" (239). In discussing other regions, he speculates more freely about the long time scale required for environmental causes and use-inheritance to produce human varieties (e.g., 235).

Modern scholars have noted repeatedly that Herder is one of the first writers who used the term "anthropology" to envision a more scientific approach to these questions.[51] In this portion of *Ideas* he expresses the "anthropological wish" for a "gallery" of accurate ethnographic representations—a wish already anticipated in the critical preface to *Volkslieder* quoted earlier. The world map that he also envisions here is inspired by the cosmopolitan voyage narrative of the Forsters, whom Herder considers to be anthropologists genuinely concerned with accurate understanding of indigenous peoples. The negative outcome of this Enlightenment gesture is plain to see: non-European peoples are consigned to prehistory or "the time of the other," as Fabian calls it. Even after it is established as a separate archaeological domain, the prejudice against prehistory is enshrined in the cursory treatment given to it in volume after volume of world history, as Smail has shown. From the point of view of conceptual history, though, Herder's thought experiments with ethnography and prehistory take on another aspect: they demonstrate the intellectual ferment and the generative uncertainty surrounding the space we now designate as prehistory, which became gradually closed off by disciplinary paradigms.

Herder, then, would have agreed with Ritson that sung oral traditions provide a glimpse of "the simplicity of the remotest periods." He was unique, however, in pursuing these glimpses into a fully worked out history of humanity. Ritson and his British colleagues confined their explorations of "simplicity and nature" to the field of native ballads; they offer a sort of applied conjectural history in the context of the ballad revival. This applied conjectural history nonetheless underscores a set of concerns shared with philosophical historians of humanity and with others who were more apt to pursue the question of human origins and stages of development. Ritson appeals to predecessors who had applied these questions to the history of music (*HE* ii), and Scott relies heavily on stock phrases from conjectural history, such as "the childhood of society," to represent the ballad tradition as a source of primary material for the conjectural historian (*MSB* 504).

Macpherson's Ossian poems were tailor-made for this kind of analysis. In the preface to his first volume of supposed translations of third-century Gaelic epic, *Fragments of Ancient Poetry* (1760), Macpherson offers the following proof of their origins in "an era of most remote antiquity": "the poems . . . abound with those ideas, and paint those manners, that belong to the most early state of society."[52] It was immediately apparent to some skeptics, including Samuel Johnson, that these works confirmed the stadial theory of society because the author, or supposed translator, had derived the relevant details from the theories themselves as well as from genuine epics, both ancient and modern. The Ossian poems, published in four volumes over the next five years, became a *succès de scandale* because the enthusiasts outnumbered these critics.[53] Goethe and Herder were among the major enthusiasts; Herder even included his own translations of this material in his anthology, *Volkslieder*. He invented a striking new title for one of his Ossianic adaptations: "Memory of Prehistoric Song" (*Erinnerung des Gesanges der Vorzeit*). The German *Vorzeit* gained currency rapidly after 1770 and is now generally translated as "prehistory," though the *OED*'s earliest example of the English word comes much later, through Edward Burnett Tylor's *Primitive Culture* (1871), which is discussed in the next chapter.[54] In the original, which Macpherson presents as his prose version of a "beautiful" and ancient "lyric ode," Ossian apostrophizes his bardic ancestors, who are prehistoric only in the sense that they are remembered through oral tradition. By adding a title that refers to prehistory—one of his numerous alterations to Macpherson's original—Herder signals the double sense of the Ossianic past as that of both the individual and the species. Gidal has identified Herder as the first of a long line of readers who respond to the trope of "proleptic self-eulogy" in Macpherson's Ossian, the prophetic vision of the land as a mute witness to the disappearance of a people (or even the whole species) who will soon be consigned to the realm of prehistory.[55]

Hugh Blair's *Critical Dissertation on the Poems of Ossian* (1763) stands out among many scholarly defenses of their authenticity for being especially well versed in conjectural history.[56] The theorists and anthologists of the ballad revival, despite similar commitments to the stadial paradigm of Adam Smith and his followers, tended to be skeptics about Macpherson's Ossian: Percy wished to claim a Nordic origin for Britain, as opposed to a Celtic one; Ritson, as we have seen, rejected oral tradition as a plausible means of transmission for these poems; and Scott, addressing Scottish partisans diplomatically in an 1805 review of the Highland Society's *Report on Ossian*, nonetheless confirmed that Macpherson was "guilty as charged," in Nigel Leask's apt phrase.[57] Though all three of them were equally invested in a prehistoric origin for song, these Anglophone theorists would have rejected Herder's definition of prehistory as Ossianic times.

Burns and Beattie were among the Scottish poets who adopted the premise that oral tradition speaks of prehistory, but without claiming to be translators of ancient poetry. Beattie's *The Minstrel* (1771) tells "the story of Man" as a long verse narrative, tracing the path of its titular hero from his birth "in a rude age" through a literary education that recapitulates the progress of society. Burns compresses a similar kind of story into the space of a lyric poem using the native Habbie stanza form, "The Vision" (1786). Unlike Beattie's poem, Burns's is loosely autobiographical and mock-heroic, but his vision of a Scottish national muse in this poem also participates in the discussion that Macpherson initiated, a critical and philosophical engagement with the long survival of oral tradition in Scotland, as opposed to England and Europe. "The Vision" alludes to Beattie and Macpherson, first of all by using the Gaelic *duan* to designate subsections in the poem (Duan First, Duan Second), as Macpherson did in "Cath-Loda" and other poems. With characteristic irony, Burns's muse deflates his vision of national glory when she begins to speak (in English rather than Scots) in the course of her visitation. "In me thy native Muse regard," she instructs the poet, but it turns out that she belongs to the numerous "lower orders" of spirits under the aegis of Scotland's national Genius. She describes her class as follows: "Some, bounded to a district-space, / Explore at large Man's infant race, / To mark the embryotic trace, / Of rustic Bard" (*BLS* 44). Deliberately assigning "harmonious Beattie" to a higher class of muse, Burns still lays claim to a national idiom by using the native Habbie stanza for this serio-comic ode. "Man's infant race" has a strictly local connotation here, referring to his muse Coila's survey of potential poetic gifts among the children of Ayrshire—a modest arena aptly confirmed by her act of crowning the poet with holly at the end. But this language adopted from conjectural history also retains its philosophical connotation, anchoring Burns's song in the fabric of the land itself. Modesty topos notwithstanding, the Muse's reward validates the poem's use of language from conjectural history to authenticate the poet's tradition (as humble, but ancient) as well as his oral-traditional project of crafting "rhyme / for fools to sing."

Burns's irony registers the impact of late eighteenth-century print culture on the integrity of oral tradition. The optimism of the 1760s—witness Percy's faith, ingenuous or not, in the authenticity of his idealized recreation of oral tradition and Herder's enthusiastic pursuit of an even more authentic remediation, inspired by his reading of Percy—soon gave way to skepticism, and to ever-more-elaborate efforts in print to certify the authenticity of oral-traditional material gathered by editors and collectors. In this section, I have focused attention on Herder because of his theoretical ambition, which elevated the survivals embedded in oral tradition into evidence for a fully developed paradigm of deep human time, an account that traces the boundary between human and prehuman time on a global scale. Whereas Ritson capitalized on Percy's equivocations about his source material to expose ballads as artifacts of print culture, Herder privileged oral transmission more unequivocally than Percy had. Anticipating the degenerationist account that gained force later in the century, Percy argued that "as the old minstrels gradually wore out, a new race of ballad-writers succeeded, an inferior sort of minor poets, who wrote narrative songs merely for the press" (*Reliques* I.380). Percy claimed that his seventeenth-century manuscript (the Percy Folio) gave him privileged access to a transitional stage of the material, before it had become corrupted by print culture, but he supplemented these versions with early printed broadside ballads as well. Herder rejected print categorically, citing Shakespeare's dismissal of "meter ballad-mongers" in 1 Henry IV.[58] Herder's *Volkslieder* is as much a miscellany as Percy's *Reliques*, but he envisions an improved ethnographic practice (his "anthropological wish") that one day will capture oral traditions authentically with all their prehistoric freight—the ballad as fossil.[59] Ritson shares some of this faith in ethnographic collecting elsewhere, but within the European context he rejects current oral tradition as even less reliable than early modern print—the ballad as artifact. In the two "Introductions" to *Songs of Innocence and Experience* and in "Earth's Answer," treated at the beginning of this chapter, Blake echoes this narrative of textuality as corruption, seeking to recover antiquity instead through his poetic practice.

Unlike Blake, who rejected the skepticism and the empiricism associated with conjectural history and with much ballad scholarship, Wordsworth and Scott incorporated an anthropological treatment of human origins into their poetic projects. In his preface to *Lyrical Ballads* (1800), Wordsworth disclaimed any systematic intentions, noting that a full defense of his innovative poetic style would require him to "retrac[e] the revolutions not of literature alone but of society itself."[60] This phrasing acknowledges that reinstating the "real language of men" implicitly presupposes an account of prehistory, even when the reform is concerned with recovering the natural language of passion rather than with recreating oral tradition. Scott takes up this challenge in *Minstrelsy of the Scottish Border* (1802), an anthology of traditional material that also included adaptations and original poems by Scott himself, along with a substantial portion of critical prose that increased with subsequent editions. Possibly in the spirit of emulation, Wordsworth allowed himself to be tempted into more ambitious conjectures in his revisions of 1802: "The earliest Poets of all nations generally wrote from passion excited by real events; they wrote naturally, and as men: feeling powerfully as they did, their language was daring, and figurative. In succeeding times, Poets, and men ambitious of the fame of Poets . . . set themselves to a mechanical adoption of those figures of speech."[61] The appeal to degeneration here as well as Wordsworth's anxiety about the "savage torpor" of modern reading audiences both seem to echo Ritson's accounts of the degeneracy of modern urban taste, which is anchored in a conjectural history paradigm. Wordsworth's project of making "the incidents of common life interesting by tracing in them . . . the primary laws of our nature" moots a deep history of human nature that not only partakes of conjectural history but also anticipates, roughly at least, the evolutionary anthropology of Bernard Chapais, who traces altruism biologically to a common ancestor we share with other primates.[62]

In the "Remarks on Popular Poetry" added to *Minstrelsy* in 1830, Scott rewrites Scottish history as conjectural history, carrying Macpherson's and Burns's projects forward on a much larger scale. The Scottish "natives are passionately attached" to warlike Celtic remains, he writes, and so make a good subject (*MSB* 509), and their Gaelic "jargon" too has a native energy to recommend it (510). The center of ballad culture, for Scott, was located on the Scottish borders (511), which provided "an atmosphere of danger, the excitation of which is particularly favourable to the encouragement of poetry" (511–12). This piece of applied conjectural history is supported by an explicit theoretical statement:

> The investigation of the *early poetry of every nation*, even the rudest, carries with it an object of curiosity and interest. It is a chapter in the history of the childhood of society, and its resemblance to, or dissimilarity from, the popular rhymes of other nations *in the same stage* must needs illustrate the ancient history of states; their slower or swifter progress towards civilization; their gradual or more rapid adoption of manners, sentiments, and religion. The study, therefore, of lays rescued from the gulf of oblivion must

in every case possess general interest for the moral philosopher and *general historian*. (*MSB* 504, emphasis added)

When Scott turns from this ambitious premise to the mechanics of ballad transmission and the history of the ballad revival, the ballad's evidentiary links to "ancient states" quickly become tenuous. Surviving ballads, he concedes, tend to be unreliable at best because of alterations prompted by "the desire of the reciter to be intelligible," often compounded over many generations (506). As an Ossian skeptic who interpolated avowedly original poetry as well as cavalier alterations to traditional texts he nonetheless describes as "very ancient," such as "Tam Lin" (326), Scott was especially aware of the limitations of ballads as a source for conjectural history. His defense of this view in 1830 is further conditioned by the self-deprecating irony and ambivalence of the antiquarian persona he cultivated during the intervening years as author and "editor" of the Waverley novels. There is, nonetheless, a serious urgency in the image of "lays rescued from the gulf of oblivion" that makes Scott a key witness for the "oral-ethnographic turn" defined by McLane as driven by an awareness of the "extinction" that threatened oral-traditional culture. This awareness is made especially keen by the ambivalent status of Scott and some other collectors as both "natives" and "sophisticated fieldworkers."[63]

The quest for human origins readily converts to a nostalgia for the recent past in Scott, and ballads appear by turns as ancient oral tradition, as 200-year-old print ephemera, or as ethnographic fiction. All these elements are present among the proponents of the ballad revival, but their relative weight differs greatly for Scott, Ritson, Herder, and the other revivalists brought together in these pages. This uncertainty itself draws attention to the potential ability of oral tradition—hard to realize in practice—to shed light on events predating the written record, on what is now called prehistory. Ballads also interact in complex ways with the historical record, and these interactions are at issue in the section that follows.

"Thus Spake on That Ancyent Man": History by Other Means

To celebrate the release of the animated film *Early Man* (2018), English Heritage organized an "early man adventure quest" that culminated in a screening of the film at Stonehenge. Although their dedicated web page for this monument specifies its Neolithic age, the page created to advertise the "quest" locates the action of the film "at the dawn of time," an expression that marks the continuity between eighteenth- and twenty-first-century discourses about prehistory and may be equally meaningful for some visitors to the monument today. "Early man" is early in much the same sense as "the childhood of society," the term that Scott inherited from conjectural history. Though it is commonly replaced by more inclusive terms, "early man" still evokes an undatable, absolute past that persists in the chronological imagination in spite of the hard dates now associated with deep time. Lacking an archaeological paradigm for prehistory, Scott and his predecessors had little choice but to overleap relative dating and establish an absolute antiquity as the origin for what they regarded as primitive culture. While prehistory remained obscure, eighteenth-century collectors and antiquaries gained an increasingly sophisticated understanding of the print and manuscript archive and its chronology. The resulting juxtaposition of the uncertain time scale of oral tradition and the local time scale of the archive generated a dialectical movement between historical and "early" times, along with a contradictory effect I will designate as time parallax. This movement gave a wider scope to the prehistoric imagination but also generated new approaches to recorded history.

Philips's *Collection of Old Ballads*—the first of its kind—already cultivates this simultaneous awareness of the distant and more recent past. On the one hand, Philips extols ballads as "the first flight of a lively imagination," an ebullition of "poetic fire" older than Aristotle and his rule-bound poetics (*COB* I.67). On the other, this history of "ancient liberty" is immensely topical for England in the 1720s, literally a Whig history aligned with the party politics of Philips's literary circle. Thus Philips declares, "our Poets of old. . . . had no tedious comments upon Aristotle to consult" but "were look'd upon like other Englishmen, born to live and write with Freedom." But an allusion to the contemporary critic "[John] D[e]nn[i]s" as a proponent of "dull rules" reveals the modern character of this discourse about antiquity. Philips's stated motives are to elevate the literary status of ballads and "shew their antiquity" (I.iii), two closely related goals. Although Philips declines the test of Hebrew antiquity on religious grounds (I.iii), it should be noted that Herder enters fully into a correlation between ballads and Hebrew antiquity, both in *Ideas* and in a dedicated work entitled *On the Spirit of Hebrew Poetry* (1782/83), and Burns touches on the same connection in his mock-heroic address to Francis Grose. Philips proceeds directly to the ancient Greeks, developing the analogy to Homer (I.v) that became standard in the literature of the ballad revival. His account of "poetic fire" gives a hint of the progress of poetry since ancient times, but the continuity between Homer's "ballads" and the "rough music" of "English Historical Ballads" remains fragmentary at best. Even as he advances a strong priority claim for these "antique pieces," Philips also apologizes for their style: "I have even omitted some of the most antique," he writes, because they are "written in so old and obsolete a stile that few or none of my Readers wou'd have understood 'em" (III.vii, cf. II.vi).[64]

The "rude times" in which Percy locates his ballads (*Reliques* I.345)—showing even more ambivalence a generation later—seems to coincide roughly with the time of chivalry. But his distinction between the original minstrels, or liberal artists, and the modern, mechanical ones (I.347) also invokes the civic humanism associated with ancient Rome. The reader who ventures into this uncertain antiquity encounters numerous contemporary poems interspersed with the ballads to satisfy the polite taste of Percy's own times. In Ritson, who insists with equal intensity that song is "coeval with mankind" but that virtually no English ballads predate 1600, this time parallax is especially pronounced: the antiquity of the object under discussion varies greatly with the scale on which it is being viewed.

Philips's early contribution to the ballad revival also clarifies the role of ballads as potential historical source material. Writing before the idiom of conjectural history became established, Philips relies largely on the written record to establish his frame of reference for antiquity but also points out the ballad's potential to fill gaps in the more recent historical record. He offers the texts of ballads that he deems early as supplements to the comparatively scanty record of the reigns of early English monarchs including Henry II, Alfred, and even Brutus, the first "King of Albion" (*COB* II.1–5). The texts do not always bear out Philips's claims that they were written to "transmit . . . glorious actions" to posterity or that they provided substantial evidence for "several fine historians" (I.viii). He also refers to the balladeers themselves as the "historians" in "times of old" (I.vii) and suggests more plausibly that ordinary people still derive all their history from "Old Songs" (I.101). This multifaceted defense of ballads as historical material, to be promoted at a time when English history is "too much neglected" (III.iii–iv), even extends to events that may be recorded only in ballad texts, such as the "Excellent Ballad" of an anonymous "Prince of England's Courtship" from a very uncertain period. "Very few of these venerable ancient Song Editors were wholly indebted to Invention for their Poetic Productions," observes Philips half-facetiously in the headnote to this ballad, concluding that if this ballad cannot be firmly located in history, this is simply because it refers to circumstances "which through length of time have been forgotten" (I.181). "Lord Thomas and Fair Ellinor" (Child #73) and "The Famous Flower of Serving-Men" (Child #106) are among those referring to real events that have, in Philips's view, "escaped the Historians" (I.216). Percy also speaks of an event that "has escaped the notice of historians" as the foundation of one strand of the plot in *Chevy Chase* (*Reliques* I.21), presented in a version that he claims is the "true original" (I.19).[65]

The ballad's historicity, as established by the ballad revival, promoted renewed engagement with literary history alongside efforts to renovate national literatures—by Ritson, Aikin, Wordsworth, and others in Britain; by Herder and Goethe in Germany; and by Walt Whitman in the United States, according to Michael C. Cohen. Cohen argues that in *Song of Myself*, Whitman aspired to the "indexical power" that scholars such as Francis James Child claimed for the national canon of ballads. Whitman actually achieved this aim, Cohen adds, only in poems that were recognizable as ballads, such as "O My Captain," which succeeded in "remaking history in the shape of a ballad."[66] Lincoln's assassination, the event in this case, escapes the historical record only because it is too recent, a topical subject like those of the news ballads mocked by Shakespeare in *A Winter's Tale* (IV.iv), though Whitman invests his ballad with the gravity of a national occasion. For Ritson and Aikin, however, the historical significance of ballads, whether topical or ancient, is not a sufficient criterion of literary merit. Hence Ritson sets the ballad apart as "meer narrative" from the more authentic song forms, and Aikin concedes that ballads display "the character of the nation" but objects that "the real character of a Scottish or English shepherd is by much too coarse for poetry."[67] Aikin therefore advocates a new kind of deliberately aestheticizing "pastoral song." As against this diachronic and progressive orientation, Herder takes a more synchronic view, reclaiming the historicity of the ballad in the name of a "history of the human spirit," an aesthetic approach that nonetheless accommodates the documentary sense of ballads. In his first work of literary theory, Herder identified the history of the human spirit as the "highest aspiration" of all literary history; one editor observes that Herder's engagement with ballads (*Volkslieder*) was initially motivated by the conviction that they offered the "best documents" available for this history. For Herder, then, ballads become a worthy object not only of literary history but also of imitation, a rich "mine" of source material (*Fundgrube*) in the present "for the poet and orator of his nation."[68]

The eighteenth-century convention of tracing the ballad tradition back to Homer further complicates the question of its historicity. In making this link, poets and critics locate the origin of the ballad at the intersection of prehistory and history, or of orality and literacy, as they were being reconfigured by literary scholarship. Philips, among others, offers a more or less mock-heroic version of this genealogy. For critics such as Aikin and Ritson, the ballad's descent from Homer is out of the question. Herder, however, is one of those who takes this genealogy seriously. Drawing an analogy to work songs collected from non-German farmworkers in the Prussian countryside, Herder praises Homer as "the greatest folk poet," redefining his epics as *Märchen* and "living popular history [*lebendige Volksgeschichte*]."[69] Responding as well to Percy's nationalism and to Macpherson's epic recoding of the ballad tradition, Herder influentially locates the material for a modern national literature in folksong. As Gidal has shown, all these contributors to the ballad revival were deeply influenced by the new Homeric criticism initiated by Thomas Blackwell's *Enquiry into the Life and Writings of Homer* (1735). Blackwell's ethnographic and geographic approach was extended by (among others) Robert Wood, who traveled extensively through the terrain in which Homer's epics are set and developed a comparison between Homer and Ossian to underscore the importance of both physical geography and living folk traditions for "authenticating" ancient poetry.[70]

In the British context, the Homeric pedigree of the ballad predates the discoveries of Blackwell and Wood, reflecting a predisposition to understand origins, national and otherwise, as the special domain of the ballad. Writing at a time when traditional ballads were unfamiliar to many readers, Addison devotes two of his *Spectator* essays to "the favourite Ballad of the Common People of England," *Chevy Chase*. He develops a classical and epic pedigree for this ballad by arguing that Homer's motive was to promote national unity and then showing the same motive at work in the ballad. Addison elaborates the point with many further analogies to Homer and to Virgil.[71] Philips includes *Chevy Chase* in his collection a dozen years later, citing the authority of Addison as a "Great Man" (*COB* I.110), and develops the analogy to ancient epic, albeit while injecting a hint of irony. Philips introduces Homer as "nothing more than a blind ballad-singer, who writ songs" (I.iii). There would be no epic, in this view, if "somebody" had not "thought fit to collect all [Homer's] ballads." The remark seems opportunistic for the editor of a ballad collection, but it also anticipates Wood's argument, over a generation later, that later editors in classical Athens assembled the Homeric fragments into a text—much as Macpherson did (in Wood's view) with the fragments of Ossian.[72]

The high seriousness of the second example illustrates the powerful effect of oral tradition as an explanatory model. Philips in 1723 makes no appeal to oral tradition and quite openly uses the association between ballads and antiquity to bolster the canonical status of recent English poets such as Abraham Cowley and John Suckling (*COB* I.v–vi). He even contradicts Addison, claiming that the "authors" of *Chevy Chase* knew the *Iliad* and the *Aeneid* and imitated them deliberately (I.110). Philips's appeal to "the best and most ancient copies extant" on his title page provides evidence predating McLane's "oral-ethnographic turn" that efforts to accommodate traditional material within established canons of history and literary history paved the way for ever longer chronologies that ultimately established oral tradition as the location of a deep past for literature. Philips also points to the predilection for "old songs; at least written upon old subjects" within epic narratives themselves, including such origin myths as "the combat of the Gods against the Giants" and "the amours of Jupiter and Semele" (iv–v). Each of Philips's three volumes includes ballads that describe or invoke the ancient origins of Albion, by way of St. George (I.23–27), Brutus (the first "King of Albion"), Lear, Arthur (II.1–24), and the illustration of Stonehenge (Fig. 3.2), among other examples. An inset tale of origins also occurs near the beginning of Blake's *Milton*: "Terrific among the Sons of Albion in chorus solemn & loud / A Bard broke forth!" (*M* 2:23–24/E 96). This song of 350 lines differs from the songs of the *Experience* bard in its epic scale, indicating Blake's trajectory from lyric to narrative forms.

The Romantic poets, too, invoke the association between Homer and the ballad tradition in a variety of registers to establish a deep past for poetry. Burns, writing in a mock-heroic vein that recalls Philips's, deploys the image of Homer as a ballad singer in "Love and Liberty: A Cantata" (1785), also known as "The Jolly Beggars." In this burlesque mini-opera, as a modern editor explains it, "Burns set out to capture, in all their glory, the lives of a set of Ayrshire vagrants drinking one night in Poosie Nansie's tavern in Mauchline." Burns introduces the balladeer who sings the last two numbers as "a wight of Homer's craft." Uncharacteristically, Burns added a footnote to this line in manuscript, explaining that "Homer is allowed to be the oldest Ballad singer on record."[73] After serenading his wife as she couples with another man, Burns's Homeric "wight" introduces himself to the audience: "I am a bard of no regard / Wi' gentle folks and a' that; / But Homer-like the glowran byke [gazing swarm], / Frae town to town I draw that" (*BLS* 88). The same singer's concluding praise of "liberty," set deliberately against courts and churches, grounds his appeal for inalienable natural rights on the animal nature of human beings, and hence "I am a bard" seems to enlist Homer and Greek antiquity in this timeless cause. Scott is more circumspect in "Remarks on Popular Poetry," crediting the Athenians (as Philips and Wood did) for their "pious care" in collecting Homer's works—without which, he writes, they would "have appeared in the humble state of a collection of detached ballads" (*MSB* 503)—and stipulating that the "early poetry" of Greece has a special place in world literature because Homer's genius was fortuitously preserved in this way. Scott pointedly excludes "sacred and inspired authors" from this assessment, but Wordsworth—less cautious than Scott, if also less democratic than Burns—makes no scruple about including the prophetic "voice / which roars along the bed of Jewish song" along with "Homer the great thunderer" and the "wren-like warblings" of "ballad-tunes" in a genealogy intended to illustrate the power of literature as akin to that of nature, composed for *The Prelude* in 1804.[74]

By incorporating oral tradition into literary history, these authors reach beyond recorded history to affirm the continuity of human nature with the "archives of nature." This is not to say that they would uniformly have endorsed Buffon's term for the emerging geological record, with its prospect of a prehuman past and (especially) a posthuman future. The genealogy of verse-making from Homer to the ballad constitutes deep human time, nonetheless, by extending the reach of history beyond the written record through the survival of an ancient practice. The ballad revivalists used conjectural history to establish a kind of prehistory through the ballad, as the case of Herder makes especially clear. They also used the written record, including Homeric epic, to situate oral tradition within recorded history, thus recalibrating the historical time scale. The scrutiny of archives, whether literary or geological, promotes the escalation and acceleration associated with temporalization (*Verzeitlichung*) in the modern era by Reinhart Koselleck's conceptual history and by Hans Blumenberg's genealogy of the difference between scales of time.[75] Even as the historicity of particular ballads becomes more firmly established, the antiquity of oral tradition itself becomes more extended and uncertain.

In literary history as in natural history, the depth increasingly revealed by the archive has a qualitative as well as a quantitative dimension. Bakhtin's term for epic time, "absolute past," captures something of this qualitative dimension: "because it is walled off from all subsequent times, the epic past is absolute and complete," regardless of the number of centuries or millennia that may have elapsed.[76] In Bakhtin's account, the stasis of the epic gives way to the more dynamic novel form, but the closure characteristic of epic action also endows it with iterability. Hence oral tradition in the present manifests the same patterns as ancient epic, and ballads speak of origins in a fashion similar to the epic's inset tales, which may themselves be considered as survivals from a previous era or indeed from "the dawn of time." This familiar location, invoked by English Heritage in their account of "Early Man," preserves the qualitative sense of high antiquity that predates and informs the more quantitative senses of deep time appearing in the nineteenth century. "The dawn of time," especially in this context, also expresses a romantic nostalgia perpetuated by the ballad revival. As late as 1861, Henry Mayhew negotiates a balance between "Ancient and Modern Street Ballad Minstrelsy"—as he titles one section of *London Labour and the London Poor*—by assigning different ages to the three parts of a ballad: the verses may refer to current events, while the chorus must have some more timeless quality to "sell the ballad," but the tune "is a matter of deep deliberation" for the ballad maker because it needs to act on the listener on a deep affective level.[77]

Mayhew's nostalgia for the primitive sensibilities of London's poor is ambivalent, much like the nostalgia of his eighteenth-century predecessors, but it obtains a kind of scientific purchase by framing the ancient components of balladry as survivals, a thesis developed more broadly by Victorian anthropologists and revived by twentieth-century critics such as Friedman, who argues that the ballad tradition "preserved a primitive mode of composition." The aesthetic value attached to these survivals depends on whether traditional ballads are seen as the antidote to cultural degeneration (as in Herder or Wordsworth) or a symptom of it (as in Ritson or Scott). McLane points to the idea that oral traditions were facing extinction, both as a motive for collectors and as a major factor behind the heterochronicity of ballad histories. She also draws out the cyclical implications of this multiscalar view of human time: "The Stone Age may be a name for our human future as much as far our imagined past."[78]

Blake offers a more synchronic model of bardic time in which there is no contradiction between art and primitive culture. Commenting on Bishop Watson's *Apology for the Bible*, Blake defends the intellectual capacity of "man in a savage state" by remarking that if Native Americans were "savages," then Homer was too, "& yet he was no fool" (E 615). He is more critical in his late anticlassical tract, "On Homer's Poetry," but the visual grammar of the illustration on this plate marks the continuity between Homer and the bardic figures in Blake's own poetry, including *Songs of Experience*. The illustration (Fig. 3.3) shows a bard performing with his harp for an audience (one more sedate than Burns's "glowran byke") gathered on the right side of the composition, which corresponds closely to the design accompanying "The Voice of the Ancient Bard"—the concluding poem in many copies of *Songs of Innocence and*

FIGURE 3.3. William Blake, *On Homer's Poetry/On Virgil* (1822), relief etching (detail). © The Fitzwilliam Museum, Cambridge, UK.

Experience. At the beginning of *Songs of Experience,* the earth replies to the Bard in "Earth's Answer." Instead of bearing witness to a nonhuman temporality, this personification seems to confirm that all time is human time in Blake's cosmology. Yet the nonhuman speaker of this poem, like its verse form, owes something as well to the ballad tradition, which includes animals among other nonhuman speakers. "The Three Ravens," with its avian speakers, provides one of the best-known examples of a multispecies environment in traditional ballads, evoking "natural man" as one species among others.[79] Later in his career, Blake devoted his energies to the epic form, at least in part because of the special temporality associated with it. This heroic or eternal present allowed Blake to develop a mythology that depicts both human and cosmic origins as a continual process, unfolding in a deep time that he considers the native element of human consciousness.

Blake's Epic Time

Blake imagines his dialogue between the Bard and the earth through the medium of oral tradition. From the Anthropocene perspective now adopted by many scholars, the earth would appear as the initiator of this dialogue, and it would appear as a dialogue initiated in writing rather than in speech. Anthropogenic inscriptions of carbon, plutonium, and technofossils are added to the existing stratigraphic record, which predates our species by billions of years. Two major twenty-first-century studies also look to inscriptions—to the earliest written texts instead of to oral tradition—to establish a deep past for literature: Wai-Chee Dimock's *Through Other Continents* (2006) and Martin Puchner's *The Written World* (2017). In his critique of Dimock's efforts to trace American literature "across deep time," as promised by the subtitle of her book, Mark McGurl rightly notes that deep time in literature can exist only by analogy, that the history of human inscription operates on an entirely different scale from the geological processes that are the domain of deep time in a strict sense.[80] The deep time of geology, however, presupposes the bardic time that evolved at the intersection of literature and conjectural history in the course of the eighteenth century. The exclusive association between deep time and geology is at least partially the product of a retrospective disciplinary purification enterprise, as may be readily seen in Cuvier's dismissal of the Hindu conception of Great Time as deriving from "nothing more than a poem."[81]

In the last section, I set out to show that poets and ballad revivalists used the figure of the bard and its uncertain origins in different ways to consolidate their histories of literature and of the human species. As Blake turns his energies to epic composition after 1797, he begins to focus on the arts and sciences as a way to measure human antiquity and thus departs from the consensus that oral tradition might be seen as antedating—symbolically at least—the written records of ancient history. Under the pressure of speculation about oral tradition, the "absolute past" of epic narration reveals an unsuspected dynamism, and the antiquity of peoples and nations becomes a site of contestation, as Katie Trumpener has shown in *Bardic Nationalism*. The fluid chronology of bardic time in the eighteenth century is more than simply analogous to the unstable chronology of the earth in the same period. The bard in his various guises figures a deep human time that is cognate with the formation of deep geological time in the sense that both create space for *longue durée* processes newly recognized as falling outside the scope of accepted chronologies.[82] Unlike *Songs of Innocence,* which begins with a strict chronology that places song before writing, the second poem in *Songs of Experience* ("Earth's Answer") already begins to question "the voice of the Bard," its command over "Present, Past, & Future," and the primacy of the "Holy Word" itself.[83] In Blake's epics, the voice of the bard engages with ancient bodies of written knowledge. Poems such as *Milton* (1804–11) and *Jerusalem* (1804–c.1820) relocate the deep past of literature from oral tradition to the process of inscription.

In these epic poems, Blake maps the human body and the earth itself as bodies of knowledge, deeply inscribed by processes of artistic creation, including writing, that he considers native to the species. Blake's approach to the problem of deep human time, then, mediates between the vehicles of writing and oral tradition while anticipating in some ways the emphasis on inscription that now appears crucial to understanding the relationship between human history and the deeper past of the earth. By making literal the ancient figure of the body (*corpus*) of knowledge, Blake achieves a higher resolution for the process of time by scaling down, as opposed to scaling up in the manner of geology. In *Milton,* his epic-scale narrative of the poet's afterlife, Blake figures the bardic time of inspired composition as a moment of intense compression:

> Every Time less than a pulsation of the artery
> Is equal in its period & value to Six Thousand Years.
> For in this Period the Poets Work is Done: and all the Great
> Events of Time start forth & are concievd in such a Period
> Within a Moment: a Pulsation of the Artery.[84]

Blake telescopes the entire span of sacred history into a single heartbeat, reaching a dramatic conclusion for this description of Los and his sons at work, building the architecture of time. Los is a major character in Blake's mythology, broadly associated with art and craft, and presented here as a personification of time. The architecture of time is a product of the arts and sciences as they are practiced in "Eternity" (*M* 27:55–56/E 125); human and cosmic time are synchronized through the bodily rhythm of pulsation. In this way, the world's body and the body of Albion, the "Ancient Man," become coterminous with the body of knowledge.

Blake's innovative approach to the problem of human antiquity sheds light on the importance of compression as a figure associated with the GTS, more commonly associated with expansion and escalation.[85] From Carl Sagan's melodious "bill-yuns" to John McPhee's geological "violence on an epic scale," deep or cosmic time has often been conjured by rhetorical feats of expansion and slowness. In its early iterations, however, deep time is explicitly a product of reflection on the embodied experience of temporality, and this act of reflection links Buffon's polemical expansion of the time scale in *Epochs of Nature* with Blake's meditations on "the moment of time" in *Milton* and *Jerusalem.* Convinced that "the human mind loses itself in the expanse of duration much more readily than in that of space or number," Buffon presents geological moments in which only a retrospect, condensed into a manageable time frame, can make imaginatively available processes for which geochemistry dictates a duration ten or a hundred or even a thousand times greater than the canonical limit of 6,000 years (*EN* 40–41). Blake makes the same period equivalent to "a pulsation of the artery" in *Milton,* and in *Jerusalem* he compresses the geological catastrophes or "many sorrows" of 8,500 years into a rainbow woven by Erin's "awful hands." A glimpse of the expanse of time transpires in a cognitive instant. The epochal histories of Buffon and Blake, as different as they otherwise are, both use retrospect to compress the epochal into the momentary, thus reflecting on the embodied experience of thinking duration.

Naturalists exploring an expanded time scale, such as Buffon, also stress the acts of reflection that produce a time scale and the figures on which it depends. Blake's reliance on the 6,000-year time scale prescribed by sacred history carries an irony that becomes more apparent when we compare it to Buffon's transformation of the six days of creation into the first six epochs of nature. In making this equation, Buffon is not so much adopting as reflecting on the tradition, already well established within natural theology, of interpreting the days of creation with some latitude as more extended periods. When he added a seventh epoch in a later revision of *Epochs of Nature*—his final epoch of human intervention, which (as detailed in the previous chapter) has become newly salient in the Anthropocene—he injected at least a hint of heterodox irony by comparing this epoch implicitly to the seventh day, the day of rest. Unlike Blake, Buffon is deliberately advocating for a much longer earthly time scale, but in doing so he capitalizes on the ambivalence of temporal units such as days and moments, much as Blake does in describing the architecture of time. Moreover, he too is reflecting on the momentary acts of mind, including retrospection, that make the contemplation of deep time possible in the first place. By conceding that "the human mind loses itself in the expanse of duration," Buffon implicitly recognizes the importance of tropes such as compression and analogy for the expansion of the time scale. Contemplating a series of shale and limestone strata on the Norman coast in his first extended treatment of temporal apperception, he shows how the momentary view of a landscape may suggest a vast expanse of geological time that is at least commensurate with "the limited power of our intelligence" and may authorize further expansions of the time scale (*EN* 40n.).

For Blake in *Jerusalem*, the separation of continents provides the geological occasion for a mythological treatment of scale and the perception of time. At the midpoint of the epic, when the full catastrophe of Albion's separation from Jerusalem has been revealed, Blake returns to a cataclysmic image with which he had been preoccupied at least since the composition of *America: A Prophecy* (1793): the sinking of the Atlantic continent. To elaborate this catastrophic analogy, he introduces a new prophetic figure, Erin, who carries political resonances of Irish revolution as well as allusions to antiquarian discourse in which Ireland figures as a fragment of a lost ancient continent and/or civilization. "With awful hands" Erin "prepares a space" for Jerusalem after her "terrible separation." This space is a rainbow, "Eight thousand and five hundred years / In its extension," drawn out of a Moment of Time across the Atlantic Vale (*J* 48:30–37). An ephemeron lasts a day, but a rainbow only lasts a moment. As an image of temporal extension, it poses a challenge to Locke's categorical distinction between extension in space and duration in time, as does the later coinage of "deep time" itself.[86] The rainbow indirectly recalls the Deluge and hence the promise of redemption that also informs Los's visionary construction of the architecture of time. In this context, however, its evanescence seems deliberately juxtaposed against the scale of geological upheaval. In the long speech by Erin that follows, the Atlantic Continent and Atlantic Mountains feature prominently in the Eternal geography from which Jerusalem is now shut out. Though Blake was almost surely exposed to "the new science of geology," as his friend George Cumberland called it, during the period of *Jerusalem*'s composition, it seems unlikely that he would have been persuaded by naturalistic accounts of the gradual separation of the continents such as Buffon's.[87] The difference of scale is nonetheless clear in the figure of the rainbow, which vividly compresses these large-scale events while elevating Erin's "labours / Of sublime Mercy."

Elsewhere in Blake the visionary compression of time—generally assigned the more conventional span of 6,000 years—is the work of Los rather than a female artist, but curiously the anomalous period of 8,500 years resurfaces later in *Jerusalem* as the age of Los (83:52/E 242). Since this variation on the standard figure for the Ages of the World "has not been satisfactorily explained," according to Morton Paley, it seems worth noting that Plato dates the sinking of Atlantis to 9,000 years before his present in the *Timaeus*. Blake's attachment to the myth of Atlantis dates to the period of his involvement with antiquarian authors such as William Stukeley and Jacob Bryant during his time as an apprentice engraver, as Vincent de Luca has shown in detail.[88] The myth of Atlantis also had a strong hold on naturalists from Francis Bacon to John Whitehurst and beyond, and Whitehurst's version of it is especially relevant both to Blake and to the prehistory of deep time.

Whitehurst's *Inquiry into the Original State and Formation of the Earth*, first published, like *Epochs of Nature*, in 1778, has both general and local affinities to Blake's mythological treatment of the earth and time. The British Isles represent vast continents for Whitehurst in the sense that he derives his theory of the earth by metonymy from his detailed fieldwork in the county of Derbyshire. Whitehurst cites Bacon as a precedent in noting the agreement between ancient creation myths and his own observations of the strata exposed in the Derbyshire mines: "the fragments of ancient learning and philosophy, supposed to have been handed down to the Phenicians and Egyptians, from the memorials of more ancient nations, concerning the primitive state of the earth," may now "be seen to have some foundation in nature and truth." Ancient and modern traditions converge in another locale that also has particular significance for Blake's myth of "the Atlantic mountains, where Giants [such as Albion] dwelt in intellect" (*J* 50: 1/E 199). Whitehurst infers that since no evidence of a crater remains near the basaltic Giant's Causeway in northern Ireland,

> the crater from which the melted matter flowed, together with an immense track of land toward the north, have been absolutely sunk and swallowed up into the earth, at some remote period of time, and became the bottom of the Atlantic ocean. . . . The consideration of such disasters, together with that of the cause still subsisting under the bottom of that immense ocean, almost persuade me to conclude that Ireland was originally a part of the island Atlantis, which, according to Plato's Timaeus, was totally swallowed up by a prodigious earthquake.[89]

Blake would have seen Erasmus Darwin's extensive footnote references to Whitehurst's theory in *The Botanic Garden* when he was at work on engravings for that poem. This exposure does not amount to influence, and the main affinity between Blake and Whitehurst in any case has less to do with geology than with their mutual reflection on the figures and the acts of mind that make vast periods of time intelligible. Whitehurst, a clockmaker by trade, also differs deeply from Buffon, whose concern is to establish a vast expanse of prehuman time; Whitehurst instead uses geological evidence to increase the antiquity of "arts and civilization."[90] Blake's epics, too, are engaged in tracing bodies of knowledge through geological time.

The image of Erin's rainbow woven across the Atlantic Vale compresses geological events associated with Ireland and Atlantis while expanding the rhythmic actions and affections of the body. Her embodied "labour of sublime mercy" inhabits the time of pulsation that frames the work of Los in Blake's *Milton*. Throughout Blake's prophetic books the continual creation of the natural and social world is typically superintended by Los, summed up as "the World of Los the labour of six thousand years" (*M* 29:64/E 128). The more atypical creative-redemptive performance of Erin reveals the dispersed, collaborative nature of this labor while offering an alternative time frame of 8,500 years and a more integrated cosmology decoupled from the prevailing binary of masculine time and feminine space. As she draws her rainbow out of a moment, she also opens an atom of space, compressing the vast liminal domain of Beulah into its center. In her long subsequent speech—something of a Jeremiad—she explicates this action in two ways, reconciling the biological clock of the body with the geological clock implied by the imagery of Atlantis. First, the continent's disappearance is merely an illusion created by "narrowed perceptions" (*J* 49:19/E 198), a major theme in this speech, as elsewhere in Blake, that accounts for the compression of time and space. Second, within the rainbow's arc "every two hundred years has a door to Eden" (48:34/E 197), permitting access to eternity through time. In this sense, "Time is the mercy of Eternity" (*M* 24:72/E 121).[91]

In *Milton* the same reflections on temporality and scale are brought to bear on the constantly recurring moments of walking, breathing, and pulsation. The "pulsation of the artery" marks the period in which "the Poet's work is done," according to the *Milton* narrator; curiously, the word *pulsation* is used only once in *Jerusalem*, where it describes the creation of Erin out of the furnaces of Los (*J* 11:2/E 154). This word is repeated in the climactic passage on the poet's work in *Milton* and helps to underscore the work of compression as rooted in bodily functions. This poem explicitly identifies Los with time itself: "Los is by mortals named Time and Enitharmon is named Space" (*M* 24:68/E 121). Thus it stands to reason that the construction of time is a major part of the "Labours of the Sons of Los" that coincide with the Great Harvest and Vintage, Blake's recurring figure of apocalypse: one group of the sons builds the whole suite of temporal units from moments up to periods, while others build containers such as human and animal bodies; porches of thought or verbal structures; and the human optic nerve.[92] Here again there is a cycle of revelations occurring every two hundred years—or seven ages—but the longest period of all, paradoxically, is less than a heartbeat, "the period / Within a moment" during which "all the great events of Time . . . are conciev'd" (29:1–3). The creation of the GTS now appears as such an event, but in its early iterations, such as Buffon's *Epochs of Nature*, the role of inspiration is readily apparent.

For geologists in the field, bipedalism is another vital faculty for scaling down cosmic time. Los, constantly in motion, "in Six Thousand Years walks up and down continually, / That not one Moment of Time be lost" (*J* 75:8–9/E 230).[93] Just before this, Blake's narrator performs the same feat himself, and the corresponding passage in *Milton* explains that "Los had enter'd into my soul" (*M* 22:13/E 116). This passage not only develops more fully the position of the moment relative to eternity, it also reflects on poetic inspiration and on the heroic present of epic action by equating the pulsation of the artery with the metrical pulse of the line. For just before Los enters his soul, the poet Milton enters his left foot, allowing Blake as narrator to bind on the "bright sandal" of inspiration and "walk forward thro' Eternity" (21:14/E 115). The poet's step and his heartbeat bookend this description of Milton's descent that concludes the first book. As Blake puts it here, "man cannot know / What passes in his members till periods of Space and Time / Reveal the secrets of Eternity: for more extensive / Than any other earthly things, are Mans earthly lineaments" (21:8–11/E 115). This declaration smacks of anthropocentrism, but in the context of deep time it should also strike us as a recognition that temporal scale, and hence the freedom to scale both up and down, forms the basis for any reckoning in time.

From a geological point of view, Blake overstates the case. Beginning with Buffon, if not before, geological observation made increasingly clear how many moments of time were lost to eternity in the geological record.[94] Even a local instance, such as Buffon's glimpse of limestone strata overlying a hill of shale on the Norman coast, threatens to exceed any scale commensurate with "our too brief existence." Scaling down even further, Buffon invites us to consider the oyster: "are not many years required for oysters that pile up in some parts of the sea to multiply there in sufficient quantity to form a kind of rock?" (Z 61). Surveying the processes of marine transport and deposition and consolidation, he enlarges the scope of this humble example from "many years" to "centuries of centuries" within a few sentences. Even this scale, Buffon considers, is probably abridged by at least a factor of ten to make it graspable (*E* 40–41). Inevitably, many moments are lost in the construction of geological periods, as Darwin also noted when he referred to the geological record as "a history of the world imperfectly kept." Hence Darwin turns to geohistory unfolding in the present. If we simply "watch the sea at work, grinding down old rocks," we are able to capture whole epochs in a moment of visual perception.[95] Buffon responded to the challenge of aggregating these kinds of observations by changing his mind repeatedly about the scale of geological time. The physiologist Marie-Jean-Pierre Flourens, also a critic of Darwin, dismissed these vacillations as "conjuring tricks" in his 1860 edition of Buffon's manuscripts.[96] Although it is pejorative, this metaphor effectively captures the rhetorical aspect of scale as it manifests itself across three very different visions of time compressed into a moment: Darwin's purely relative positioning of weathering in the context of long-term geological processes; a similar observation in Buffon, rendered unstable by his pursuit of absolute dating; and Blake's eternal present adopted as a vantage point on the unfolding of cosmic time.

Though incorporating geological and biological metaphors of embodiment, Blake's poetic history of the human species lacks the strong naturalistic emphasis characterizing the deep time narratives treated in the first half of this book, those of Buffon and the Forsters. Like these authors, however, Blake cultivates a sense of the deep past, of species time in this case, by reflecting on the scalar qualities of time. These critical reflections on the reciprocal workings of scale and human cognition, with the attendant risk of "losing oneself in the expanse of duration," as Buffon put it, deserve a place in the conceptual history of deep time. Blake's depiction of cosmology through the practice of craft knowledge (weaving, blacksmithing, poetry) likewise calls attention to embodiment as the limiting condition for perceptions of time. As discussed in my introduction, the *OED* locates the first instance of the term "deep time" in Thomas Carlyle's 1832 review of Croker's new edition of Boswell's *Life of Johnson*. Carlyle uses this figure, catachrestically giving time extension in space, to anticipate the enduring power of Johnson's example, over and above that of his works: a *longue durée* to be sure, but one very much within the scope of historical time and specifically of literary history. Rather than separate early uses of the term such as this from authentic geological usage, a conceptual history must account for the continuities that link the two. As Stephen Jay Gould argued so memorably in *Time's Arrow, Time's Cycle*, the rhetoric of deep time in its geological sense owes much to ancient patterns of narrative time.

Blake's revisionist approach to epic has been a fruitful source for literary scholars thinking about these patterns, which include not only arrows and cycles but also *chronos* and *kairos*, the Greek concepts of quantitative and qualitative time, with their rich rhetorical and Biblical history.[97] Dimock identifies the architecture of "unnatural time" in Blake's *Milton* as "something that takes the entire history of the species to emerge," an "environment" engineered to "save human beings from biology." Steven Goldsmith points out that even though Blake's account is far from naturalistic, his notion of the "dense instant" or "time of pulsation" as the period of the poet's work in *Milton* still depends on the physiology of the heart to convey the reciprocal action of extension and compression. According to Goldsmith, this passage registers Blake's recognition that "the exceedingly small and the vast share the quality of being beyond and before narratable time." Pulsation, being neither active nor passive, suggests an "auto-poetic" quality in time itself that Blake uses to establish critical distance from the idea of the heart as the seat of sympathy or compassion, according to this reading.[98] Though Goldsmith does not refer to it here, poetic meter provides a relevant example of what he calls "inorganic pulsation," an inhuman temporality mirrored in space by Blake's analogy between the "red Globule of Mans blood" and the "unwearied Sun" (*M* 29:19–23). My concern in linking Blake with geological thought is not the geological content but the reflections on time, scale, and embodiment that shape the eternal present of his epics and similarly inform geohistorical speculations, including Buffon's polemical expansion of the time scale in *Epochs of Nature.* Blake's vision of Los as a geological agent who "displace[s] continents" and "alter[s] the poles of the world" (*M* 9:16–17) has a special poignancy in the Anthropocene, the ostensibly geological epoch in which the vastness of natural time may prove to be no more adequate than sacred history for securing the future of the species.

Like Herder, Blake uses myth to create a contact zone between geology and prehistory. Whereas Herder, also a poet in his own right, chooses the philosophical idiom of conjectural history for this endeavor, Blake settles on epic poetry as a performative medium for his antiquarian syncretism, incorporating the speculations he gleaned from the ballad revival and from other learned works that he read or illustrated. Together with mythological figures of geological agency, such as Los, Erin, and Albion, Blake also develops a paradigm of primitive arts and sciences throughout *The Four Zoas, Milton,* and *Jerusalem,* themselves "poems of the highest antiquity," as Blake assured visitors to his 1809 exhibition (E 542–43). In all these works he seeks to establish that "the Primeval State of Man, was Wisdom, Art, and Science" (*J* 3:48/E 146). In the synchronic dimension of this myth, the "Ancient Man," or Albion, "convers[es]" routinely "with the animal forms of wisdom" (*FZ* 138:31/E 406), evoking the pastoral state of society associated with oral tradition and effectively recovered in *The Four Zoas* as the "ancient golden age renewed" (126:29/E 395). Shortly after writing these prophetic lines, Blake abandoned this poem and returned from Sussex to London when he experienced an epiphanic renewal of his own artistic vocation: "I have . . . resumed my primitive & original ways of Execution in both painting & Engraving," he announced in a letter of 1803, reflecting on the historicity of his own method in relation to the larger narrative of the arts and sciences (E 724). Many other accounts of ancient knowledge, both in Blake and elsewhere, give special attention to the useful arts, foregrounded as artisanal practice in this letter and in Blake's poet-prophet-blacksmith character Los as well as in his theoretical writings, which insist on the inseparability of "invention" from skilled "execution." Useful arts also featured in the new synthesis promoted by holistic institutions (such as the Royal Institution) that departed from the Enlightenment model of learned societies (such as Blake's quondam nemesis, the Royal

Academy) and contributed to what Jon Klancher terms a "fundamental redefinition" of the arts and sciences themselves in Blake's time.[99]

In his writing about the ambitious paintings of this period, especially *The Ancient Britons,* Blake explicitly connects his own artistic process with the prehistoric origins of art. Blake exhibited this painting, now lost, together with fifteen others in 1809 and published *A Descriptive Catalogue,* containing notes on each picture, for visitors to the exhibition. In his long commentary on *The Ancient Britons,* he announces that "the British Antiquities are now in the Artist's hands" (E 542). Among many other "antiquities" in this syncretic account of Albion as equivalent to "the Atlas of the Greeks," King Arthur, and the Biblical patriarchs, Blake cites both his own "voluminous" poem in progress on "the ancient history of Britain," *Jerusalem,* and an antiquarian compendium he had illustrated during his time as an apprentice engraver, Bryant's *New System*: "the antiquities of every Nation under Heaven, is no less sacred than that of the Jews. They are the same thing as Jacob Bryant, and all antiquaries have proved" (E 543). According to Leslie Tannenbaum's reading of this passage, "Blake is here distinguishing history as *kairos* from history as *chronos,* that is, history as a number of moments containing divine and 'intemporal significance' from history based upon clock time." Tannenbaum adds that Blake's "concept of history juxtaposes events, things, or persons from different time periods in order to relate them to a central theme or paradigm that transcends the world of time and causality."[100] Blake's disruption of sacred history therefore proceeds from a very different motive than the revisionism of Buffon. In deliberately synchronizing his artistic process with the "naked civilized men" depicted in this painting, he also establishes the continuity beyond history of an ancient and native body of human practices that guarantees the survival of images in the same way that metrical verse preserves the rhythm of pulsation.

Calling on Albion to "return" to his prelapsarian state of unity, Blake's narrator in *Jerusalem* reminds him that "Mountains are also men; everything is human. Mighty! Sublime!" (*Jerusalem* 34:48/E 178). When Blake insists repeatedly on the humanity of landforms such as "the Atlantic mountains, where Giants dwelt in intellect" (*Jerusalem* 50:1/E 199), his literalized metonymy is authorized, as he tells us, by Jewish tradition and by other ancient humanized cosmologies. Blake's contention extends to scripture the claims of Herder, Buffon, Reinhold Forster, and other contemporaries that oral traditions contained recoverable traces of ancient natural events. At least since Isaac La Peyrère's *Men Before Adam* (1656), chronologers and theologians had used naturalistic accounts of the Biblical Deluge to situate human antiquity in relation to observable changes in the natural world. Natural philosophers in the later eighteenth century changed the character of this discourse by relating local oral traditions to the genesis of particular landforms. Examples include Herder's geological origin story of Prassarama's arrow (discussed earlier in this chapter) along with Forster's story of Maui dragging a net through the South Pacific and Buffon's account of northern Asia, examined in previous chapters of this study. Analogously, Blake invokes the "rude monuments" of "Persian, Hindoo, and Egyptian Antiquity" in his *Descriptive Catalogue* (E 530) to authenticate his appeal to the "stupendous originals" on which he bases his vision of public art as an ancient/native human practice. The story of Atlantis, as Blake incorporates it into *Jerusalem,* shows that geology is part of the allegorical framework he uses to establish the epic scale of these origins. More explicitly than his account of "the poet's work" in *Milton,* Blake's allegory of Erin drawing the rainbow out of "a Moment of Time" translates cosmic time into the practice of a craft (*J* 48:31/E 197). The practice of weaving here reiterates or instantiates ancient wisdom concerning a catastrophe of the geological past, the sinking of Atlantis.

Blake's abiding interest in the scholarly literature in which he was immersed during his apprenticeship, ranging from Bryant's *A New System: or, an Analysis of Antient Mythology* to Cook's *A Voyage toward the South Pole,* allowed him to integrate ethnographic accounts of ancient oral tradition with philological accounts of ancient writings. Blake's idea of the original earth and original man as one body (Albion) has been identified with the kabbalistic figure Adam Kadmon by W. H. Stevenson, following Blake's own cue in "To the Jews." George Gilpin makes another relevant identification in his reading of *The Book of Urizen,* arguing that "Los attempts to restore the beautiful metaphor of the 'world's body' of classical science." Gilpin points out that Bacon in 1605 had mocked "the ancient opinion that man was Microcosmus," but in *Wisdom of the Ancients* (1609) Bacon also put forward the influential argument that ancient myth contains encoded or mediated allegories of scientific knowledge.[101] Whitehurst and Erasmus Darwin were among the contemporaries who preserved this esoteric dimension of Baconian science. Whitehurst's physico-theological thesis concerning the "primitive islands" that appeared out of the first chaos finds an echo in Forster's more secular claim that the volcanic islands of the Pacific signify the land's original rising out of the sea and therefore approximate the original "seat of the human race."[102] Blake's narrative of art and science is indebted to scholarship on ancient texts, but its affinity to Forster's field notes on the "ancient systems" of "primitive" peoples shows that this narrative has an ethnographic side as well. His depiction of Albion, the "Ancient Man," as tattooed with images of the sun, moon, and stars (*J* 25), combines ingredients from the Cook voyage accounts of tattooing among Pacific islanders, from illustrated histories of the ancient Britons, and from antiquarian accounts of astronomical knowledge among preliterate peoples as a survival from antediluvian times.

Blake's innovative synthesis of antiquarian approaches translates the heroic present of epic tradition into a kind of species time, a "moment" in which the history and primitive condition of the species could be instantiated in weaving or engraving as much as in poetry or song. Blake considered the arts and sciences in some form to be original human attributes or faculties, as did Herder and some of the other ballad revivalists considered in this chapter. Blake's particular emphasis on embodiment helps to create a unified perspective on this view of human antiquity. Taken together, Blake's poetic project, Herder's philosophy of history, and the ballad revival that inspired them present a pre-evolutionary history of the human species. Influenced in a variety of ways by the emerging GTS, these authors begin to bridge the distance between conjectural history and prehistory by collecting materials on the origins of human knowledge from both written and oral tradition, from science and from art.[103] Blake's focus on embodiment as the shared condition of the species and the planet, by way of pulsation and related images, opens an avenue to the rhetoric of compression that enables him to reflect on the scale of time. The GTS in its modern form (Fig. 1b) provides an overarching diachronic paradigm that strongly influences the narrative of human evolution. Species time, or deep human time, a qualitative prerequisite for the emergence of deep time in this form, has a more synchronic character. This sensation of "depth" in time becomes palpable in the synchronic paradigm of ancient and modern cultural practices promoted by Blake and Herder, as against the more diachronic thrust of Scottish conjectural history and its influence on other strands of the ballad revival.

Positioning themselves in the widening gap between sacred and secular time, "Earth's Answer" and *Jerusalem* both present images of bodies originally marked by knowledge. In "Earth's Answer," Blake extends this claim to the world's body by drawing on the nonhuman speaker as well as other conventions associated with the ballad tradition. The systematic opposition to the doctrine of original sin, or "Eating of the Tree of Knowledge" (E 564), in his later works provides a useful context for the idea of knowledge as the original condition, not only of humanity but of the rotating planet. Blake develops this hint of a

FIGURE 3.4. *A Woman of the Island of Tanna.* Engraving by James Basire (probably assisted by William Blake) from a drawing by William Hodges. James Cook, *A Voyage Toward the South Pole* (1777), vol. 2, Plate 45. Courtesy of the Linda Hall Library of Science, Technology, and Engineering, Kansas City, Missouri.

deeper past for the world's body by invoking indigenous craft knowledge, such as weaving and tattooing, in his depiction of the "Ancient Man" in his epics. The strong antiquarian component in these later poems, itself the product of Blake's apprenticeship as an engraver, inflects their imagery of art and science with a sense of practice and process. Art historians have attributed to Blake one particular image in Cook's *Voyage toward the South Pole,* published toward the end of his apprenticeship, partly on the basis of its strong resemblance to the frontispiece of *Songs of Experience.* This image of a Tannese woman and child (Fig. 3.4, from an encounter discussed at length in Chapter 1) marks the first in a series of ethnographic portraits engraved by Blake in the course of his career, including subjects from Surinam and from New South Wales in the 1790s.[104] This aspect of Blake's professional work, spanning the transition from lyric to epic in his poetic career, illustrates the importance of "knowledgeable practices" for his understanding of the human condition. Amanda J. Goldstein uses this term to designate Blake's engagement with scientific thought, arguing that he uses images of artistic practice to locate "livable form and its science . . . between fact and construct," as opposed to the objective mastery promoted by Baconian science.[105]

Conclusion

Blake elevates the embodied practice of knowledge by placing it in the heroic present generally reserved for heroic action in the epic tradition. In so doing, he recovers a deep human past obscured by the written record, as proponents of the ballad revival also set out to do by exploring oral tradition. These cultural recovery projects rendered dynamic the "absolute past" long associated with the written record of human antiquity, especially the Hebrew Bible. Buffon's experimental revision of geological time during the same historical period mobilizes the qualitative experience recorded by many naturalists in the field who were forced to contemplate an indefinite "abyss of time" by the sheer scale of the phenomena that they were documenting, often for the first time. Like geological time, deep human time or species time has a qualitative dimension that is prerequisite to a concrete grasp of hard dates in human evolution, such as the divergence of proto-humans and chimpanzees from their most recent common ancestor 4.6 million years ago. In the first two chapters of this book I concentrated on naturalists who positioned human origins against the backdrop of geological time and on the reciprocal effects of these two horizons. This incipient distinction between "human" and "natural" time later comes to bear on modern evolutionary accounts of human behavior, including the arts—to be explored further in the next chapter.

The literary approach to cultural origins addressed in this chapter also has something in common with evolutionary paradigms, but the affinity is more symbolic and less direct. By picturing early humans and the origins of art through language, authors associated with the ballad revival established a metaphorical register that lingers in scientific discourse about stratigraphic signatures and genetic transcription and translation. Hans Blumenberg argues that Enlightenment expansions of the time scale served to compensate for the acceleration of history and the increasing sense of civilizational belatedness.[106] Blake, Herder, and their contemporaries may be understood as reacting to modern temporality in a similar fashion, but they also cultivated a new kind of attention to the cultural present by looking to oral performance and artistic practice for survivals from prehistory. When Ritson argues that singing and dancing are "coeval with mankind" or when Blake declares that "the Primeval State of Man, was Wisdom, Art, and Science," they are not setting human origins explicitly against a prehuman past but are nonetheless producing permanent and distinguishing attributes of human species being. The power of memory to access this deep human past was subject to debate, and Blake famously rejected memory altogether in favor of inspiration. He nonetheless shares with Herder and with many of the actors in the ballad revival a synchronic understanding of culture, of art as primitive culture, that merged into the idea of evolutionary memory later deployed in some evolutionary accounts of the arts as well as by science fiction writers such as J. G. Ballard, who argues that "the central nervous system is a coded time scale," enabling individuals to activate "the ancient organic memory . . . laid down in your cytoplasm millions of years ago."[107]

4

The Descent into Deep Time in Darwin and Lubbock

VOYAGE NARRATIVE, COMPARATIVE METHOD, AND HUMAN ANIMALITY

[T]he new views with reference to the Antiquity of Man . . . will, I doubt not, in a few years, be regarded with as little disquietude as are now those discoveries in astronomy and geology, which at one time excited even greater opposition.

—JOHN LUBBOCK, *PRE-HISTORIC TIMES* (1865)

AS A RESPECTED archaeologist and successful politician, John Lubbock (1834–1913) was in a good position to reassure his Victorian readers that the radical expansion of anthropological time was merely a temporary disruption occasioned by the progress of science. Lubbock's declaration signals the completion of the prehistoric turn defined in the first three chapters of this book.[1] Having argued that colonial natural history and geological fieldwork in the mid-eighteenth century began this disruption of human temporality, I addressed a parallel body of evidence in the previous chapter, looking to the ballad revival of the eighteenth and early nineteenth centuries for its poetic expansion of the deep human past. Ballads and oral tradition became a powerful linguistic medium for focalizing species time, the point at which the prehistoric human past recedes into the deep time of geology. This chapter shifts attention from the form and content of oral tradition to the physical capacity for song and other evolved human traits that helped to define species time for Lubbock, Charles Darwin, and other evolutionary thinkers in the mid- to late nineteenth century. My purpose in juxtaposing human evolution and the ballad revival is not to anticipate evolutionary explanations for the arts (though the art of song will feature again in the ninth section of this chapter) but rather to historicize the geologically inflected narrative of origins consolidated by Darwin's *Descent of Man* (1871). This narrative retains key conjectural and figural elements inherited from its Enlightenment versions.

Lubbock's prediction that the "new views" supporting a very long human prehistory would cause "little disquietude" now seems optimistic. By making these views a consequence of revolutions in astronomy and geology, Lubbock rewrites them as a chapter in the true history of nature, thus effecting what Pratik Chakrabarti terms a "naturalization of antiquity."[2] The naturalness of human origins and the historicity of nature had both been very much in question in earlier explorations of the deep human past. The chronological gap in this book between the ballad revival and evolutionary theory accentuates the distance between antiquity as an area of uncertainty and antiquity as an object of scientific consensus. The career of Lubbock's mentor Darwin, extending from colonial voyaging to the era of social science, partially bridges this distance. Darwin, Lubbock, and their contemporaries are more apt to be hailed as discoverers of deep time than are Buffon, Herder, or Blake; yet the form of the Victorian narratives, including the "descent" into deep time, shows their indebtedness to the literary tradition traced earlier in this study. To a large extent, Darwin's and Lubbock's consensus around human evolution still depended on the uncertainty and conjecture surrounding the abyss of time. This consensus also depended on the conflation of geological and ethnographic time, claiming paleontological authority for the age of "modern savages" and scientific validity for the racism that grew out of the ethnocentrism inherited from earlier conjectures in this maximally colonial era. Lubbock's belief in the superiority of Europeans over "savages" inspires the confidence of his prediction.

Deep Time and Human Prehistory

By the early nineteenth century, scholars and ballad enthusiasts had traced oral traditions in many parts of the world back to ancient times. Their claims are intrinsically hard to prove, but they firmly established the idea of oral tradition as a gateway to the deep human past. In the last chapter, I borrowed the words of a twentieth-century editor who captured this scenario vividly: "doubtless the cave-man made up a ballad after an outstanding adventure with a sabre-tooth tiger or a neighboring tribe."[3] Unlike the Romantics, Roger Lancelyn

Green had a century's worth of archaeological findings at his disposal, as well as the benefit of ironic distance, when he used the ballad as an index of species time. By contrast, Buffon was one of only a few naturalists in his time who were familiar with the fossil evidence of Pleistocene megafauna—the mammoth and the mastodon or "Ohio animal" rather than the saber-tooth tiger, in this case—and who also entertained the possibility that they overlapped with early hominins (as presupposed by Green's anecdote). As we have seen, Buffon was then ridiculed by Cuvier and others for hazarding such romantic conjectures. But at the same time that Cuvier was forging a more scientific paleontology that excluded the possibility of overlap between ancestral humans and the megafauna (or so he thought), very early hand axes made of flint were being found in increasing numbers in England and France. Some of these were found in caves, and by 1865 Lubbock was able to conclude decisively that "cave men" had indeed hunted the megafauna with these weapons. The image of the Pleistocene or Paleolithic hunter, which contrasts so sharply with the ancestral humans conjured up by Blake and Herder, nonetheless evolved from the same Enlightenment matrix: antiquarianism, voyage narrative, and conjectural history.

This chapter traces the expansion of human species time through Darwin's writings over a period of almost four decades. As a youthful voyager in the 1830s, Darwin was naturally influenced by earlier voyage narratives in his ideas about human origins. His later work, and Lubbock's, even begin to treat Darwin's early voyage account as part of this archive. By the time he wrote *The Descent of Man*, Darwin was able to incorporate a great wealth of material produced by the new disciplines of anthropology and archaeology. In this work, Darwin adapts the "comparative method" that informed these human sciences, which allowed an archaeologist like Lubbock to compare "modern savages" with Stone Age humans. On the one hand, this method itself owes much to the Enlightenment traditions of voyage narrative and conjectural history, as I shall argue in a reading of Lubbock's *Pre-Historic Times*, one of Darwin's major sources for *The Descent of Man*. On the other hand, Darwin became the first to extend the comparative method in a substantive way to nonhuman species. The human capacity for making art—if not the ballad per se—is one of several that Darwin presents in *The Descent of Man* as shared not only with ancestral humans but with other species. In this historical context, *The Descent of Man* formulates a new method—neither archaeological nor geological, though partaking of both sciences—for exploring the deep past of the human species via trans-species comparison.[4] Darwin's ambivalence toward "savages," a central topic of this chapter and of much previous scholarship, is conditioned by his reliance on ethnographic material for examples of human-animal traits.

At the very end of the book, titled in full *The Descent of Man and Selection in Relation to Sex,* Darwin remembers a scene that took place almost four decades earlier, one year into his *Beagle* voyage: "The astonishment which I felt on first seeing a party of Fuegians on a wild, broken shore will never be forgotten by me, for the reflection at once rushed into my mind—such were our ancestors" (*DM* 689). Although his visit to the Galápagos Islands later on this voyage is more widely remembered today, constituting a Darwinian origin myth in its own right, Darwin's encounter with the Fuegians (here the Yaghan people) reveals him as engaged with a very traditional form of speculation about human origins. This encounter situates him deeply within the *longue durée* of European exploration. From early modern reports of Patagonian giants and other "savages" to exhaustive studies of ethnography and natural history performed in the era of Cook and Bougainville (and cited throughout Darwin's *Beagle* narratives) through to twentieth-century anthropological texts inspired by *The Descent of Man* itself, European encounters with indigenous peoples have been imagined as forays into humanity's deep past. The peoples of Tierra del Fuego (now southern Chile and Argentina) have held a privileged place as "the most wretched savages," not only since John Reinhold Forster gave them this ranking in 1778, but since the sixteenth century.

Although *The Descent of Man,* a late work, was the first in which Darwin situated human beings within the deep evolutionary time that he promoted in his earlier books, it remains indebted for its scale-shift to the pre-Darwinian natural histories of "primitive" peoples and environments in which the possibility of deep time first came into view. Darwin's career marks the point of intersection between natural history and the sciences of evolution, including anthropology. The literary trope of recognition "rushing into [his] mind" like an inspiration points beyond his own early fieldwork to acknowledge the earlier literature of discovery, if in a more figural way than the many conscientious citations of James Cook, Louis-Antoine de Bougainville, Alexander von Humboldt, and other explorers that appear in Darwin's *Beagle* narratives. Perhaps no other literary moment did so much to establish a place for humans in evolutionary time. Johannes Fabian demonstrated clearly that this recognition comes at a cost, that of backdating modern humans who happen to be non-European. Fabian also points out the extent to which this "allochronism" is an inherited Enlightenment tendency, but Darwin's debt to the Enlightenment in this multilayered scene of recognition is complex and warrants detailed analysis.[5] His first report on the encounter owes something to the European voyagers' inclusion of ethnography within natural history and to the special role of the "Fuegians" (Yaghan, Selk'nam, and others) as a nexus between the two—a role already established by the time of Cook's second voyage, as detailed in Chapter 1. The younger Darwin, like Joseph Banks and the Forsters, was especially interested in geology on his voyage, and geology helps to set the stage for this primal scene of human evolution, the deeper history found by him within this "wild, broken shore."

Deep time is hard to reconcile with literary history, as illustrated by the 2013 debate between Americanists Mark McGurl and Wai-Chee Dimock.[6] My contention here is that the shock of recognition, given form by a literary trope that may be traced to Aristotle's discussion of anagnorisis in the *Poetics*, played a role in the formation of deep time as a category, even as it expanded beyond prehistory to the prehuman time of geology. By 1871, this trope of recognition had been cemented by two new specialist literatures for which it was a premise. The disgust that the Forsters experienced at Tierra del Fuego was exceptional and not typical of their experience as voyagers. With Darwin, this response becomes more typical as a response to human difference—we might call it "ancestor revulsion"—and conditions the otherness attributed to deep time in a different way. Moreover, "geological time" was a more familiar concept by the time of the *Beagle* voyage—volume two of Charles Lyell's *Principles of Geology* reached Darwin en route—and an object of scientific consensus by the time he wrote *The Descent of Man*. Just as important, Darwin returned near-obsessively to his encounter with the Fuegians throughout his writing career, invoking it as a "moment of truth" and "ethnographic authority," as Ian Duncan has argued. The affective experience of human difference thus became more fully incorporated with the concept of long-scale time during a period when terms such as "geological time" and even "deep time" were just beginning to appear in print.[7]

During the same period, the term *primitive* became less current in geology and more current in the human sciences. When Darwin returned to the Fuegians in 1871, he situated them and the problem of human origins in an evolutionary context, referring the ethnographic considerations that informed his *Beagle* narratives to a new group of specialists. In fact, *The Descent of Man* repeatedly cites the anthropologist Edward B. Tylor, whose *On Primitive Culture* appeared in the same year, and relies even more heavily on the pioneering prehistoric archaeology of Lubbock, Darwin's friend and neighbor in the Kentish village of Downe.[8] Tylor still believed, as the Forsters did, that conjecture had to play a role in cases where there was no written record, and Lubbock recognized gaps in the archaeological record, different perhaps only in degree from Darwin's account of the incompleteness of the "geological record," which he described as a "history of the world imperfectly kept."[9] These new sciences left behind the progressive framework of four stages of society that still provided the basis for the Forsters' fusion of human and natural history. They employed new kinds of conjecture, inferring more ancient and more gradual developments from modern "vestiges" (Darwin) or "survivals" (Tylor).[10] In principle, Darwin's theory decoupled development from teleology, but traces of teleology remain in this anthropological context, especially in Lubbock. For Tylor, deep time resolves into cultural difference, but Darwin takes up the behavioral survivals that Tylor studied as closely related to the anatomical vestiges of prehuman origins enumerated in *The Descent of Man*.

This slowly unfolding process of specialization (and temporalization) poses a challenge to received accounts of deep time as a concept triggered in spectacular fashion by improved geological understanding. Stephen Jay Gould and John McPhee, who popularized the term *deep time* in the 1980s, consider human history chiefly as a foil to set off the magnitude of geological time. In *Deep History* (2011), Daniel Lord Smail and Andrew Shryock incorporate archaeological time more fully into their account of the emerging distinction between historical and natural time, which they call "the time revolution." The present chapter engages with some of the same materials that Smail and Shryock use to establish 1859 as their date for this revolution, including Darwinian theory and the confirmation by numerous researchers that the gravels of the Somme in France truly contained man-made artifacts mingled together with the fossil bones of extinct Pleistocene fauna.[11] This indisputable eruption of the human past into a geological period well beyond the reach of Scripture and the written record was essential for opening the question of where human beings fit into the history of the planet. Nevertheless, it is an Enlightenment question, raised repeatedly, if in vague and experimental terms, by works on natural history and antiquities such as Isaac La Peyrère's *Men Before Adam* (1656) and Nicolas-Antoine Boulanger's *Anecdotes of Nature* (c. 1750). Whether or not we consider this period as a "prerevolutionary" or (with Thomas Kuhn) "pre-paradigm" phase, I hope to put the issues in a different light by placing evolutionary theory and prehistoric archaeology at a late stage of the formation of deep time as a concept.

As I have argued previously, the vocabulary of deep time developed slowly and depended in part on the recognition of new and different human histories outside Europe and outside Scripture. La Peyrère was one of a long line of thinkers engaged in reconciling New World discoveries with the chronology implied by Scripture, and Reinhold Forster joined this conversation 120 years later with his own thesis on the peopling of the Americas. Boulanger, the author of explicitly anticlerical works such as *Antiquity Unveiled*, took a strictly naturalistic approach to the history of the earth, which in turn inspired Buffon (as detailed in Ch. 2). By the time of Cook's voyages, a European antiquary such as Giuseppe Giovene was in a position to compare stone tools brought back from Polynesia with European Stone Age finds, though these were still undatable.[12] In the 1770s, the Forsters grounded their speculations about the history and prehistory of Pacific Islanders on their developing knowledge of how the islands formed. Inspired by their new observations as well as by the radical Enlightenment tradition, they challenged a set of stereotypes that kept the idea of human prehistory tethered to the written record. They identified human and natural processes that could only occur over "a series of many ages," in George Forster's words (*V* II.592). Among Darwin's predecessors, the Forsters are especially important for recognizing that human time was deeper than they thought and that the uneasy kinship they felt with non-Europeans was mediated by large-scale processes of environmental history that were beyond their comprehension. Writing nearly a century later, Lubbock gave a systematic account of formal similarities between modern stone tools and hundreds of implements like those found by Giovene, to which he gave the name "Paleolithic" (*PT* 2). Lubbock also identified climate change as the key environmental factor that needed to be understood in order to integrate human origins into geological time.

In turning to Darwin and Victorian prehistory, I wish to emphasize the importance of these framing questions and other legacies of the Enlightenment not only for the early evolutionary sciences but also for concepts of deep time and the primitive as they continue to unfold today. In the Anthropocene—an epoch that is both human and geological at once—primal or "Paleo" approaches to diet, exercise, and lifestyle are routinely supported by appeals to the science of human evolution, and this is one reason why Darwin's forty-year labor of integrating ancestral humans is so important to the conceptual history of deep time. In the context of Tierra del Fuego, the word *primitive* emerges as a key player in this story of converging and diverging time scales. Darwin locates his "savages of the lowest grade" in this archipelago "strewed with great fragments of primitive rocks," as I noted earlier.[13] Tylor correlates human and geological time in a more literal sense, situating the primitive artifacts beneath the surface: "Where now is the district of the earth that can be pointed to as the 'Primeval Home' of Man, and that does not show by rude stone implements buried in its soil the savage condition of its former inhabitants?"[14] Alongside the psychological shock of recognition, which epitomizes human ancestry in an individual or population to produce a biological sense of kinship, the encyclopedic approach exemplified by Tylor offers a stratigraphic (or sociological) certainty that all human populations are more or less primitive. For Lubbock, however, the shock of recognition is accompanied by a complete reversal, for he concludes that all "modern savages" are morally and intellectually "inferior . . . to the more civilised races" (*PT* 560) and that their cultures, however seemingly diverse, are of value primarily because they "throw light on the ancient remains found in Europe" (*PT* 417). This distinction between Tylor's more cosmopolitan theory of "survivals" and Lubbock's racialized hierarchy may be overdrawn, but it illustrates the horns of the dilemma on which Darwin often finds himself caught as he integrates ethnographic material on "savages" into his account of human origins.

It is easy to forget that Darwin is closer in time to Cook and the Forsters than he is to us. Over the ninety-three years between Reinhold Forster's *Observations Made During a Voyage Round the World* and *The Descent of Man,* fossilized human bones were discovered for the first time in France and Belgium; hundreds of flint implements were found mingled with the fossil bones of extinct animals; and the geologist Charles Lyell devoted a whole book to the problem of human origins, *Geological Evidences of the Antiquity of Man* (1863).[15] Developments after Darwin, including radiometric dating, have provided a broad empirical foundation, but deep time could not have solidified as a concept without the developments to which Darwin, Lubbock, and their contemporaries were responding. In grappling with the evidence that human antiquity must be calculated on the same scale as geological time, they resisted many of its implications. The qualitative measures that made the sense of deep time unsettling were increasingly supplanted by a data-driven approach, including statistical tables and global surveys, making the new chronology more acceptable to some of their readers—as Lubbock hoped it would. The need to allay readers' anxiety (as well as their own) about secular time, human marginality, and human animality led them to suppress the synchronic dimension of human species time, lending credibility to what is now called "social Darwinism."[16] In taking these steps to make deep time acceptable, Darwin infused new prejudices into the concept while also opening new imaginative possibilities. Above all he transformed the comparative method into a new means for exploring the deep past: not through geology, archaeology, or ethnography but by deriving prehuman and ultimately human traits from other, "lower" animal species.

Darwin and the Voyaging Intertext

Like the classic conjectural histories of Adam Ferguson, John Millar, and many others, Lubbock's *Pre-Historic Times* draws extensively on travel narrative for its evidence concerning "primitive" culture. By the late nineteenth century, however, the literature was much more extensive, encompassing detailed studies of the languages and customs of places colonized by Europeans in addition to narratives by explorers. Even a generation before, the young Darwin on Tierra del Fuego was conscious that "travel as science" had a history. Reinhold Forster's 1778 study of differing "varieties of the human species" was included in Darwin's shipboard library and provided a model for "exploring the past," a method formalized by Joseph-Marie Degérando in 1800: "the philosophical traveller, sailing to the ends of the earth, is in fact traveling in time." Fabian critiques this substitution of time for space in *Time and the Other*, locating these philosophical travelers of the Enlightenment near the beginning of a process that generated "naturalized Time" out of secular time. Anthropology became a modern discipline, Fabian argues, by deriving a strict teleology of progress from evolutionary theory, a development that he terms "intellectually regressive" as compared to experimental Enlightenment practices of "temporalization." "*Primitive*[,] being essentially a temporal concept," implied that the traveler's point of origin was not only geographically central but historically advanced. Even much later, Fabian argues, this "Typological Time" continued to enable the distinction between traditional and modern societies: "savagery exists . . . in *their* Time, not ours."[17] The "comparative method" of Victorian anthropologists, as emphasized by George Stocking, preserves some of the Enlightenment commonplaces of cultural contact, perhaps especially the analogy between modern natives and ancient Europeans. Tylor's more encyclopedic, less teleological approach, anticipating the cultural anthropology of the mid-twentieth century, shows that this analogy in its less racialized forms was not just a negative limit but also an enabling condition for the study of human evolution and divergence.[18]

Darwin's *Beagle* voyage (1831–36) falls roughly in the middle of this development of scientific discourse out of travel narrative, which could also be framed, following Frank Palmeri, as the emergence of a modern form of conjectural history out of the eighteenth-century genre.[19] Darwin's *Beagle* narrative still shows very clearly the geological inflection of the word *primitive,* a specific sense that is neglected in Fabian's account. (There are actually multiple narratives, including the diary and the letters, but my main point of reference is the *Journal of Researches* commonly known as *The Voyage of the Beagle.*) From the perspective of Lubbock and the later perspective of Darwin himself, who was writing *The Descent of Man* under Lubbock's influence, the *Beagle* narrative had become a part of the same archive as the Enlightenment accounts. Darwin's initial encounter with the Fuegians is something of a palimpsest, incorporating both the wonder and the philosophical reflections of the earlier accounts to produce an intertext that remains an active inheritance for the late Victorian science of human origins. This affectively charged encounter partakes not only of Fabian's "horizontal stratigraphy"—the premise that "dispersal in space reflects . . . sequence in time"—but also of two vertical stratigraphies, the geological research that provides the context for the encounter and the deep textual history of these "savages of the lowest grade."[20]

Because of the lasting influence of Darwin's *Beagle* narrative, it can be hard to disentangle the early assumptions of Victorian social science reflected in it from the older paradigms of conjectural and natural history. Cannon Schmitt calls Darwin's idea of the Fuegians as ancestors a "quintessentially Victorian speculation," but in fact it is as old as exploration and may well extend across cultures, as suggested by the recurring hypothesis that indigenous peoples in some Pacific islands might have regarded European explorers as returning ancestors themselves. Darwin's success in "opening up a biological time as profound as the 'deep' geological time" of Lyell, Schmitt argues, depends on the idea of species or race memory. Schmitt notes perceptively that "the assertion that the Fuegians cause Darwin to recall having been a savage himself . . . depends on a slippage in meaning between the individual-biographical and the species-historical." While the evolutionary result is new, the anxiety is not, for this "slippage in meaning" depends on another slippage or elision between European "ancestors" and indigenous people supposedly "very like them."[21] This elision, endemic to voyage narrative and stadial philosophies of history, is broadly modern and European rather than quintessentially Victorian. As Schmitt himself points out, Darwin borrows one of his most famous savage tropes, the savage who looks at a ship as something without a history, from the eighteenth-century voyage narrative of Bougainville, whose account of the Fuegians is also quoted in *Voyage of the* Beagle.[22]

Darwin and the Fuegians, Revisited

The powerful afterlife of Darwin's ambivalent notion of primitive humans became clear to me during a visit to my son's third-grade classroom. A teacher in his public school enrichment program had assembled several age-appropriate books on the history of science for use in a course entitled "Art and Science." I was startled to find a chapter on the Fuegians entitled "The Civilized Savages" in the book *Darwin and Evolution for Kids* by Kristan Lawson. According to this chapter, both *Beagle* voyages "came into conflict with the local Indians, a very primitive tribe called the Fuegians." Lawson reproduces in very uncritical terms what she takes to be Darwin's "astonishment" at the failure "of their experiment in civilization," the attempt to repatriate the Yaghan o'run-del'lico (Jemmy Button) and the Alakaluf yok'cushlu (Fuegia Basket) and el'leparu (York Minster) together with an English missionary.[23] Lawson's version shows not only the strong persistence of Darwin's myth of primitive humans but also, more to my present purpose, his strong bias in relation to the Enlightenment antinomy of the "noble" and "ignoble savage."

If so-called savages, and the Fuegians in particular, take their place as "utterly central to Darwin's aims in the *Descent* . . . as stand-ins for prehistoric forms of human life," as Schmitt argues, then this is due to the tropes of comparativism, degeneration, animality, and temporal displacement developed by Bougainville, the Forsters, and their contemporaries.[24] Challenging the authority of armchair historians of civil society, Reinhold Forster mobilized his ethnographic fieldwork on the voyage of the *Resolution* (1772–75) to claim that he was the first author to have "had the opportunity of contemplating mankind in the state of original simplicity, and in its various stages from that of the most wretched savages, removed but in the first degree from absolute animality, to the more polished and civilized inhabitants of the Friendly and Society Isles" (*O* 9–10). This contrast between civilized Tahitians and wretched Fuegians—discussed extensively in Chapter 1—shows that Forster was more than a little indebted to Bougainville and previous voyagers, if not to armchair philosophers, but it also encapsulates the model of degeneration that is Forster's own contribution. The Fuegians serve throughout as his example of ancestral animality because Tierra del Fuego was the most extreme latitude that he visited, and he was convinced that "the warm tropical climates seem to have been originally the seat of the human race" (342) and its "ancient systems" of arts and sciences (196). As populations migrate toward the poles in this biogeographic scheme, the harshness of the environment "entirely alters their mode of living; their habits, their language, and I might almost say their nature. They must therefore of course, by degrees, degenerate into a debased forlorn condition" (197). Forster's view of the Fuegians' "stupidity" is recycled by Lubbock in a catalogue of references that covers more than two centuries' worth of encounter narratives (*PT* 526), including Darwin's, in order to "illustrate" the state of culture in "prehistoric times."

The imprint of the Forsters' observations is also apparent in the *Beagle* narrative itself. Gregory Radick has argued that Darwin similarly viewed the Fuegians as "degenerate descendants of more advanced ancestors," never changing his mind on this point.[25] On Tierra del Fuego, George Forster comments that "no man can be more like an animal, and hence unhappier, than he who is so deficient in understanding and reflection that he cannot devise any means to protect himself against the most unpleasant bodily sensations of cold and exposure" (*R* 923). Darwin, similarly, observes that "here 5 or 6 human beings, naked & uncovered from the wind, rain & snow in this tempestuous climate sleep on the wet ground, coiled up like animals," and calls the Fuegians "the most abject and miserable creatures I anywhere beheld."[26] If the Fuegians have degenerated, however, they cannot be taken as "living exemplars of primeval stone age ways of life," as Nicholas Thomas defines the conventional view of twentieth-century anthropology. The difference helps to explain Fabian's remark that this early evolutionary paradigm is, paradoxically, "intellectually regressive" because less open-ended.[27] Both before and after evolution, the European idea of the antipodes had a powerful influence on perceptions of people from extreme latitudes as the most primitive, as Peter Hulme has shown. In other cases, however, no theoretical justification is given for the primitive character of the Fuegians, and the issue is further complicated both by ethnic differences among the four populations lumped together as "Fuegian"—Yaghan (Yàmana), Selk'nam (Ona), Mannenkenk (Haush), and Alakaluf—and by pervasive philosophical ambiguities noted by Graham Burnett, among others.[28]

To some extent, the model of the primitive Fuegian conforms to early modern encounter narratives from across the New World. Darwin cites at least one of these in *Voyage of the* Beagle, the narrative of Sir Francis Drake, and the name *Tierra del Fuego* itself derives from the first circumnavigator of the globe, Ferdinand Magellan. In 1520, Magellan became the first voyager to make the transit from the Atlantic into the Pacific via the strait that now bears his name. Magellan only noted in passing the watch fires of the Yaghan along the strait—hence the name *Tierra del Fuego*—but his expedition spent many months in Patagonia, not far to the north. Magellan's companion Antonio Pigafetta recorded many ethnographic observations that would become commonplaces, with at best varying degrees of accuracy, in the description of places judged to be remote or inhospitable, such as the nakedness of the Patagonians, their lack of cooked food, and the gendered division of labor: "their women, laden like asses, carried everything," he notes. Christopher Columbus, describing the first native people he encountered, on San Salvador, calls them "a people very short of everything," emphasizing the lack of everything from clothing to iron.[29] Thomas Jefferson, writing in the context of the early Republic, had political as well as intellectual reasons for rejecting some of the inherited tropes associated with so-called savages, including the idea of degeneration that still had a lingering hold over Darwin two generations later. Jefferson shows more willingness to incorporate indigenous peoples into civil history than does Darwin, who insists on their naturalness. Writing in the tradition of voyage narrative, Darwin is more apt to place them in geological time and thus closer to the point of human origin.[30]

By the time Darwin writes *The Descent of Man*, the Fuegians' ostensible proximity to animals has acquired a new evolutionary importance. Alexis Harley, noting that this proximity is a two-way street, points to the tradition of highly anthropomorphic descriptions of New World animals that Darwin inherited from Humboldt, whose influence he acknowledges explicitly in the *Beagle* narrative. The assumption of animals' proximity to humans is so implicit, Harley argues, that it may be viewed as "the [pretheoretical] ground from which sprouts his readiness to conjecture human/nonhuman common descent." Even before his evolutionary turn, then, Darwin may have attached a more biological meaning to the early modern convention of brutish humans, as when he prefaces his remarks on the Fuegians' exposure to the elements by reflecting that "it is a common subject of conjecture what pleasure in life some of the lower animals can enjoy: how much more reasonably the same question may be asked with respect to these barbarians!"[31] In addition to numerous firsthand observations of the constraints imposed by their hunter-gatherer existence, Darwin often uses the observations of other voyagers to support his picture of animal existence. When he writes, for example, that "their skill in some respects may be compared to the instinct of animals," the remark is sandwiched between anecdotes from John Byron and Francis Drake that illustrate supposed deficiencies in moral reasoning and in the use of tools among these peoples.[32]

Darwin scholars rightly emphasize the resiliency of Darwin's first impression of the Fuegians and its lingering effect in his work over time, including the idea of the savage as it eventually informed his model of human evolution. Radick finds evidence in *The Descent of Man* that Darwin did not change his mind about the Fuegians in the course of more than forty years, and Schmitt seems to assume a quasi-evolutionary motive for his initial treatment of the Fuegians: "Darwin required that Fuegians be brutish and abject, and so plausibly close to animals, but also educable, and so like the rest of humanity."[33] Darwin's ambivalence about "savages"—the word itself carries this intellectual legacy—nonetheless betrays investments in the Enlightenment discourse of savagery and civilization, as well as in the affective experience of deep time in its inchoate, preevolutionary form. These elements stand out even in one of the most historically specific features of Darwin's encounter, the presence on board of the Alakaluf and Yaghan people mentioned earlier, three captives taken by FitzRoy on the *Beagle*'s first voyage and partly "civilized" over three years in England. (This is the notion repeated by Lawson's expression "civilized savages.") While on the mission to repatriate these individuals, Darwin reports in a letter to J. S. Henslow: "I had an excellent opportunity of geologising and seeing much of the savages."[34] The contrast made him doubt that the repatriates would retain their new "European habits" and fueled his Enlightenment-style conjectures about the divergence of populations over long periods of time.

The formative influence of Lyell's *Principles of Geology* on Darwin at this time has been well documented by scholars.[35] Adelene Buckland argues that the geological imagination itself, as Darwin adapted it from Lyell, was "characterized . . . as the exclusive property of what was imagined to be the 'modern' human."[36] Buckland demonstrates the influence on Darwin's *Beagle* narrative of Lyell's declaration that "we may restore in imagination the appearance of the ancient continents which have passed away," and a very similar expression surfaces in Darwin's prose toward the end of *The Descent of Man* (678). Buckland makes the point that for Lyell, this faculty is "unavailable to humans 'in an early stage of advancement,'" though I would add that this category also excludes European thought under Catholicism, to which Lyell devotes the majority of his attention in seeking to establish a geology appropriate for "more enlightened ages."[37] On the *Beagle* voyage, the sublime scale of geological forces and the abyss of time opened by Darwin's geological research requires an "imaginative recovery." This recovery, Buckland argues, is enabled by his encounter with the Fuegians but is achieved "at the expense of those same indigenous peoples." She cites passages from Humboldt, Lyell, and Darwin in which "the senses are replaced by the power to range across hemispheres and millennia," a power that is explicitly denied to "savages" by Darwin in particular.[38] In Buckland's reading, "the ability to see a 'Fuegian' as human" is presented as "a mark of [Darwin's] enlightened imagination," and his *Beagle* narrative makes visible the "radical racial consequences" of any attempt to correlate human and geological time. In her reparative reconstruction of what the imaginative life of the Yaghan might have been, Buckland rightly notes that "other scales of imagining—human-human, human-animal, intercultural, and interspecies forms of imagining—are easily lost at the planetary and geological scale."[39]

Lyell constructs his history of geology on the model of the conjectural histories of civil society. His project of clearing away "the fabulous tales of former ages" therefore not only makes geology Eurocentric, as Buckland shows, but also makes it vanishingly modern, a manifestation so new as to set off more effectively the abyssal antiquity of the earth. Lyell's innovation, which expanded the scope of geological time, grew out of a careful reckoning with the geological thought of the Enlightenment, particularly that of James Hutton, and Palmeri has shown how deeply Darwin too remained indebted to Enlightenment thought even later, in *The Descent of Man*.[40] Darwin's observations on the voyage, still prior to his reckoning of evolutionary time and to evolutionary theory as such, occupy a transitional place in the long conceptual history of deep time belied by Lyell's assertion of a revolutionary change.[41]

In his foundational paradigm of conceptual history, Reinhart Koselleck designates the transitional period between premodern and modern concepts as *Sattelzeit*, a "saddle time" (analogous to a mountain pass) between two paradigms. Even so, "modernity" for Koselleck designates a rupture that is arguably more characteristic of the history of democracy, revolution, and other sociopolitical concepts that illustrate his account.[42] The concept of deep time is a product of continuous formation during the period of middle modernity and beyond, incorporating non-European influences and responding to paradigm shifts in the sciences, as we see in Darwin's evolving engagement with temporality. Part of my concern here is to distinguish this process from the product of "temporalization" (Koselleck's *Verzeitlichung*). Buckland's powerful analysis of the colonialism inherent in the geological imagination reveals Darwin's time on Tierra del Fuego as a transformational moment in the history of geological time. Like the "time revolution" associated with Darwin and prehistory in 1859, however, this transformation is also the result of a long process manifest in the history of voyage narrative, which informed this encounter as deeply as Lyell's new geology or the *Beagle*'s particular mission. Earlier voyagers' use of geological evidence to gauge the antiquity of ostensibly primitive indigenous cultures initiated a break in chronology that continued to shape both evolutionary theory and the new science of prehistory.

Recognition and Revulsion

The Fuegians are depicted in both noble and ignoble guises in the visual archive of the Cook voyages, but Conrad Martens's illustrations of hirsute Fuegians on the *Beagle* voyage (Fig. 4.1) echo Darwin's stronger embrace of the "ignoble savage" topos.[43] Ann Chapman comments: "This man looks like Jemmy Button in 1834 when Fitz-Roy and Darwin returned to say good-bye to him." Ethnic and cultural differences among populations near Cape Horn, which the Europeans understood poorly in any case, are only one factor accounting for the differences among early images of "Fuegians," but Chapman's surmise is warranted by the date of Martens's drawing, March 1, 1834—four days before the Beagle encountered "Jemmy" (o'run-del'lico) again. The subject seems to be another Yaghan man; as Chapman explains, FitzRoy mistakenly believed that the people in the area called themselves "Tekeenica," probably a corruption of a native expression meaning "I don't understand what you mean."[44] The lack in Darwin of any honorific dimension to "savage," as we find it in Jefferson and other Enlightenment thinkers, suggests that revulsion precedes what Darwin calls the "recogni[tion of] our parentage" in *The Descent of Man* (193). By 1871, he has come to justify his initial view of the Fuegians as the "lowest grade" of savage, the most primitive, by arguing that they are ancestral in some nearly literal way. His memory of a purely subjective experience of recognition in Tierra del Fuego seems to inform his certainty in this work that "we can partly recall in imagination the former condition of our early progenitors; and can approximately place them in their proper position in the zoological series" (*DM* 678). The word "series" evokes the geological context of Darwin's shipboard observations, and the entanglement of memory and imagination attests to the peculiarly literary quality of this argument. The scene of recognition also evokes a form of evolutionary memory, to be theorized much later as a genetic imprinting that survives across many generations.[45]

Ancestor revulsion played an important part in the development of a deep, geologically grounded human past out of the relatively shallow conceptions of indigenous peoples as noble or ignoble savages, or of modern primitives

FIGURE 4.1. "Fuegian (Yapoo Tekeenica)." Engraving by Thomas Landseer after Conrad Martens. Robert FitzRoy, *Narrative of the Surveying Voyages of His Majesty's Ships* Adventure *and* Beagle *Between the Years 1826 and 1836* (1839), vol. 2 (frontispiece). Courtesy of HathiTrust.

recapitulating ancient European history, put forward by many earlier voyagers as well as conjectural historians of civil society. Reinhold Forster's close attention to environmental factors did not suffice to decouple his natural history of the human species entirely from notions of progress, and his decision to place the Fuegians lowest on the scale, "just one degree removed from animality," clearly influenced Darwin before he had the opportunity to make extensive comparisons of his own (*O* 205). In one typical example, Forster finds luxury articles made of "bone, shark's teeth, &c." together with "delicate . . . victuals" on Tahiti; in New Zealand, some "conveniences," together with palatable food; and at Tierra del Fuego only the "bare necessaries of life," "bit[s] of seal skin" for clothing and meat that he (like Darwin after him) considers "putrid" and "disgustful" (*O* 140–41). Forster offers parallel accounts of the increasing impoverishment and "deformity" of the plant and animal kingdoms as he moves away from the tropics. Here too, Tierra del Fuego is at the bottom of the scale, consisting mostly of "black rocks, perfectly naked," with the flora at one location consisting of only two species (*O* 118–19). As noted earlier, George Forster's contempt for the "animality" of the Fuegians stands in stark contrast to his relative open-mindedness toward Pacific peoples. He claims, for example, that here even the sailors are disgusted by the native women. Although the Fuegians present "the most loathsome picture of misery and wretchedness" (*V* II.628), he notes that they nevertheless wear paint and jewelry and concludes that "ideas of ornament are of more ancient date with mankind, than those of shame and modesty" (*V* II.631).

This inherited narrative, more than other accounts of human origins, shapes Darwin's own encounter and his eventual evolutionary argument. Ancestor revulsion in the form of sexual disgust triggers the implicit claim that the Fuegians are in fact more "ancient" and not, as the elder Forster argues, degenerated from a better ancestral condition. Darwin's disgust with the Fuegians in 1834, though prompted by food-sharing efforts rather than sexuality—as he recalled vividly in 1872—links his argument for Fuegian ancestrality historically and tropologically to George Forster's century-old narrative.[46] This long conceptual history informs Darwin's vision of evolutionary time at more than one critical juncture in his book-length argument for human descent from a common "hairy, tailed quadruped" ancestor species and underwrites that argument more fully than twenty-first-century readers might be inclined to suspect. Lubbock's influence on Darwin's later work is well documented, and his extensive use of Darwin's *Beagle* material on Tierra del Fuego in *Pre-Historic Times*—explicitly incorporating Darwin's observations with those of earlier voyagers—must have prompted Darwin's renewed reflection on the encounter throughout *The Descent of Man* as well as in *The Expression of the Emotions in Man and Animals* (1872), originally planned as the third part of *The Descent of Man*.[47]

For Darwin, as for the earlier voyagers, the harsh geological setting of Tierra del Fuego, its primitive rocks and seemingly meager resources, informed the recognition and revulsion that prompted him to place its people in prehuman time. Again and again, this temporalization manifests itself through the intertextuality of voyage narrative—here we see the Forsters' narrative form of stadial environmental history together with a direct allusion to the narrative of John Byron, who circumnavigated the globe shortly before Cook:

> Their country is a broken mass of wild rocks, lofty hills, and useless forests: and these are viewed through mists and endless storms. The habitable land is reduced to the stones on the beach; in search of food they are compelled to wander unceasingly from spot to spot, and so steep is the coast, that they can only move about in their wretched canoes. They cannot know the feeling of having a home. . . . Was a more horrid deed ever perpetrated, than that witnessed on the west coast by Byron, who saw a wretched mother pick up her bleeding dying infant-boy, whom her husband had mercilessly dashed on the stones for dropping a basket of sea-eggs![48]

The voyaging intertext works to repress the shock of recognition inspired in Darwin by this encounter, which continues to reverberate, along with its geological context, forty years later in his theory of "the former condition of our early progenitors" (*DM* 678). Burnett points to the importance of the Fuegians for the larger argument of *The Descent of Man* by coining the term "savage selection": because they are "barely human anyway," as Burnett describes the view of the time, they illustrate something nearly identical to natural selection operating on the human species, and thus allow Darwin to "elide the question of agency" implied by his notion of selection.[49] New developments in archaeology and anthropology in the intervening years, beginning with the consolidation of those sciences themselves, finally make it possible for Darwin to incorporate his many cross-cultural encounters on the voyage into the theory of evolution.

Before turning directly to *The Descent of Man*, it will be useful to consider the ambitious expansion of the "comparative method" by Darwin's younger contemporaries in the emerging social sciences, particularly Lubbock's *Pre-Historic Times*, the second edition of which (1869) is cited extensively in *The*

Descent of Man. Lubbock makes Darwin's *Beagle* account of the Fuegians a central piece of his chapter on the Danish shell-mound builders and returns to it again, along with many quotations from Darwin's predecessors, in one of his chapters on the "manners and customs of modern savages." Lubbock relies on Darwin to support his conjectures concerning the seminomadic lifeways of the Stone Age people who built shell mounds or middens on the coast of what is now Denmark and evidently moved from place to place "like the Fuegians, who lead, even now, a very similar life" (*PT* 229). A two-page quotation from Darwin's *Journal of Researches* then provides descriptions of dress, built environment, and family structure, "illustrating" those features of the mound builders' culture that were not preserved in the archaeological record along with the shell mounds. In the later chapter, Lubbock dedicates a separate section to the Fuegians as "the lowest of mankind" and cites no fewer than seven authorities, including Reinhold Forster and Darwin, to support this categorization (525–32). Darwin found his own early fieldwork reflected back to him in the form of evidence for the comparative method; in what follows I shall argue that this reflection enabled him to take the next step and expand the comparative method to nonhuman animals, buoyed by the new consensus that some human groups were indeed more animal than others. Lubbock's version of this method was rich enough in geological material to encourage this approach. According to Lubbock in *Pre-Historic Times*, "the Van Diemaner [Tasmanian] and South American are to the antiquary what the opossum and the sloth are to the geologist" (416).

It is tempting to attribute a prophetic glimpse of human evolutionary history even to the Darwin of the *Beagle* voyage, who was steeped in Enlightenment voyage narrative, yet was already vexed by the species question and was strangely ambivalent about the Fuegians from the start. By returning in 1871 to Enlightenment commonplaces about ancestral humans in primitive settings, Darwin seems to invite this attribution himself. If his reaction was not exactly unconscious ancestor worship, it was nonetheless a recalibration of deep and human time. The startling contrast between FitzRoy's Anglicized Fuegians and their "wild brothers," in Darwin's phrase, suggested a potential for change in human populations that must have seemed staggering in the new context of geological time. According to Aristotle in the *Poetics*, "a recognition is finest when it happens at the same time as a reversal, as does the one in the Oedipus. . . . for such a recognition and reversal will contain pity or terror . . . and in addition misfortune or good fortune will come about in the case of such events."[50] Aristotle adds that in some cases the effect is heightened by a mutual recognition. We can only speculate about what the Alakaluf, Yaghan, and Selk'nam recognized in Darwin and other European explorers, but misfortune certainly came about for them in the form of genocide motivated in part by nineteenth-century social Darwinist arguments. Darwin's recognition, like that of his late Enlightenment predecessors, was indeed accompanied by a reversal in the sense that it placed all members of the human species closer together in time, undermining presumptions of European superiority—even though Darwin himself resisted the leveling implications of this insight.

Comparative Method in Lubbock's Pre-Historic Times

When Darwin compared the Fuegians to ancient "barbarians" during the *Beagle* voyage, he was following earlier scientific voyage narratives by the Forsters and others. When Victorian social scientists such as Henry Maine and John McLennan began systematically comparing "native" and "ancient" populations, they transformed this eighteenth-century tactic into a method. In *Victorian Anthropology*, George Stocking shows that Maine, Lubbock, and Tylor responded to Darwin's evolutionary theory with increasing sophistication as they elaborated this "comparative method" in the course of the 1860s. Stocking argues that McLennan—whose account of "primitive marriage" would have a significant impact on *The Descent of Man*—was the first to consolidate the comparative method into a "full sociocultural evolutionary argument" in 1865. Stocking suggests that at this time Tylor was still weighing the "relative merits of the historical and comparative methods," while Lubbock was "still documenting the antiquity of man," though both were strongly invested in the comparative method.[51] Lubbock is of particular interest for my account of deep time because his "prehistoric archaeology" involved the most substantial appeal to geological evidence. In addition to having (arguably) the strongest influence on *The Descent of Man*, Lubbock developed a rich geological vocabulary to figure the unsuspected antiquity of the "primitive savages" depicted by his comparisons. For Lubbock, time's arrow points backward.

Lubbock's approach is analogous to the uniformitarian method of actualism, which traces ancient phenomena "by reference to causes now in operation," as Lyell calls them in the subtitle of *Principles of Geology*. Lubbock establishes the descent into deep time as a narrative form, one subsequently used by Darwin to reach an even deeper prehuman past. Where Reinhold Forster had been content to trace his ethnographic analogies no farther than the barbarians recorded by ancient writers such as Strabo, Lubbock's *Pre-Historic Times* takes the reader down through Bronze Age burials, Neolithic flint quarries, and newly uncovered Swiss lake dwellings to the cave dwellings and glacial "Drift gravels" in which "savage" implements made of flint were found mingled with unmistakable remains of extinct megafauna. Lubbock distinguishes these finds from previously recognized Stone Age (Neolithic) sites as "Paleolithic" (*PT* 74). He attaches hard dates of 240–100 kya—albeit tentatively—to this new epoch, and specifically to the gravels of the river Somme (404), where the first widely recognized specimen of a flint axe mingled with the fossil bones of large mammals was found by Jacques Boucher de Perthes in 1841. In a more speculative vein, Lubbock suggests, contra Lyell, that even the possibility of ancestral humans in the Miocene period (>5 mya) cannot be ruled out (412). Lubbock's signal contribution, however, was to expand the range of qualitative measures for deep time. Lubbock calls the geologist as a "witness," concluding that "all the facts of geology tend to indicate an antiquity of which we are but beginning to form a dim idea" (408). But he questions "whether even geologists yet realize the great antiquity of our race" (410). His exhaustive account of the prehistoric Somme valley culminates in a sweeping panorama that folds in large-scale ecological factors including erosion and climate change:

> What date then are we to ascribe to the men who lived when the Somme was but beginning its great task? No one can properly appreciate the lapse of time indicated, who has not stood on the heights of Liercourt, Picquigny, or on one of the other points overlooking the valley; nor, I am sure, could any geologist return from such a visit without an overpowering sense of the changes which have taken place, and the length of time which must have elapsed since the first appearance of man in Western Europe. (375; cf. Fig. 4.3)

Lubbock's appeal to an embodied imaginative experience recalls Darwin's injunction to "watch the sea at work" in *The Origin of Species* and even Buffon's much earlier drama of observing the inch-by-inch accumulation of a hill of shale in Normandy (see Ch. 2). Lubbock's more specialized focus on human antiquity reflects the rapid development of geology and other sciences during the intervening century. The controversy around human antiquity in Victorian England was, if anything, more intense than the debate provoked by Buffon's proposals concerning the age of the earth, and therefore many of Lubbock's narrative strategies, both inherited and new, are conservative. Like Buffon, he draws attention to the difficulties of absolute dating and positions himself at the end of almost every chapter as erring on the side of caution with his proposed dates. I have been arguing that the comparative method of "illustrating" prehistoric times by means of "the manners and customs of modern savages" is itself inherited. Many of Lubbock's illustrations draw directly on much earlier voyage narratives, and the method not only recalls past ethnographic practice but also counterbalances the new assertion of a deep past with a strong synchronic dimension, a claim for evolutionary proximity between "savages" of whatever date and "civilized" Europeans. As Hans Blumenberg points out, European efforts to expand the domain of prehistory simultaneously diminish the sense of belatedness associated with modernity, instead opening new vistas for the future flourishing of European civilization.[52] Compared with earlier champions of deep time, Lubbock has the added advantage of being able to confirm accumulated evidence of "men among the mammoths" that had so far been rejected or suppressed, including that of Boucher de Perthes and even earlier discoveries dating back to the early eighteenth century.[53]

These considerations allow Lubbock to draw his reader ever deeper into the Paleolithic. Regrettably, his concluding chapters on "modern savages"—the last to be written, possibly added to give the work more popular appeal—make it very clear that this descent into deep time is purchased at the cost of a racialized hierarchy, an ideology seeking to establish the construct of "lower races" as an evolutionary fact.[54] Lubbock's approach to prehistory is superseded now due to the "appalling ignorance" revealed by this prejudice and other methodological flaws, but this judgment rendered by the archaeologist Steven Mithen is tempered by a striking testament to Lubbock's power as a storyteller: Mithen invents a narrator for his own prehistory who is also named John Lubbock, a modern "prehistoric travel[er]" who journeys into the past with *Pre-Historic Times* under his arm, marking corrections as he goes.[55]

In light of the voyaging intertext, it is unsurprising that the Fuegians are the first group cited by Lubbock as living "even now, or . . . very lately, in an Age of Stone" (*PT* 3). The first detailed instance that he offers illustrates a discovery of Iron Age arrows in Denmark by comparing the owners' marks on the shafts to "those on the modern Esquimaux arrows" (11). Lubbock's line engraving here provides the first of many visual depictions that support his typological approach to artifact comparisons, which extends in later chapters to Maori and Neolithic adzes; a huge variety of flint implements (mainly prehistoric); designs etched in bone ("Esquimaux" and Paleolithic); and larger built works, among others. He does not always cite specific sources on the "Esquimaux"—a vague umbrella term, like "Fuegians," used indiscriminately of different Arctic populations—but he returns to them often and finds their artifacts useful for illustrating the full range of prehistoric remains, including even the very early designs of animals etched in reindeer horn that were discovered in caves in the Dordogne region of France in the 1860s (323). Having proceeded with caution in assigning many types of archaeological sites to the Neolithic period throughout the first half of the book, Lubbock appeals to many recent discoveries in caves and Drift gravels to declare finally that "we can no longer doubt that our ancestors . . . co-existed in England with the mammoth, which they no doubt hunted, as the wildest tribes of Africa and India do now" (287). This extremely vague analogy provides at best a vague illustration of Paleolithic practice—given that mammoths are not elephants, that in the case of India at least the refined *keddah* system for capturing elephants is already documented in ancient texts, and that the ecological settings are drastically different. Lubbock points out that more than one extinct pachyderm species roamed northern Europe during the Pleistocene, but the real object of his broad claim about mammoth-hunting is to diminish resistance to the idea of an early human antiquity and not (as elsewhere) more scientific documentation of Stone Age technologies.

Lubbock's version of the comparative method, built on analogies between archaeology and paleontology, serves more than one purpose. The vivid image of English mammoth hunters makes a deep human past more plausible and acceptable—in part by putting "primitive savages" in their place (*PT* 347)—while more precise analogies support the archaeologist's work of reconstruction:

> Our fossil pachyderms . . . would be almost unintelligible but for the species which still inhabit some parts of Asia and Africa . . . and in the same manner, if we wish clearly to understand the antiquities of Europe, we must compare them with the rude implements and weapons still, or until lately, used by the savage races in other parts of the world. (416)

Lubbock concedes that "the present habits of savage races are not to be regarded as representing exactly those that characterized the first men," but argues that modern societies differ in their "degree" of progress, just as Stone Age peoples did (539). He therefore presents a table comparing the rates of progress made by various non-European populations by the time of their first European contact in order to facilitate more precise analogies between modern groups that (for example) used dogs for hunting (Aboriginal Australians) and Stone Age peoples that appear to have done so (Danish shell-mound builders). The same graphic also tabulates the presence or absence of navigation, pottery, agriculture, and a variety of tools and weapons among fifteen population groups (541). The volume is replete with detailed accounts of tool-making in particular, both from travel narratives and from inferences that could be drawn from fieldwork on Stone Age sites. In this way, Lubbock situates the student of human origins in geological time. "Archaeology," he declares, "forms . . . the link between geology and history" (2), including even the very recent history of non-European peoples.

As Darwin does not long afterward in *The Descent of Man*, Lubbock concludes with modern history to illustrate human descent from ancestral forms—in Darwin, we find "musical powers," some "sexual differences," and other features that can be traced to earlier animal species; in Lubbock, patterns of tool use or subsistence that may be traced to Stone Age human or hominin ancestors. Up to this point, however, Lubbock's descent is more deliberate and gradual, beginning with recent prehistory and moving down through the strata. He begins with the Bronze Age and then devotes five chapters to Neolithic sites, pointing out continually that there is insufficient evidence to assign an earlier date to these sites. Throughout most of *Pre-Historic Times*, Lubbock places his emphasis on relative dating before finally committing himself to a range of absolute dates (240–100 KYA) for the Paleolithic. The system of "three ages" (Stone, Bronze, Iron), introduced by C. J. Thomsen and modified by Lubbock's subdivision of the Stone Age, is itself a system of relative dating. Lubbock's reliance on materials and manufactures for his system of classification sometimes results in a chronology without dates, as in his treatment of Scottish "beehive houses" and other "rude dwellings" that have persisted from the Stone Age "down to the present day" (*PT* 53). Since there is no written record of their origin, these architectural forms belong to prehistory in the qualitative sense outlined by Herder in 1784: there may be vast expanses of time prior to the written record, but no chronology in a strict sense. Lubbock's scrutiny of the production of flint flakes at Stone Age sites, illustrated with a traveler's image of contemporary Aboriginal tool-making (88), has a timeless, typological character, as does his richly illustrated treatment of the form and function of implements across place and time. These examples present a cautious approach to chronology, which avoids any commitment to a long time scale or drastic change over time.

Lubbock's Deep Time Vocabulary

Lubbock's repertoire of figurative language multiplies qualitative approaches to the problem of deep time, often expressing some form of uncertainty. Uncertainty about the lapse of time allows Lubbock, like his predecessors, to dwell in the abyss of time while also promoting caution (or seeming caution) with regard to unwarranted quantitative measures. Artifacts of uncertain date are described by him repeatedly as "extremely rude" (*PT* 145), a term often invoked in earlier treatments of primitive rocks as well as cultural products by naturalists and poets alike. The Swiss lake dwellers, Lubbock reports, began as "rude savages," but the chronology of their progress remains uncertain because "the progress of the human mind is but slow" (213). Like George Forster on New Caledonia, Lubbock is "baffled" in his effort to date the rise and fall of this culture. His turn to geological evidence in the second half of *Pre-Historic Times* finally lends some support to quantitative measures, but geology too is above all a source of "facts which impress us with a vague and overpowering sense of antiquity" (410). Of a piece with this strategic conservatism is Lubbock's deferred treatment of the cave dwellings (302 and passim), which, he says, provide the only secure evidence of a Paleolithic period, yet cannot be engaged until a comparable antiquity has been ruled out for every other type of site. To underscore his caution, Lubbock returns repeatedly to the figure of insufficient evidence—a literal claim, to be sure, but deployed rhetorically by him—rejecting early dates when these seem insecure (315) and accepting long established evidence only when it has been thoroughly weighed (306–7). Like Buffon in *Epochs of Nature*, he draws attention to his conservative choice of dates over earlier ones that *might* be warranted by the evidence (330), often presenting a *terminus ad quem* for a given monument or site, such as Stonehenge. Here Lubbock mistakenly points to the surrounding Bronze Age burials to reject a Neolithic date for Stonehenge and perhaps (by analogy) for Avebury too (113–16).[56] In promoting his ideological view of the "mental inferiority" of "savages," Lubbock insists that the presence of grave goods in prehistoric burials does not imply "a definite belief in a future state" (142), and an inherited prejudice about the New World leads him grossly to underestimate the age of "American semi-civilization," which, he claims, "will not require an antiquity of more than three thousand years" (278).

On his home ground of Europe, Lubbock places a more appropriate emphasis on the slowness of ecological processes, including climate change. Lubbock cites the abundant evidence of third-growth forests to support his overly cautious dating of North American prehistory (*PT* 277) but assigns larger chronological implications to the succession of three different forest types in Denmark, where more paleo-ecological data are available, here inferring an immeasurable lapse of time (380). He is inclined to be skeptical about catastrophic extinction events (285) and limits to twelve the class of mammal species that disappear from the record between the Paleolithic and the Neolithic (291). At the same time, he attaches due weight to these extinctions witnessed by prehistoric peoples (today widely held to be caused by them). These comparatively recent extinctions, he argues, not only answer the charges leveled by "opponents of Mr. Darwin's theory"; they also displace early humans temporarily from their central role as the protagonists of prehistory: "Great as is the interest attaching to the existence of man at a period so much more ancient than that hitherto assigned to him, there is something which, to many minds, will appear even more fascinating in the presence of such a fauna" (299).[57] Lubbock closes this chapter on Quaternary mammals with a panorama of "gigantic Irish elk" and other megafauna persisting into the "human period," a vista no less epic than his geological panorama of the Somme valley. The mammalian extinctions are "more fascinating" at this point in the story because they provide clear evidence of global climate "fluctuations." Lubbock develops this theme more fully than his contemporaries, including Lyell, because he recognizes the qualitative potential of this large-scale process for amplifying the depth of human antiquity.[58] Both the Ice Age and the Somme panoramas manipulate the scale of representation to situate early humans securely in deep time.

Personification may be the most vivid figure in Lubbock's deep-time repertoire. Although contextual evidence concerning their discovery helps to support the early date of the flint axes in the Somme valley, Lubbock notes that it is not strictly necessary, for "the flints speak for themselves" (*PT* 344). In his extended prosopopoeia, the flints speak of the "time revolution" later named by Smail and Shryock, and they speak in English, conveying the support of the English scientific establishment for propositions that had seemed vague or visionary when uttered in French or Acheulian. Geologists, for Lubbock, are critical "witnesses" (343–44) who can corroborate the testimony of the flints, and Boucher de Perthes, their French discoverer, had failed to recruit these expert witnesses to his cause. Lubbock points out that part of the issue was Boucher de Perthes's catastrophism, his preference for a time scale compatible with sacred history. As Lubbock tells the story, a series of English antiquaries and geologists visited Boucher de Perthes and his collection in Abbeville in 1859 and alerted the Geological Society of London, which duly appointed a committee, consisting of Roderick Murchison, Lyell, and Lubbock himself, to investigate (333). Although they are not "judges" of human prehistory, Lubbock here argues that geologists are indispensable as witnesses because they are never "hasty" in accepting new evidence—hence Boucher de Perthes's eighteen-year wait for recognition.

Lubbock does not state explicitly what I take to be the geologists' main credential as interlocutors of the flints: their authority over the time scale needed to make human prehistory fully a part of geochronology. To the expert eye, the weathering of the flints speaks unmistakably of their age; geology in Lubbock's time has the benefit of more accumulated evidence to illustrate the rate of weathering and other gradual processes than Buffon did when he guessed that the shale in Normandy might have accumulated at the rate of five inches a year. When John Evans rediscovered a hand axe similar to the French examples, one originally found by John Frere in 1797, he found that this flint, too, spoke a language that was only now becoming intelligible, its age "irreconcilable with even the greatest antiquity until lately ascribed to the human race" (*PT* 332). Lubbock's illustrations here include a color lithograph (Fig. 4.2) that renders one of the hand axes as a monument, recalling the visual style of the mastodon teeth engraved for Buffon's *Epochs of Nature* (Fig. 2.2). This worked but weathered stone shows that geology has now caught up with human antiquity and has finally established the proximity between human origins and past epochs in the earth's history—a proximity strongly suspected by the Forsters, Buffon, Herder, and other Enlightenment thinkers, as I have argued, and evoked by them in qualitative and figurative terms that remain essential to Lubbock's effort to make the flints speak. This language also owes something to ideas about human antiquity from non-Western sources, a legacy which the earlier thinkers still acknowledged in various ways. Lubbock also invokes directly Enlightenment precursors such as William Stukeley and John Bagford, who discovered a similar hand axe entombed with a whole mammoth skeleton in the heart of London in 1715. The evidence, Lubbock exclaims, has been accumulating unrecognized for "nearly a hundred and fifty years" (336).[59]

Notwithstanding this *longue durée* of half-unwitting prehistoric discovery, acceptance of human antiquity in Lubbock's own time was far from general, and he engages energetically with recent sources. Just as the flints are made to speak of human antiquity, the beds in which they are found must yield the true age of the hunters who used them, and of their quarry: "we must solve this question by examining the drift gravels themselves," Lubbock declares (*PT* 353). For a qualitative measure, he borrows the eloquence of a writer for *Blackwood's Magazine*—almost certainly Henry Darwin Rogers—who depicts the "fascinating grandeur" of the Ice Age hypothesis while maintaining that the "other side" of the question is equally well supported by the evidence (356–57).[60] Lubbock answers Rogers's counterargument by designating the sandstone boulders in the Somme valley, with Joseph Prestwich, as "glacial

FIGURE 4.2. "A Flint Implement Found near Abbeville" (color lithograph). John Lubbock, *Pre-Historic Times*, 2nd ed. (1869), Plate III. Courtesy of the University of Missouri, MU Libraries.

erratics" and providing a series of sections through the valley (Figs. 201–3; Fig 4.3 herein). This sentence from the "visual language" of geology, as Martin Rudwick aptly calls it, provides the raw material for Lubbock's closing panorama of "a savage race of hunters and fishermen" (370) as well as a strong support for the emerging quantitative approach.[61] In his chapter on "the antiquity of man," Lubbock continues to rely on the geology of his time, noting that Lyell's estimate for the age of the Somme valley coincides with the near end of his own range at 100

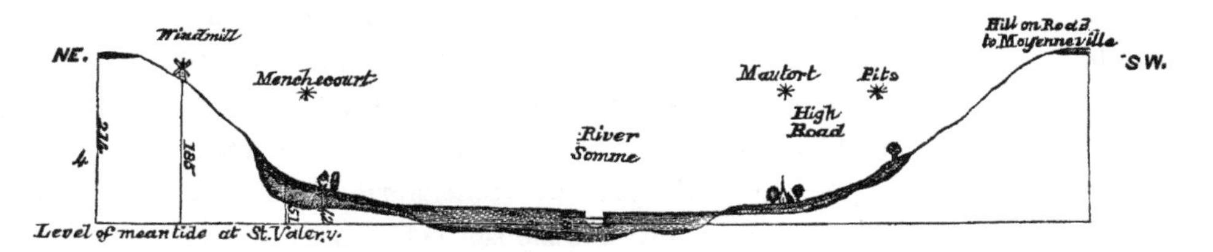

FIGURE 4.3. "Section across the Valley of the Somme at Abbeville." John Lubbock, *Pre-Historic Times*, 2nd ed. (1869), Fig. 201. Courtesy of the University of Missouri, MU Libraries.

KY (390). He also challenges the global generalizations of geology from the local and ecological vantage point of archaeological excavation, suggesting that Lyell underestimates the rate of geological change but that Archibald Geikie overestimates it when he predicts the complete erosion of Europe within 4 MY (402–3). In this way, Lubbock generates his own broad date range for the Paleolithic, greatly enlarging the scale of human antiquity while maintaining that the true extent of geological time is still unknown (407–8). He stipulates, however, that these quantitative estimates are "brought forward not as a proof, but as a measure of antiquity" (409–10). The proof lies in a qualitative sense of the large-scale changes—climate change, extinction, reconfiguration of land masses—that can now be seen to postdate human origins.

Race and Comparative Method

Lubbock's intervention places ancient and modern "savages" firmly in deep time. Thanks to its enduring quality of conchoidal fracture, flint continues to flake the same way, whether in Neolithic France, nineteenth-century Australia, or a sunny parking lot behind the University of Missouri anthropology building today. Ethnocentrism is probably the main source of Lubbock's confidence when he states that "the new views with reference to the Antiquity of Man . . . will, I doubt not, in a few years, be regarded with as little disquietude as are now those discoveries in astronomy and geology, which at one time excited even greater opposition" (*PT* ix). These new views, as corroborated by Lubbock's concluding chapters on "modern savages," increase the purchase of traditional colonial ideology by setting Europeans apart from people who can now be "proved" to have remained unchanged for a quarter of a million years. The qualitative and affective registers of deep time, however, are synchronic as much as diachronic, and Darwin's response to the Fuegians shows how unsettling the proximity between scientists and "modern savages" can actually be. Darwin is more apt to see civilizational refinement as a thin veneer overlaid on the deep strata of human nature—not only prehistoric but prehuman. Lubbock is more confident that racial difference clinches the distance between white and nonwhite people, in biology if not in time, and this confidence makes him an important transitional figure. The slippage between science and ideology so apparent in Lubbock's social science provided an essential impetus for *The Descent of Man,* which could not have been written in the same way without the precedent of Lubbock. On the one hand, Lubbock throws his geological caution to the winds when his racialized "modern savages" enter the equation, as when he uses the belief that Melanesians are "Negroes" to support a claim that the Australian continent separated from Africa within the span of human prehistory (378–79). On the other hand, the scientific values that structure his narrative are themselves held up as evidence for the proposition that only European civilization stands outside deep time in its synchronic sense.

Lubbock frames his narrative retrospectively as a story about the triumph of science, archaeological science in particular. He traces the system of Stone, Bronze, and Iron Ages back to Lucretius, and from there through the conjectural histories of the Enlightenment to his Danish contemporaries C. J. Thomsen and Sven Nilsson, to whom "must be ascribed the merit of having raised these suggestions to the rank of a scientific classification" (*PT* 6). Archaeology allows the study of antiquity to cast off tradition and be guided by geology alone. In the case of Europe, he argues, tradition offers no guidance on the history of the megafauna, a comparison that proves the methodological strength of "examining the drift gravels themselves" (353). Outside Europe, history is still necessarily excluded as a source of evidence concerning "pre-historic times," but Lubbock briefly acknowledges a few "traditions and myths" from various parts of the world that provide some "indirect" evidence. These are ultimately "untrustworthy" because of their "tendency to the marvelous," and so the archaeologist is "relieved . . . from the embarrassing interference of tradition" and "free to follow the methods which have been so successfully pursued in geology" (416). This declaration comes at the beginning of the first of Lubbock's three chapters on "modern savages" and rules out any further discussion of traditional knowledge possessed by non-European people themselves.

This framing makes the comparative method especially visible as ideology. The anachronism of construing "modern savages" as "early" makes it clear that living non-Europeans are primarily of antiquarian interest for Lubbock. Here, deep time functions as a mechanism for exclusion in the manner portrayed by Fabian and Buckland in their critiques of voyage narrative, but with an added archaeological dimension: placed side by side with evidence from Stone Age sites, "savage" customs and manners are reduced to corroborating evidence with no ethnographic importance of their own. The voyaging intertext plays an important role in these three chapters because all their material is taken from voyage narratives and replicates the hierarchy established by these narratives (and examined closely in Ch. 1), with Fuegians at the bottom and Tahitians at the top. The comparative function of voyage narrative is brought home by an anecdote about the archaeologist Édouard Lartet, who proved that bone needles like those found in caves in Dordogne could be made using flint implements. Lubbock calls this experiment "ingenious" but notes that Lartet instead "might have referred to the fact stated by Cook in his first voyage" that a group of Maori used stone to drill holes in a piece of glass (*PT* 537). Lubbock here adds that "the fortifications of New Zealand, as well as the large morais [*marae*, meeting grounds] of the South Sea Islands, are arguments in favour of the theory which ascribes" certain European remains "to the later part of the Stone Age." Here the utility of "savage" customs as corroborating evidence is especially clear. The Cook voyages are also the main source for Lubbock's derivative claim that Tahiti is "the Queen of Islands" (458). He quotes admiringly Cook's measurements of the "morai" [*marae*] of Oberea [Purea], attaching a new archaeological significance to the scale of this "heathen pile," as Joseph Banks called it at the time (472). As an exemplar of "the highest stage in civilization" without access to metal, Tahiti counts for Lubbock as more "advanced" than the most refined Neolithic societies (458). Their rapid adoption of iron after European contact clinches this point for Lubbock. Tahiti stands as an exception for him—as it did for the voyagers—whereas more typical "modern savages" offer the archaeologist "glimpses" of the origin of culture, a process that he traces both to the imitation of prey animals and to trial and error (572–73).

As noted in my earlier discussion of the voyaging intertext, the Fuegians most clearly exemplify this "early" civilizational stage for Lubbock, who attributes to them a racially marked proximity to nonhuman animals. Lubbock takes a monogenist view of human origins, like Darwin, but places racial differentiation at a very early stage in human prehistory, making nonwhite races "early" in a double sense. Noting marked differences among skulls found in caves at Cro-Magnon (southwestern France), in Belgium, and elsewhere, Lubbock concludes that "even at this early period, Europe was already occupied by more than one race of man" (*PT* 330). This example suggests the instability of race as a construct, since the Cro-Magnon specimens are now held to predate the dominance of *Homo sapiens*; it seems more accurate to say that *only* at this early period was Europe occupied by multiple species of hominins. Modern ethnic groups such as "Esquimaux" and Fiji Islanders are also designated as "prehistoric races of men" (162–63), but the clearest evidence of Lubbock's racial essentialism may be his decision to update (by some 250 MY) the separation of the Australian continent from Africa on the basis of a perceived racial affinity (378–79). Darwin's *Beagle* account of the Fuegians is quoted at great length in the service of "carrying our imagination back into the past" from the site of a Stone Age shell mound on the Danish coast (233). Although the Fuegians do not share the "heavy overhanging brows" attributed by Lubbock to the shell-mound builders, the pathos of Darwin's encounter with them moves Lubbock to call them into service for his own prehistoric romance.[62] The synchronic dimension of deep time crystallizes in Darwin's rendition of his disgust and revulsion, as much spontaneous as inherited from earlier voyagers. In the later chapter, Lubbock offers a lurid survey of the ethnographic material provided by Darwin and other voyagers in support of classifying the Fuegians as the "lowest" modern "savages" (525–26), adding as an afterthought that the Danish shell-mound builders were "more advanced" (532).

Lubbock calls in the Fuegians one more time to support his closing argument that all "savages" are "inferior . . . to the more civilized races" (*PT* 560) and that therefore European dominance (the subject of a short final chapter) is fully justified. In this respect Lubbock parts company with Darwin, whose problematic ideas about race are finally motivated by a principled opposition to slavery, as Adrian Desmond and James Moore have argued.[63] Nevertheless, his argument culminates with the same anecdote from Byron's voyage, of a Fuegian man "dashing" his three-year-old son "against the stones," that also plays a central role in Darwin's *Beagle* account. Lubbock provides an exhaustive catalogue of supposed evidence for the "extreme mental inferiority" of "savages," surveying systems of counting, inability to grasp "abstract ideas," and beliefs that fail to qualify as religion (563–64).[64] While this outpouring of prejudice is not entirely without its promised application to prehistory—it explains Lubbock's contention that Neolithic grave goods imply no belief in an afterlife, for instance—the main thrust of these chapters seems to be that only the European mind is evolving (590). He presents this justification for the impending extinction of "savages" in the final chapter, citing Tylor, Alfred Russel Wallace, Thomas Henry Huxley, Francis Galton, and Herbert Spencer—a range of Victorian authorities now generally regarded as having disparate views of social Darwinism.

These thinkers cited by Lubbock were, however, united in their acceptance of the long time scale required by evolutionary theory. Wallace is of particular interest here for his cyclical vision of a human future in deep time. Wallace argued in 1864 that the action of natural selection had "become transferred from physical to mental variations," but that the early prehistoric epoch predating this "transfer" could still be understood "by a powerful effort of imagination": "it is just possible to perceive [man] at that early epoch existing as a single homogenous race without the faculty of speech, and probably inhabiting some tropical region."[65] Wallace's investment in this mind/body dualism leads him to allow natural selection an even wider scope in time to bring the species to the point of racial divergence, and he casually posits an "Eocene or Miocene" origin for the species in this passage. It also leads him to place a strong emphasis on the "grand revolution" occurring when civilization set in, which he associates with domestication and human control on an even larger scale than that envisioned by Buffon in his seventh Epoch, when "Man assisted the operations of Nature." In Wallace's version, humanity has finally "escaped . . . natural selection" and appropriated much of its power. The "anticipation of the future" is one of the key mental faculties which began to be selected for as adaptive, according to Wallace, and he displays it in this paper by anticipating that the species will "continue to improve and advance till the world is again inhabited by a single homogeneous race."[66] Wallace projects deep time into the future in the manner of some conjectural historians, and in the manner of Anthropocene thought—though the outcome he imagines is drastically different. For Lubbock, prehistory is entirely a thing of the past.

Lest we imagine that primitive cultures might be of more than antiquarian interest or might deserve a place outside the deep past, Lubbock candidly reveals his rhetorical strategy toward the end of his long survey of "savage" peoples. The narrative form of a descent into deep time, adapted and modified again by Darwin for *The Descent of Man*, takes on a moral inflection here as Lubbock concludes each survey of a culture's accomplishments with reassurances of its rudeness, a narrative descent into barbarism. "Savages," he assures us, are far from being diminished by his comparison between them and Paleolithic ancestors: "their real condition is even worse and more abject than I have endeavoured to depict" (*PT* 571). Even the "advanced" Tahitians practice infanticide, he shows, and almost every account of a given society ends with a similar depiction of inhumanity or animality, all instances of which are summed up in the closing argument for "savage inferiority." The movement of these chapters therefore replicates the movement of Lubbock's prehistoric narrative from the Bronze Age down to the lower Paleolithic. The three chapters on "modern savages" may be compared with popular compilations of voyages, such as Richard Hakluyt's *Principall Navigations, Voiages, and Discoveries of the English Nation* (1589) and John Hawkesworth's *Account of the Voyages Undertaken by the Order of His Present Majesty* (1773), because they synthesize existing voyage narratives to offer the reader an experience of recognition in their detailed accounts of other cultures. In Lubbock's version, this recognition is accompanied by a drastic reversal. By comparing them to Stone Age societies, Lubbock transports non-European others into deep time, using his expanded geochronology to magnify the distance between "early" men and their evolved successors.

"In the Dim Recesses of Time"

Darwin, however, calls on the same geological scale of time to allow for an evolutionary process that more genuinely requires it: the descent of the human species from nonhuman mammals and even more remote ancestors. The recognition that he recalls from his voyaging days—"such were our ancestors"—partakes of the same revulsion, the same desire for distance expressed in Lubbock, but *The Descent of Man* harnesses that revulsion to a different end. Darwin moves beyond the human "ancestor" to a "more humble creature," using an invidious comparison to make descent from other primates—and hence an even deeper evolutionary past—acceptable to his readers: "For my own part I would as soon be descended from that heroic little monkey, who braved his dreaded enemy in order to save the life of his keeper . . . as from a savage" (*DM* 689).[67] Darwin is able to use Lubbock's evidence that human evolution occurred in geological time to leverage an even longer time scale for human descent from nonhuman mammals by way of "semi-human progenitors." In the *Beagle* narrative he asked rhetorically what reason the Fuegians have to exercise their "higher powers of the mind," and *The Descent of Man* solves this problem by showing that the so-called higher powers are inherited from nonhuman animals over such a long period that the importance of cultural or racial differences is minimized.[68] In this way, Darwin extends the comparative method across the boundary between human and nonhuman mammals and even—most notably in the eleven chapters devoted to sexual selection—across the boundary between mammals and other classes of animals, both vertebrate and invertebrate.

I have argued that *Pre-Historic Times* is of independent interest for its contributions to the vocabulary of deep time, and Darwin's approach to the same representational challenge develops in a different direction, as dictated by his more holistic view of evolution. Nevertheless, Lubbock's influence on *The Descent of Man* was considerable, and it is worth summing up their common ground before pursuing the descent into deep time in its Darwinian form. In "The Teacher Taught?" Alison Pearn points out that when Lubbock made descent from an apelike ancestor consistent with human perfectibility, he produced "valuable propaganda that made broadly Darwinian ideas far more widely palatable." Pearn argues that this tactic by "Darwin's diplomat" is just as important as the influence that Darwin acknowledged explicitly by "refer[ring] to Lubbock for evidence of the unity of the human species and for a much longer timescale for human [evolutionary] history than previously accepted."[69] Pearn thus credits Lubbock for making "Darwin—both the man and his science—acceptable and respectable" intellectually as well as personally and politically (most visibly by leading the effort to have him interred in Westminster Abbey). Pearn traces the influence of *Pre-Historic Times* on *The Descent of Man* in particular, presenting substantial evidence to support her hypothesis that "Lubbock's and Darwin's shared thinking on the most fundamental questions of human origins may in fact have been arrived at through discussion," that "neither man could claim ownership."[70]

Darwin cites Lubbock dozens of times in *The Descent of Man*, returning the compliment that Lubbock paid him in *Pre-Historic Times*, while also departing from Lubbock on some key points of early human evolution and maintaining the larger biological and geological scale of analysis established with *The Origin of Species*. Lubbock's defiance of "the opponents of Mr. Darwin's theory," noted above, may be traced to an exchange that occurred almost ten years before. Pearn shows that Darwin had already quoted Lubbock's declaration that "every species is a link between allied forms," from an article published in 1860, in the third edition of *The Origin of Species* (1861). In his revisions for the second edition of *Pre-Historic Times*, Lubbock reiterates this claim in support of Darwin. The scrutiny of "allied forms" is more intense in *The Descent of Man* than in the earlier instances, since human descent is at stake, and there seems to be an echo of Lubbock in Darwin's suggestion that fossil remains of a Miocene human ancestor might be found in Africa, though characteristically Darwin develops the idea in greater detail. Lubbock speculates that remains of ancestral humans predating the European Stone Age might be found in "hot . . . tropical climates" and might be found to date to the Miocene (*PT* 412). Darwin specifies a region of the African continent that has "not as yet been searched by geologists" (*DM* 184) and hazards an even earlier (Eocene) point of divergence for the stock of the Miocene ancestor (182–83).

Lubbock's influence is pervasive in *The Descent of Man,* and Darwin's casual mention of Lubbock's "well-known terms, . . . paleolithic and neolithic" (171) suggests that he expected *Pre-Historic Times* to be widely known among his readers. There are also some references to Lubbock's *The Origins of Civilisation* (1870), but as Pearn suggests these were "tacked on to discussions in *Descent* at a late stage," just before it was sent to the printer. Even though Lubbock (following McLennan and others) here argued that humans were originally promiscuous, Darwin decided not to depart from his defense of pair bonding in early humans.[71] The difference is a fundamental one. For Darwin, pair bonding is an evolutionary inheritance, indispensable to his argument that sexual selection at a remote period produced many key traits of the human species: "At a very early period, . . . [man] would have been monogamous, like the other animals" (*DM* 662), with the result that "sexual selection has been the most efficient" cause of "differences in external appearance" (675). Darwin's extensive evidence on birds (Ch. 13–16) clinches this point for him; "it would be a strange anomaly" (585), he notes repeatedly, if "the same cause," especially female choice, had not "acted on" the "higher" mammals as well (605). Here, then, Darwin situates human emergence in a much larger expanse of evolutionary time, deliberately crossing species lines and even larger divisions in an expanding series of comparisons signaled by the anthropomorphic extravagance of his depictions of "marriage arrangements" among "reptiles and fishes" and other non-mammalian species (255).

Near the beginning of *The Descent of Man,* Darwin notes that "the high antiquity of man has recently been demonstrated by the labours of a host of eminent men" (18), and by 1871 there was even more reason to feel confident that that human antiquity would soon be uncontroversial, as Lubbock had predicted. Darwin was well aware, nonetheless, that human antiquity carried a different set of implications for him than for Lubbock or Lyell or even Ernst Haeckel and the other evolutionary naturalists who are also cited in this passage. In *The Origin of Species,* Darwin strongly emphasized the length of time required for marked evolutionary change, drawing his support from Lyell's expansion of geological time in *Principles of Geology.* The consensus on human antiquity that emerged afterward encouraged Darwin to trace the common ancestry of humans and other mammals across the full extent of evolutionary time, and this radical step is signaled by his deceptively lighthearted references to love and marriage among even more ancient classes of animals. These references might be considered as anthropomorphic stowaways concealed on his voyages into deep time. Desmond and Moore point out that Darwin exaggerated in declaring that there was ample precedent or consensus concerning human descent and sexual selection (*DM* 703). He parted ways with Lyell over human evolution and nurtured a very different conception of "primitive marriage" from that of Lubbock, McLennan, and the other social scientists who became his authorities on the human species. For Darwin, human variety depends on sexual selection, and sexual selection is an evolutionary inheritance as ancient as the animal kingdom itself.

With this expanded reach in mind, the long middle section of *The Descent of Man* on sexual selection (Ch. 9–16) may be understood as Darwin's intervention in the comparative method. Lubbock and others derived their notions of primitive marriage from a comparison between "savage" and "civilized" societies, extrapolating an ancestral species-wide condition from this difference in the usual fashion. Darwin opposes the "communal marriage" implied by the comparative method through a longer, more deliberate series of comparisons—not only between societies but also between humans and other animals and even between living and extinct species:

> From the strength of the feeling of jealousy all through the animal kingdom [and the corresponding "special weapons"], as well as from the analogy of the lower animals, more particularly those which come nearest to man, I cannot believe that absolutely promiscuous intercourse prevailed in times past, shortly before Man attained to his present rank in the zoological scale. . . . Thus, looking far back enough in the stream of time, and from the social habits of man as he now exists, the most probable view is that he aboriginally lived in small communities, each with a single wife, or if powerful with several, whom he guarded against all other men. (*DM* 658–59)

Darwin chooses the unusual figure of the "stream of time" to establish the authority of biology (as we would call it today) over anthropology but achieves a diplomatic formulation by adding the clause about "social habits." Darwin arrives at his rebuttal to "promiscuous intercourse" by stages, navigating the stream of time from invertebrates to birds to mammals. His objection to this thesis rests on the same principle as his defense of sexual selection in other mammals. Just as is as "improbable that unions of quadrupeds in a state of nature should be left to mere chance" (588), so too "the pairing of man" cannot be the product of "mere chance" (655), because in both cases at least some traits of the species seem most likely to have been produced by sexual selection.

Because the original promiscuity thesis leaves everything to chance, Darwin here parts ways with the anthropological approach to prehistory, arguing instead that "men and women, like many of the lower animals, might formerly have entered into strict though temporary unions for each birth," which is all that is required for choice to be exerted (*DM* 657).[72] In the first of his chapters on mammals, Darwin reminds us that the most substantial evidence "shewing that the female selects her partner" comes from birds and insists that "it would be a strange anomaly if female quadrupeds, which stand higher in the scale and have higher mental powers, did not . . . exert some choice" (585). He concludes that "the same cause . . . has acted on mammals and birds," and specific cases including the pervasive "taste for the beautiful" help to consolidate the analogy with humans (604–5). The prevalence of patterns of selection among "reptiles and fishes" and even among invertebrates, including shellfish and insects (the subject of two full chapters, 10–11), sufficiently shows that these are universal: "the exertion of some choice on the part of the female seems almost as universal a law as the eagerness of the male" (257). While conceding that "sexual selection acts in a less rigorous manner than natural selection" in his first chapter on the subject, Darwin makes the point that its effects are stronger over time, a critical point for my argument in this chapter: "the advantages" derived from sexual selection "are in the long run greater than those derived from rather more perfect adaption to their conditions of life" (262). The new consensus on human antiquity creates the opportunity for Darwin to show that sexual selection operated early enough to have a decisive effect on human variety: "Of all the causes which have led to the differences in external appearance between the races of man, and to a certain extent between man and the lower animals, sexual selection has been the most efficient" (675). Racial differences thus emerged gradually as an expression of the "taste for the beautiful" specific to early human populations as they migrated to different locales (664–65).

Ascending the stream of time even further, Darwin locates nonadaptive traits such as the loss of body hair "at an extremely remote period" (*DM* 670) in a domain that falls outside the new province of prehistory and escapes ethnographic analysis. Still, he does not hesitate to employ the comparative method in its ordinary sense when it suits him; the Fuegians are called into service once again when an existing society is needed to confirm a zoological principle such as female choice, here as one of the "utterly barbarous tribes [in which] the women have more power in choosing, . . . than might have been expected" (666). There is at least a suggestion here that female choice in these societies may constitute a survival from a distinct evolutionary epoch.

Similarly, Darwin argues that "the greater size and strength" of the human male are "due in chief part to inheritance from his half-human male ancestors . . . preserved or even augmented during the long ages of man's savagery" (628). He argues further that ethnographic evidence militates against any racial or Lamarckian explanation of these differences. In these instances, Darwin is offering a novel variation on anthropological theories such as Tylor's doctrine of survivals, which does depend on ethnographic evidence. In attributing traits to sexual selection, Darwin draws from living and fossil animal species rather than customs and manners or prehistoric artifacts. The traces of counting by hand preserved in written systems of enumeration, among other survivals identified by Tylor, support the universality of a prehistoric "barbarism" of which living peoples "retain no tradition" (170–71), but this barbarism belongs to the more recent "ages of man's savagery" as measured by evolutionary time. Rather than stating explicitly that the evidence of Tylor, Lubbock, and their colleagues is insufficient for tracing the descent of man fully, he juxtaposes their findings with his own analysis of more ancient inherited traits. Whereas Lubbock proposed some absolute dates—however tentatively—for the epochs of prehistory, Darwin's emphasis on prehuman ancestry makes him reliant on the more qualitative measures familiar from the history of deep time.

Human Descent and the Art of Song

The long indefinite period intervening between "half-human" sexual selection and rudimentary counting—the "middle ages of man's savagery," to adapt Darwin's language—is of special interest here for two reasons. Evolutionary social science in the twenty-first century has established a much more nuanced transition between ostensibly natural traits such as the loss of body hair and cultural acquisitions such as counting.[73] Darwin's pioneering map of evolutionary time depends on a firmer distinction between these two categories, but he is not unwilling to explore areas of uncertainty in between, including the origin of the arts and the "taste for the beautiful." The second point of interest in this very early phase of prehistory is that it preserves an intellectual inheritance from the Enlightenment, which offers many conjectural histories of the origin of art. Some of these histories are associated with the ballad revival, treated at length in the previous chapter of the present work, and Darwin's speculations on "the capacity and love for singing and music" (*DM* 632) share several points in common with eighteenth-century accounts of the ballad revival. Ian Duncan reads *The Descent of Man* as "resuming the ambitions of the Enlightenment natural history of man" and traces to Friedrich Schiller "Darwin's key idea, that the aesthetic sense is the medium of a specifically human evolution."[74] On the question of the origin of art, Darwin is also indebted to a tradition in French thought that extends from conjectural histories such as Antoine-Yves Goguet's *De l'origine des lois, des arts, and des sciences* (1758), acknowledged by Lubbock as a foundational text in archaeological anthropology (*PT* 6), to Boucher de Perthes's first book-length publication on his discovery of Paleolithic flint implements in the valley of the Somme, titled *De l'industrie primitive, ou des arts à leur origine* (1846).

Like Herder and other proponents of the ballad revival, Darwin regards song as one of the most ancient and permanent natural-cultural faculties of the species. Unlike the Enlightenment thinkers, Darwin pursues the idea that song is an inheritance shared with nonhuman animals (albeit overlooking the affinity long established as a trope within poetry, relative to birdsong especially). "An instinctive tendency to acquire an art is not peculiar to man" (*DM* 109), he observes in a chapter comparing "the mental powers of man and the lower animals," which also covers imitation and elementary reasoning. With respect to song in particular, Darwin here anticipates his argument on sexual selection: "primeval man, or rather some early progenitor of man, probably first used his voice . . . in singing, as do some of the gibbon-apes of the present day" (109). The zone of indistinction between "primeval man" and "some early progenitor of man" is located (for Darwin) beyond the pale of archaeology and anthropology. The appeal to what "probably" happened in this speculative context typifies the inherited rhetoric that marks *The Descent of Man* as a conjectural history in the Enlightenment tradition, according to Palmeri.[75] By tracing music in particular to the "creaturely voice" we share with nonhuman animals (in Tobias Menely's elegant phrase), Darwin deepens the human species time invoked by those who defended the antiquity of oral tradition, ascending this stream of time even further to its origin in other species.[76] As much as they express a shared inheritance, the arts are also surprisingly important for showing species integrity, as Darwin's primary example of a peculiarity too "unimportant" or "singular" to "have been independently acquired by aboriginally distinct species or races" (207). Our uniform anatomy further proves that "all the races of man are descended from a primitive stock" (205), but ethnographic data once again provide a valuable opportunity to assimilate social science into the evolutionary argument: "the close similarity between the men of all races . . . is shewn by the pleasure which they all take in dancing, rude music, acting, painting, tattooing, and otherwise decorating themselves" (207–8). Duncan locates the origins of art in the last example, "on the sculpted and scarified surface of the savage body," where the "categorical distinction between . . . nature and culture" visibly "disintegrates."[77] Darwin's notion of arts in general as aboriginal faculties recalls Blake's treatment of craft knowledge and the arts more broadly as native endowments of the human species.

Darwin's fullest discussion of music and related arts occurs near the beginning of Part III of *The Descent of Man*, on "sexual selection in man." It occurs as a digression in the middle of a conjectural history of sex differentiation during the "long ages of man's savagery" via "characters" including bodily strength, as noted above. For Darwin, man's mental "superiority" to woman—surely the largest ideological blind spot in the whole work—is itself a product of sexual selection (*DM* 631).[78] In pivoting to "the capacity and love for singing or music," he acknowledges that this is "not a sexual character in man" (632) but integrates it into the history of sexual selection on speculative grounds; above all it is important to him as a locus of evolutionary memory. Although it is difficult to date relative to the loss of body hair or sex dimorphism, this "most mysterious" faculty (636) offers an affective experience of continuity between modern humans and their "progenitors." "Whether or not the half-human progenitors of man possessed, like the singing gibbons, the capacity of producing, and therefore no doubt of appreciating, musical notes, we know that man possessed these faculties at a very remote period," Darwin declares, appealing to both archaeological and ethnographic evidence. He adds here that "poetry, which may be considered as the offspring of song, is likewise so ancient, that many persons have felt astonished that it should have arisen during the earliest ages of which we have any record" (636). The temporal structure of this genealogy is nearly synchronous, moving as it does from ancestral species all the way down to recorded history. The claim regarding poetry, which includes a likely allusion to the ballad revival, closely mirrors Herder's reasoning concerning the preliterate origin of traditional tales like that of Prassarama's arrow: poetry survives among the earliest written documents because it preexisted writing.

Darwin follows the same pathway to prehistory to posit an even earlier branching that would have been nearly inconceivable to his Enlightenment precursors, though some of the early polygenist race theorists did posit other kinds of human-animal kinship.[79] Music as a form of evolutionary time travel creates an experience of kinship even with prehuman ancestors. The universality of a musical faculty, for Darwin, may point either "to the practice of our semi-human progenitors of some rude form of music" or "to their having acquired the proper vocal organs for a different purpose," like other animals (*DM* 637). In describing the descent into deep time via music, Darwin introduces a key term from his basic stock of evidence for human evolution, reversion: "The sensations and ideas thus excited in us by music, or expressed by the cadences of oratory, appear from their vagueness, yet depth, like mental reversions to the emotions and thoughts of a long-past age" (638). This formulation captures the affective quality of deep time, the structure of feeling without which "deep time" could not have gained its currency as a category for geological events. In the opening argument from anatomy, too, reversion offers a bridge across deep time: "The principle of reversion, by which a long-lost structure is called back into existence, might serve as the guide for its full development, even after the lapse of an enormous interval of time" (57). This principle accounts both for rudimentary features that are universal, such as the human tailbone (or poetry), and for structures that recur only in isolated cases—Darwin here cites certain muscular structures that are rare in humans but universal in some of the apes (62). Music occupies a special place between these kinds of anatomical features and the practical arts that account for human "pre-eminence." In reviewing these, Darwin again uses the same qualitative measure of time to express the slowness and fluidity of evolutionary change: "Archaeologists are convinced that an enormous interval of time elapsed before our ancestors thought of grinding chipped flints into smooth tools" (69).

The reversion experienced with music rekindles sensations associated with this gray area between evolution and prehistory. Likewise, Darwin claims, the "discovery of fire, probably the greatest ever made by man, excepting language, dates from before the dawn of history" (*DM* 68). Thus "man-like animals" as far back as the Neanderthal already had the most essential equipment of modern humans, including "capacious" skulls (75), fire, chipped flint, and presumably song. The interval of time since then, though "enormous," is still comparatively short in evolutionary terms, for "we" are still "not removed by very many generations" from "the savage state," as Darwin later clarifies (163). Notwithstanding his endorsement of Lubbock's evidence for the Stone Age mentality of "savages," this account of the emergence of the human species may be understood as an antiracist intervention on Darwin's part, since he appeals to anatomy and evolutionary time to refute Wallace's claim that the brains of "savages" are closer to those of apes than those of modern humans (68).[80] In concluding his discussion of music, Darwin reintroduces the hypothesis of sexual selection as the best available mechanism for integrating music into the evolutionary narrative. If it was "used by our half-human ancestors, during the season of courtship," it becomes possible to "understand how it is that music, dancing, song, and poetry are such very ancient arts" (638–39). This explanation further accounts for the experience of reversion through music: "The impassioned orator, bard or musician . . . little suspects that he uses the same means by which his half-human ancestors long ago aroused each other's ardent passions, during their courtship and rivalry" (639). Notably, music is one of relatively few adaptive traits that he presumes is gendered female. Although he reaches a certain degree of closure here, anticipating a future trend in evolutionary aesthetics, Darwin's treatment of music and the other arts still stands out as a space of uncertainty and conjecture. He states again that "we know not" why beauty in any form gives pleasure (651), maintaining aesthetic and affective response as sites of openness that allow subjective and synchronous engagement with the deep history of the species.[81]

Darwin, Deep Time, and the Human Animal

This affective descent into deep time undergirds a narrative form that recurs in many places in *The Descent of Man* where Darwin rapidly surveys, in the course of a few sentences, an evolutionary process that would take millions of years, even it were calibrated on the GTS as it was understood in 1871. Unlike Darwin's own figure of "the stream of time," which implies a lateral descent from the source of the stream to its mouth, the vertical descent into deep time is a geological figure that implicitly structures many aspects of *The Descent of Man*. The title "descent of man" offers yet another spatial framework, that of the genealogical tree, illustrating further the suggestive indeterminacy produced when time is translated into spatial metaphors. The notion that human beings are "descended from some lower form" (*DM* 676) conveys the "biological outrage" against human narcissism identified by Freud but also carries the association with progress that was so important to many of those who embraced Darwin's thought. The vertical descent into deep time that concerns me here is no less metaphorical than the others, but it accords with the law of superposition and gains a certain solidity from its convergence with the geological pursuit of origins in the depths of the earth. The narrative descent into the primitive past structures Darwin's argument in a fundamental way, recalling the backward direction of time's arrow in Lubbock. The narrative scaffolding of steps or stages, however compressed, generally informs these episodes. They also depend on a rich qualitative vocabulary, references to "primordial" or "primeval times [or men]," and to "semi-human progenitors" with their many variations, incantatory words that Darwin uses over and over again in *The Descent of Man* to conjure deep time. "Recalling in imagination" many scenes from deep time, he further exploits the imaginative legacy of geological concepts such as "revolution," inherited from Buffon and Cuvier among others.

Like the aesthetic experience of poetry or music, bodily form provides a subjective, virtually instantaneous vehicle for traversing geological time. The most compressed of these narratives recalls the "moment of pulsation" in Blake, the 6,000 years compressed into the poet's heartbeat. Body hair provides the rhythm for Darwin: "The direction of the hair on our own arms offers a curious record of our former state" (*DM* 178), when we had fur to ward off the rain. The present direction of these hairs, however, shows that they were "adapted to throw off the rain" before prehuman ancestors became bipedal. This local narrative of descent sets up a larger one in which Darwin famously concludes that our progenitors "properly" belong to the "Old World stock" of monkeys, "however much the conclusion may revolt our pride" (182). The affect of revulsion here recalls the explorer's trope of ancestor revulsion, the experience of recognition and reversal associated by Darwin and his predecessors with the Fuegians in particular. Invoking the lapse of time to soften this blow, he cites the best fossil evidence then available, the Miocene "ancestor" noted earlier as a key point of connection between Darwin and Lubbock, specified by name in Darwin's text as the "Dryopithecus of Lartet."[82] Recognizing that the biogeography of this extinct species would depend on a better understanding of the tectonics (as we now call it) of the Mediterranean basin, Darwin here reverts to the inherited geological vocabulary documented in the previous chapters of this study: "Since so remote a period the earth has certainly undergone many great revolutions, and there has been ample time for migration on the largest scale" (182–83). Herder's sense of the most recent revolutions of the earth as sites of contact between geology and prehistory seems especially relevant here, given his particular interest in prehistoric migrations (*I* 31). The complex trope of "revolution" provides a richer metaphorical vehicle to complement the more pedestrian descent into deep time by way of arm hair. In the late nineteenth-century context, the revolutions of the earth provide a relatively uncontroversial qualitative measure that nonetheless solidifies the geological foundation of evolutionary time.

Having produced a "record of our former state" that documents our position among the mainly quadrupedal apes, Darwin rapidly expands the evolutionary field to its full extent. The figure of geological descent is necessary to understand the movement of his narrative from the anthropoid apes through the "lower animals" to more and more primitive instances of "lowly organization." These vertical denotations are not merely, and perhaps not even primarily, evaluative, since they are loosely correlated with the principle of superposition, according to which the oldest rocks and fossils (under normal circumstances) lie nethermost.[83] Unlike the descent of man itself, which nominally begins with the ancestor uppermost, the descent into deep time begins in the present:

> We will now look to man as he exists; and we shall, I think, be able partially to restore the structure of our early progenitors, during successive periods, but not in due order of time. This can be effected by means of the rudiments which man still retains, by the characters which occasionally make their appearance in him through reversion, and by the aid of the principles of morphology and embryology. (*DM* 188)

The narrative first descends from Dryopithecus and its ilk to "a much earlier period when the uterus was double" and ultimately to "a still earlier period" when "the progenitors of man must have been aquatic in their habits; for morphology plainly tells us that the lungs consist of a modified swim-bladder." At the base of this imaginary sequence, "the heart existed as a simple pulsating vessel; and the chorda dorsalis took the place of a vertebral column. These early ancestors of man, thus seen in the dim recesses of time, must have been as simply, or even still more simply organised than the lancelet or amphioxus" (188).

In his recapitulation, Darwin specifies that this primitive marine ancestor "seems to have been more like the larvae of the existing marine Ascidians [also invertebrate chordates] than any other known form" and adds that "in the dim obscurity of the past we can see that the early progenitor of all the Vertebrata must have been an aquatic animal, provided with branchiae, with the two sexes united in the same individual" (*DM* 679). In both passages, an affiliation with the "dark abyss of time" is clear. The earlier passage cites additional support from embryology for the probable hermaphrodism of the ancestral species (189). Darwin signals the connection here using the word "hence," as he does very often in *The Descent of Man*. This pattern of reasoning from an embryological state to an ancestral one is a variation on Lyell's actualism, the practice of inferring long-past processes from present phenomena. Given that Lyell looms so large in Gould's history of geological time, it seems apt to designate this pattern of induction as deep-time reasoning. The "hence" used as a hinge between present and past enables rapid transit through time; as a form of comparison, this logic reveals its indebtedness to the comparative method by surfacing frequently in intra-species comparisons (between "savages" and ancestral humans) as well. Darwin's regression from "man as he exists" to the lancelet (188), arguably his most vertiginous descent into deep time, also stands out for concluding on a note of caution. Noting that "it seems, if we turn to geological evidence, that organisation as a whole has advanced throughout the world by slow and interrupted steps," he warns that "we must not fall into the error of looking at the existing members of any lowly-organised group as perfect representatives of their ancient predecessors" (192). Even more than Lubbock, Darwin shows an awareness of the fallacy that can result from deep-time reasoning, a conflation of the apparently primitive and the ancestral. Michael D. Crisp and Lyn G. Cook have pointed out that this fallacy persists in modern evolutionary thought.[84]

In Darwin's retrospect, he frames *The Descent of Man* as setting out to "recall in imagination," if only "partly," "the former condition of our early progenitors" (*DM* 678). The descent into deep time provides a narrative framework for this imaginative creation of evolutionary memory, typically supported in characteristic Darwinian fashion by a mass of factual detail. Like those recognized by Gillian Beer in *Darwin's Plots,* the descent into deep time is also a plot structure that appears at times without its body of facts, as in the romance of music used in prehuman "courtship" (638) or of the discovery of fire "before the dawn of history" (68).[85] Indeed, it was recognized as such by J. G. Ballard, who first named the "descent into deep time" and made it a central theme in *The Drowned World* (1962). Inspired by Freud as well as Darwin, Ballard articulates a theory of "archaeopsychic" or "neuronic time" through one of his biologist characters, who rediscovers an evolved human faculty for descending the spinal column as a strategy for coping with the climatic conditions of a superheated earth, compared by Ballard to those of the Triassic period. In the chapter entitled "Descent into Deep Time," the same scientist interprets the protagonist's nightmare of "baying reptiles" and a gigantic sun as not a dream but "an ancient organic memory millions of years old," a "cytoplasmic" recall triggered by rising temperatures and ecosystem change.[86] Ballard's "back to the future" scenario also recalls the cyclical recurrence of planetary states hypothesized by Lyell and caricatured by Henry De la Beche as "Awful Changes," in which reptiles feature prominently.[87]

The descent into deep time, often associated with reversion or regression, is a staple of scientific romance, though assigned to the future rather than the past in one of the most famous examples, H. G. Wells's *The Time Machine*

(1893). Ballard's version stands out for its vertical orientation, as it derives flora and fauna from specific points in the fossil record and evolutionary memory from the base of the human spinal column. Another scientific romance, *The Time Stream* (1931) by John Taine (Eric Temple Bell)—the title may allude to the *Descent of Man* passage discussed earlier—constructs a more synchronic paradigm in which the time traveler may access any point in time through a kind of trance. This novel is also a eugenicist fantasy that succumbs wholly to the logic of "lower" races implied by the pursuit of deep time, a logic that Darwin embraces to a large extent and that Ballard also flirts with, though not, in my view, uncritically.[88] For Darwin, however, human as well as human-animal differences are always a matter of degree, not kind, and "mental powers" as such are practically as ancient as life itself (*DM* 87).

There is a large literature on Darwin, race, and racism. Buckland has shown that this issue is not tangential to the history of geological time.[89] The comparative method inherited from the Enlightenment, which has been a major focus of this chapter, carries an alternative view of human difference that predates the pseudoscientific racial taxonomies of the late eighteenth and nineteenth centuries. Palmeri argues systematically that the eighteenth-century conjectural histories were, as a rule, Eurocentric rather than racist. This generalization does not hold in every case, but it has the merit of historical precision.[90] Even in Darwin and Lubbock, writing well after the advent of racial "science," the term "savage" connotes the civilizational stage inherited from conjectural history, while also becoming associated with nonwhite races as a group—a vague but no less ideological form of racial categorization.[91] The indefinite time horizon also inherited from conjectural history remains a factor in these late nineteenth-century accounts of human evolution and helps to account for some of the differences between Lubbock and Darwin.

I showed earlier that Lubbock moved cautiously toward absolute dates for the epochs of human prehistory, positing very early racial differentiation within this context. Darwin preserves the space between geology and prehistory by means of sexual selection, which operates on an evolutionary if not strictly geological time scale and links humans with nonhumans. Since racial differentiation postdates early human migrations and their associated history of sexual selection (*DM* 664–65), he effectively places it later in the history of the species. Because of this extended relative time scale, common descent becomes more important to species identity than it is for Lubbock (also a monogenist) and present variation becomes less important (205–7). Skin color specifically is singled out by Darwin as an "indifferent" character relative to evolutionary history (229). Darwin specifies, too, that the "inevitable" extinction of "savages" is due to intertribal competition (with Europeans among others) rather than unfitness (211–12), the more racist explanation favored by Lubbock and others. All the same, this long discussion (211–22) still amounts to an apology for colonialism, as scholars have generally argued.[92]

The youthful Darwin's perception of the Fuegians as being "like animals" provided an indirect alibi for the dispossession of the Yaghan, the Selk'nam genocide, and other atrocities in the region. Buckland resignifies this dehumanizing judgment by sketching a restorative portrait of the Yaghan, based on anthropological literature, as a resilient people whose spiritual beliefs allowed them to conceptualize their culture as part of a vibrant multispecies community.[93] As the development of Darwin's ideas shows, his act of relegating these "savages" to the deep past had a reciprocal effect on his understanding of human descent. His misrecognition of the Fuegians developed into a recognition of the evolutionary proximity of all modern humans, anticipated in part by the youthful Darwin's reflection on the rapid changes undergone by the group who had been kidnapped by FitzRoy on the previous *Beagle* voyage and taken to England (*DM* 86). Buckland argues that the "geological imagination" functions as an instrument of colonial domination at the moment of misrecognition, an index of the exclusive European command of science. This misrecognition of human otherness, however, is reversed by the otherness of geological time—consistently figured both through and against human otherness since its inception as a concept, as I have argued throughout this study. To see Darwin's pursuit of deep time merely as an expression of racism is to miss the synthesis that produced a temporal location for the human species as a whole within geological time.

Returning to the idea of the human animal after having framed his theory of evolution and engaged with the new disciplines of social science, Darwin positions the species as a whole as a recent variation on the "man-like animal" that can be traced in the fossil record (*DM* 69). By extending the comparative method across the line that divides extinct from living species, Darwin also established the descent of the human from nonhuman species. In accepting, however reluctantly, that "Jemmy Button" (o'run-del'lico) and the other Yaghan from the vicinity of the Beagle Channel "resembled us" (86), Darwin sets the stage for the recognition that differences among human groups are insignificant compared to those between modern humans and nonhuman animals. Yet these too, despite the many interposing "gradations"—including the "ape-like progenitors" who "probably lived in society" (83)—are fundamentally "like us" in many ways. Thus Darwin both relocates the human/animal boundary to a lower position on his relative time scale and makes it more permeable. The "common progenitor" of all humans—deeper in the fossil record even than Lubbock's "ancient savages"—"would probably deserve to rank as a man," Darwin urges (678). Darwin observed in Brazil that slavery "degrades" its victims below this rank; his opposition only intensified as the rank of "man" became ever more inclusive in the progress of his evolutionary thought.[94] Through the early hominin ancestor, "man is the co-descendant with other mammals of a [still more ancient] common progenitor" (676). The indefinite time horizon of "descent," inherited from conjectural history, allows the genealogist to scale up (as here) or down, recursively. Darwin therefore remains committed to the ideology of "lower races" in spite of his own extensive evidence that the analogy between non-Europeans and hominin "ancestors" is heuristic at best.

Nonetheless, he establishes an interspecies community of Quadrumana and on more than one occasion imagines the speech of nonhumans, as when he claims that "an anthropomorphous ape . . . would admit that though he could form an artful plan for plundering a garden—though he could use stones for fighting or breaking open nuts, yet that the thought of fashioning a stone into a tool was quite beyond his scope" (*DM* 150). Although evolutionary time is relative, this collective of primate "ancestors" at times lends itself to a kind of epic discourse, an absolute past that sets the scene for archetypal actions by "primeval men, or the ape-like progenitors of man" (154). Darwin acknowledges at one point that the now-standard term *primates* describes this collective more accurately (176), but he still uses the now-obsolete term *Quadrumana* throughout, perhaps to emphasize the importance of common descent over against the specific difference between "two-handed" and "four-handed" primates. As Palmeri has shown, Darwin remains especially indebted to conjectural history when hypothesizing the "origin . . . of the moral sense" during these "rude times" (140–42).[95] In his quest to "recall in imagination" the moral conduct of "primeval men," Darwin draws inferences at will from dogs, bears, elephants, "savages," and "civilized" Europeans, whose "sneering" at the theory of nonhuman descent reveals the "snarling muscles" they share with other animals (60).

Conclusion

Darwin's theory of human evolution extends human species time to a deep prehuman past that converges with the fossil record and thus with geological time. More and more ancestral hominins continue to emerge from the fossil record today, but Darwin still provides the foundation for the modern consensus concerning human species time. Although students of Darwin generally understand *The Descent of Man* as extending the argument of *The Origin of Species* to the special case of humans, I have argued that the intellectual legacy of the Enlightenment is of particular importance for the later work. Darwin reminds us of his own experience as a voyager and his immersion in Enlightenment voyage narratives throughout the work by his constant reference to the Fuegians and his liberal use of methods and rhetorical figures inherited from earlier voyagers—what I have called "the voyaging intertext" in this chapter. Together with this formative material and his own mature theory of evolution, Darwin synthesizes archaeological and anthropological principles from the social sciences established in his own time. Of Darwin's many contemporary influences, I have been most concerned with Lubbock because Lubbock adapted Enlightenment conjectural history to the newly expanded scale of geological time. In different ways, *Pre-Historic Times* and *The Descent of Man* capitalize on the associations between geochronology and human origins established by a range of Enlightenment thinkers, including those studied in the previous chapters of this book.

Darwin's readers were well acquainted with the "astonishment" that he remembered as having caused him to exclaim, "such were our ancestors." Though Darwin now seems all too eager to relegate non-European "savages" to the deep past, his ultimate concern was to establish something far more controversial, "namely that man is descended from some lowly organised form" (*DM* 689). In Lubbock the expansion of prehistory conveniently dovetails with the triumph of European civilization. In Darwin's case, the insistence on European superiority becomes visible as a kind of special pleading, as when he argues that "the mental powers of the higher animals do not differ in kind, though greatly in degree, from the corresponding powers of man, especially of the lower and barbarous races" (604–5). The emphasis ("especially") belies the marginal importance of race difference in the evolutionary context, and the widely held view of "savages" as animals is here converted to a different purpose, that of locating the human species altogether in the "zoological series" that bridges human history and deep time.

Approaching the end of this conceptual history, it becomes clear that the idea of deep time incorporates the prejudices infused into it by Darwin and his scientific contemporaries, who assumed the burden of proof. Deep time is now an indispensable concept not only for doing science but also for understanding the fraught planetary niche of what we still call "civilizations." Its long history of imaginative possibility and human exceptionalism is more relevant than ever.

AFTERWORD

Evolutionary Nostalgia and the Romance of Origins

IN 2013, one of the students in my course on deep time made an arresting declaration on his or her anonymous course evaluation: "Deep time is NOT an academic subject." I was beginning to explore the topic of this book and had a chance to create a course for a select group of science students whose education would be fully funded by a National Science Foundation grant with the twin aims of integrating mathematics into the life science curriculum and recruiting students from diverse backgrounds.[1] Not all these students were eager for a course in the humanities, and the student quoted above raised an especially trenchant objection to the course that had been foisted on them. In support of it, one might add that the literature on deep time is recent, dating only to the 1980s, and still relatively slender. The idea of deep time is widely recognized among geologists, but many of these students felt (as I learned to my surprise) that geology was not a "hard" science to begin with. Although I have a higher opinion of geology, my main interest lies with deep time before geology, with a kind of language about the very ancient past that preceded and made possible the naturalized use of this metaphor in a scientific context. I have made this argument on the basis of eighteenth- and nineteenth-century writings newly attuned to the uncertainty around questions now associated with prehistoric and prehuman times. These historically specific writings at the same time illuminate the affective experience of any human being confronted with evidence of "revolutions anterior to the memory of man" (Buffon), whatever form that evidence might take. Pascal Richet's much wider survey *A Natural History of Time* begins appropriately with a chapter on origin myths across cultures. Like many of the mythmakers, the authors under consideration in this book responded to the feeling of being thrown into a world that antedated them by who knew how long.

The Forsters, Buffon, Herder, Blake, Lubbock, and Darwin also share a historically specific awareness of the questions that were bringing the Western disciplines of geology, archaeology, and prehistory into being in their lifetimes. My endeavor to construct a history of deep time through their work takes inspiration from Paolo Rossi's remark that "the object of the history of science does not in the least coincide with the object of science."[2] The recent literature on deep time, however, does not always bear out Rossi's statement. John McPhee himself asserted no priority claim with regard to his use of "deep time" in works of popular geology. Stephen Jay Gould asserted that priority claim on his behalf, and it has been an influential one, as noted in my introduction. *Time's Arrow, Time's Cycle* is a somewhat unusual case in being a history of science written by a working scientist, but it is worth noting that Gould here reverses the trajectory suggested by Rossi and his source, Georges Canguilhem: Gould adopts the term *deep time* from a science writer describing the current practice of geology and projects it backward into the history of the science, fusing two objects into one. McPhee and Gould both offer histories of deep time, but Gould scales down the inquiry to pursue the origins of the geological perspective. An opposite instance, somewhat more in keeping with Rossi's principle, would be Henry Gee's *In Search of Deep Time* (1999), which argues that most evolutionary explanations have been constructed in defiance of deep time. In this view, the history of evolutionary narratives since Darwin's time has been at variance with the science of deep time, an object that exceeds the scope of causal explanation or narrative continuity.

I have argued that deep time is not exclusively or even primarily a geological field but is nonetheless an "object of science" that cannot be easily disentangled from its complex history. Rather than attempt a history uninfluenced by the assumptions of modern geology, I shall conclude here by surveying a proliferation of more recent histories of deep time, all of them taking the modern GTS as their starting point but diverging into the differing realms of ethnography, science fiction, climate fiction, literary scholarship, and discourse on the Anthropocene. Time is only as deep as the narratives that mark its extent, and although geology now provides a paradigmatic context for other deep time narratives, this does not mean that deep time is sealed in the black box of geological reckoning. Gee's broad critique is motivated by the insight that many modern narratives of evolution, both popular and scientific, overleap the radical uncertainty and discontinuity that geological time imposes. They pretend to be literal genealogies, he argues, rather than heuristic models, neglecting the premise of evolutionary theory itself: "Evolution is a consequence of deep time." This critique is purchased at the cost of an oversimplified history, however:

> [Darwin] saw evolution as generally slow and gradual yet the scholarship of his time viewed Earth as no older than the few thousand years allowed in the Bible. The realization, by Victorian geologists such as Darwin's geological mentor, Charles Lyell, that Earth was inconceivably, if not immeasurably old, gave Darwin's idea of natural selection time enough to change one species into another.[3]

To be fair, Gee's objective is not a history of the concept of deep time but rather a critique of the history of life, "a kind of 'anti-history' that recognizes the special properties of Deep Time."[4] Even as an anti-history, however, I would argue that Gee's narrative belongs to the broad category of histories of deep time that emerged in the late twentieth century. For his anti-history, Gee appeals to literary models from Shelley to Kafka to Borges, and in this respect his project converges with my own account of deep time by way of narrative forms, forms that tend to be self-conscious about their own fictionality.

My history traces deep time to an imaginative space between prehistory and geology. From a strictly geological point of view, prehistory may be irrelevant. Seen from the floor of the Grand Canyon, the uppermost strata along the rim are at least 270 million years old, predating the earliest mammals. The geological point of view itself, though, still bears traces of the expanding time scale built on points of contact, both observed and conjectural, between the rock record and the prehistoric origins of species. Well into the nineteenth century, when there was little real debate (pace Gee) over the principle of a long time scale, scientists such as Lyell and Lubbock still acknowledged the precedent of non-Western traditions built on vast prehistoric or prehuman expanses of time. Throughout this study, I have cited a willingness on the part of eighteenth-century thinkers to entertain these traditions as a form of evidence, in contrast to a more Eurocentric attitude associated with the nineteenth-century disciplines, beginning with Cuvier's dismissal of the *Mahabharata* and its vision of Great Time as "nothing more than a poem" (discussed in the introduction to this book). For Lyell, writing toward the middle of the century, defining the boundaries of geology, and hence of geological time, was an even more central concern.

In *Principles of Geology* (1830–33), Lyell begins with a substantial history of ideas about the earth, incorporating classical antiquity and "Oriental cosmogonies" (more and less ancient) along with medieval and early modern Europe. Lyell's engagement with these premodern and/or non-Western ideas betrays some ambivalence. To a degree, the modern scientific idea of geological time must still compete in this arena against other views that place human history farther back in cosmic time or call for quantitative measures that are excluded by science. Lyell's rationale for his long preliminary history links this large set of competing ideas to a "controverted question" of his own time as he sets out "to demonstrate that geology differs as widely from cosmogony, as speculations concerning the creation of man differ from history." The larger argument of this three-volume work, which limits geological inquiry to "causes now in operation," thus depends on an analogy stating that the modern sciences are still insufficiently distinct, "just as the limits of history, poetry, mythology were ill-defined in the infancy of civilization."[5] There is ample common ground, in Lyell's reading, between the "superstitions" of the ancients and the "unphilosophical" theories still prevalent during the "infancy of the science" of geology, as recently as a generation before his own time. These theories still have the "air of a romance" because they depend on catastrophes as a causal mechanism. Modern catastrophist theories, according to Lyell, share with ancient cosmogonies a recognition of the "great periodical revolutions in the inorganic world" divorced from any grasp of the length of time these transformations required.[6] The long time scale of Hinduism, considered in some detail through Sir William Jones's translation, *Institutes of Hindoo Law, or the Ordinances of Menù* [Manu] (1796), and likewise the Great Year of the Greeks, are no more than superstitions for Lyell because they lack an evidentiary basis in the "ancient history of the globe," the fossil record. He claims that the "exaggerated traditions" of revolutions in the natural world "merely received corroboration" from geological observations the ancient Brahmins and others may have made, and had their real origin in folk memory of local natural catastrophes.[7] All the same, Lyell remains impressed by the agreement between ancient cosmogonies and the rock record and seems to construe them, in spite of himself, as a shadowy kind of evidence for the true nature of deep time.

Without disputing the scientific merit of Gee's position, it is easy to see what his emphasis on the irremediably fragmentary nature of deep time owes to Lyell. By defining geology against cosmogony, or the pursuit of origins, Lyell engaged in a kind of exaggeration himself, an amplification of gaps in the fossil record that is echoed in Darwin's famous gloss, "a history of the world imperfectly kept." Lyell's own much later work, *Geological Evidences of the Antiquity of Man* (1863), thoroughly examines the evidence of Pleistocene humans but nonetheless limits the scope of prehistory, in this instance to foreclose a rigorous evolutionary explanation for human descent.[8] Lubbock, as we have seen, is more willing to place Pleistocene humans in the context of evolution and hence a longer prehistory, but still maintains the boundary between (European) science and (non-Western) tradition defended by Lyell. Though declaring himself freed of the "embarrassing interference of tradition" (*PT* 416), Lubbock also appeals to the earlier work of James Cowles Prichard on human antiquity. Lubbock notes Prichard's grudging acknowledgment of the "extravagant claims to a remote and almost fathomless antiquity, made by the fabulists of many ancient nations" as ultimately more reliable than the short chronology of Archbishop Ussher, the seventeenth-century theologian who is widely remembered for dating the origin of the earth to 4004 BCE (*PT* 376–77). Once again, the language of deep time is smuggled in from non-European tradition in the guise of a scientific critique of superstition. Histories of deep time after geology retain the broad scope of this poetic language, incorporating ethnography and geology, science and myth.

Lyell's conviction that "the superstitions of a savage tribe" can "exert a powerful influence on the mind of the philosopher" creates a space for prehistory in his geological argument.[9] His effort to marginalize these "superstitions" is of a piece with his effort to discard the long time scale of ancient myth and (even more decisively) the short time scale of Christian Europe as religious prejudices. These efforts constitute an acknowledgment of the imaginative space in which a consciousness of deep time slowly developed. In this large framework, the period from late antiquity to the present seems to have done more to "retard" the progress of geology than to advance it. One form of religious prejudice becomes an allegory for the others. In tenth-century Samarkand, for example, the philosopher Omar was banished (as Lyell tells it) because he required a greater time scale than the Koran allowed; similarly, in sixteenth-century Verona, the insights of Girolamo Fracastoro into the true antiquity of fossil shells were buried under the objections of his more orthodox co-religionists.[10] The eighteenth-century naturalists I have foregrounded did very little, according to Lyell's partisan narrative, to alleviate the many centuries of "profound oblivion" that preceded them, and even James Hutton's revolutionary recognition of the limitless scope of geological time was suppressed by forces of religious prejudice that Lyell implicitly compares to the Inquisition. Lyell's radical critique of catastrophism and the search for origins betrays his own rhetorical bent, and he ultimately endorses the idea of geology as a poetic science. He compares the very recent confirmation of geological principles to the Copernican Revolution as a moment of imaginative freedom: "by the geologist myriads of ages were reckoned, not by arithmetic computation, but by a train of physical events—a succession of phenomena in the animate and inanimate worlds—signs which convey to our minds more definite ideas than figures can do, of the immensity of time." In this way, the geologist experiences "a bliss like that of creating."[11]

The literary tradition banished from the science of deep time still shapes the narrative form of Lyell's geology, which contains one of the first modern iterations of the motif I have described as the "descent into deep time." In the last chapter I focused on works by Lubbock and Darwin that are primarily concerned with establishing sites of contact between prehistory and geology. Lyell is more concerned to emphasize the disproportion between the human epoch and geological time as such, but as a consequence his emphasis on qualitative measures is even more pronounced. His preference for geological "signs" over numbers recalls the long tradition of qualitative statements and affective response among eighteenth-century naturalists who observed the strata, including the Forsters, Buffon, and many others. Lyell's descent into deep time in his third volume leaves the human world rapidly behind in moving from the Recent (now Holocene) down toward the Eocene epoch in the course of this volume. His insistence on "signs" and qualitative measures vividly marks the depth of this descent while proving the uniform action of "causes now in operation" across deep time. This approach at the same time excludes any upper limit on the age of the earth. As a modern account, it takes deep time in its geological sense for a premise but remains categorically different from the twentieth-century deep time narratives in which a hard radiometric date is attached to this premise.

Radiometric dating and geophysical research on the origin of the earth create new temptations to deviate (as Lyell would see it) from geology into cosmogony. Traces of premodernity persist in more recent narratives that take the numeric age of the earth for granted, including both the wonder evoked by Lyell and the romance of origins rejected by him. The tendency to associate human origins with geological inquiry seems especially strong in the Anthropocene, with its focus on the human species. The known age of the earth, and of ancestral hominins, now makes it impossible to conflate the two, but as Lyell noted there is a strong analogy between cosmogony and "speculations concerning the creation of man." The proximity of these questions is evident today in works of climate fiction and Anthropocene popular science, which often call on prehistoric or prehuman actors to witness "scenes from deep time." Following in the wake of McPhee and Gould, Martin Rudwick, in *Scenes from Deep Time* (1992), locates this trope of virtual or proxy witnesses to events of the deep past in nineteenth-century popular science, in illustrations that did not observe the limits of geology as strictly as Lyell might have wished. Following Gould, Rudwick credits McPhee with attaching "deep time" to "the unimaginable magnitudes of the prehuman or prehistoric time scale."[12] As a study of nineteenth-century illustrations, Rudwick's account has a warrant for equating "prehuman" and "prehistoric," though the Crystal Palace exhibition of dinosaurs and popular illustrated books sometimes hinted at the chasm between the "primitive world" and human time.

Actors have appeared on the stage of deep time at least since the molluscoid, pachyderm, and hominid "giants" of Buffon. Although the human species has been reduced to a marginal role by modern geology, the desire to connect prehistory with geology that survives in Rudwick's examples remains stronger than ever in all manner of nontechnical deep time narratives. In the early twenty-first century, this desire takes the peculiar form of evolutionary nostalgia, as evident in the "Paleo diet," barefoot running, and other primal lifestyle choices. In a long historical view, these back-to-nature movements could be traced back to Romanticism or even farther. The trace of conjectural history, that early form of prehistory which offers some concrete reasons for nostalgia about human origins, is important here, as is the antecedent myth of a Golden Age. The ancestral Paleolithic hunters often invoked today play a role comparable to that of Buffon's giants or Vico's poets, while authors drawing from living indigenous traditions (such as the endurance running of the Rarámuri/Tarahumara in northern Mexico) carry on the problematic legacy of the European explorers. McPhee's account of the geological past shows that human actors are not strictly necessary (except as narrators or chronologers), but the dream of a long prehistory resurfaces in the reception of more strictly geological or evolutionary texts, in narratives that recenter the human inheritance within deep time.

The discourse on the Anthropocene provides a more historically specific context for evolutionary nostalgia in its current form. Although more typically construed as a formal recognition of anthropogenic climate and earth system change, the Anthropocene might also be seen as a reaction against the late twentieth-century popular science of deep time. McPhee's relativizing image of human history as falling prey to the first stroke of a nail file on the king's outstretched middle finger, enthusiastically echoed by Gould, compounds the geological "outrage" long associated with the marginalization of humans that resulted from the series of scientific revolutions. In diagnosing the three major outrages (*Kränkungen*) or "blow[s] to man's self-love," Freud subsumed the discovery of geological time under the second, biological outrage of Darwin's theory, though some scholars have considered the abyss of time as a blow in its own right, the ultimate realization of human marginality.[13]

The Anthropocene notion that "we are becoming players in geologic time" promises to avenge this injury.[14] The discourse on the Anthropocene also capitalizes on the clarity of modern geological timekeeping, establishing stratigraphic markers in recent history to underscore the laser-like precision that now renders the abyss of time a strictly imaginative problem. Considered as a symptom of the Anthropocene, evolutionary nostalgia further capitalizes on the increasing precision of the science of human evolution. By contrast to Darwin and Lubbock, consumers of popular science today seek a sympathetic recognition of ancestral humans without the reversal. At its best, the impulse to emulate the smaller carbon footprint of Pleistocene humans exemplifies the "timefulness" advocated by Marcia Bjornerud. At its worst, this impulse can appear as a denial of irreversible anthropogenic impacts, a form of what Bjornerud calls "time denial."[15] Catastrophe by comet, though associated with the romance of origins for Whiston, Buffon, and their critics, also has a fatal history that can be repurposed for the catastrophic erasure of anthropogenic impacts—a form of climate denial—as in the film *Don't Look Up* (2021).

Bruce Chatwin's account of Aboriginal wayfinding practices, *The Songlines* (1987), anticipates the response I am calling evolutionary nostalgia. It also preserves elements of the popular science of deep time as it existed in the 1980s and before. Chatwin's travel narrative performs the substitution that Nicholas Dames has identified with "modern" nostalgia, which he defines as "the substitution of an inaccessible *time* for a still-real place."[16] I began this book with a chapter on Pacific voyage narratives because Australia and the Pacific islands have played a special role in this translation of place into time—in the production of deep time—since at least the eighteenth century. The Aboriginal novelist Alexis Wright has reframed this legacy in her anti-nostalgic romance *The Swan Book*, in which modern technology registers as a fatal disruption of indigenous wayfinding practices.[17] In *The Songlines*, Chatwin focuses instead on what he takes to be unbroken continuity in the oral transmission of a method for navigating the landscape through song. One strand of this narrative is occupied with Chatwin's experience in Australia as he learns this method from Aboriginal informants and white Australian mediators. As he develops his thesis that substantial traces of a culture more than ten thousand years old survive in the practice of "walkabout," Chatwin introduces a second narrative strand, an evolutionary one. His experiences yield insights into the centrality of walking, of nomadism, in early human evolution. The affective and synchronic experience of deep time in the field inspires a diachronic descent into deep time through the science of human evolution, culminating in a long excursus on the discovery of *Australopithecus africanus* and the history of the human species.[18] Running occupies a similar place in the more openly nostalgic evolutionary narratives of runners, including Bernd Heinrich's *Why We Run: A Natural History* (2001) and Christopher McDougall's *Born to Run* (2009).[19]

These works, too, are histories of deep time because they situate themselves at the nexus between prehistory and geology. An area of uncertainty, aptly named the abyss of time by eighteenth-century writers, existed at this location before these two disciplines existed as such, and many living writers working under the guidance of these disciplines—an increasing number, it seems to me—continue this tradition of "witness[ing] time in a way that transcends the limits of our human experiences," in Bjornerud's pithy phrase. Other scholarly accounts of deep time and the Anthropocene have emphasized the alien or inhuman quality of the scale of geological time. Mark McGurl proposes H. P. Lovecraft's ancient alien race as a literary approximation of this quality, and Kate Marshall, defining the Anthropocene novel, places her emphasis on the stark desert settings of novels in which the protagonists perform large-scale inscriptions comparable to those made by geological processes.[20] I am arguing that the idiom of deep time has been used historically, and continues to be used, to figure both the proximity and the distance between human origins and geological time. This view is less anthropocentric than anthropic, an acknowledgment that the geological record necessarily reflects back to us a history of our own species. The mere act of seeking "to understand our place in Time" could be classified, along with wayfinding, running ("our ultimate form of locomotion"), eating, or having sex, as an act of "reasserting our kinship . . . with ancient man."[21] "The ancient man" also provides Blake's image of human ancestry and now seems both sexist and speciesist, but the twenty-first century offers a rich repertoire of alternatives, including "bipedal savanna hunting apes" as well as an "extended family of living organisms" that may be traced all the way back to 6 a.m. on the synecdochic 24-hour clock that stands for geological time.[22]

The descent into deep time is a prominent feature of the postapocalyptic narratives that are flourishing today in the age of climate change fiction. In *New Wilderness* (2020), Diane Cook offers a naturalistic depiction of climate refugees (former New Yorkers) who have reverted to a hunter-gatherer existence in very short order. Lydia Yuknavitch portrays a blasted earth in *The Book of Joan* (2017) on which survivors take refuge in caves and all remaining biodiversity is subterranean. These motifs of reversion and regression are inherited from scientific romance and science fiction. Ballard's *The Drowned World* (1962) is a special case, both because it anticipates a climate crisis in some ways like our own and because it names the "descent into deep time" as a topos inherited (as Ballard sees it) from literary modernism as well as from Darwin and Freud. Classic science fiction offers many more examples of disaster followed by return to an earlier step or stage of development, sometimes layered with other forms of regression, as in Walter M. Miller Jr.'s strange amalgam of twenty-sixth-century barbarism and medieval monasticism (*A Canticle for Leibowitz*) or Samuel R. Delany's vision of print culture without clock time (*Dhalgren*). Apocalyptic literature itself is ancient, and from Biblical times it has favored stylistic archaisms, making it difficult to claim historical specificity for any apocalyptic genre. The current concentration of apocalyptic scenarios, in cli-fi specifically, is remarkable all the same. Their affiliation with deep time (or their character as an expression of the current fascination with deep time) also comes across in explorations of posthuman or nonhuman agency, as McGurl and Marshall have noted. Alan Weisman's *Earth without People* (2007), a nonfictional but counterfactual history of the future, presses an imaginary "reset" button that reestablishes evolutionary conditions as they might have been if humans had not become dominant. In Reza Negarestani's curious theory-novel *Cyclonopedia* (2008), petroleum holds discourse with the sun.

The modern "catastrophism" diagnosed by Lyell has developed in a direction even more indebted to the romance form, and in some cases (e.g., Yuknavitch) it seems that the catastrophic agents associated with world formation are now invoked to imagine a kind of catastrophic renewal. The interrelated genres of climate fiction and postapocalyptic fiction are proliferating to such a degree that it becomes difficult to take a stand as a critic and dwell with a particular example.[23] One contemporary novel stands out, however, in relation to the conceptual history of deep time. Jeff VanderMeer's *Annihilation* (2014) offers a descent into deep time that unsettles both the distinction between human and nonhuman agency and the vertical geometry of deep time itself.

By unsettling these boundaries—indeed, the setting of the novel, Area X, has no defined boundaries—VanderMeer captures the abyssal quality that is so central to the idiom of deep time, as I have argued in this book. Lyell argued that geological (and evolutionary) change is nondirectional, not "progressive" in the sense of improvement or increasing organization over time. Yet the temporal direction expressed by the principle of superposition (the lowest strata are the oldest) remains clear and inescapably informs the imagery of deep time from the bottomless abyss of time to (more rarely) the "summit of the time scale" (as in Buffon). The central feature of Area X is a deep vertical shaft, dug by nature and/or art, that is referred to by the narrator/protagonist of this novel as a "tower," but is referred to by the other characters as a "tunnel."[24] The narrator, a biologist on a team of explorers, studies vaguely prophetic inscriptions found on the walls of this descent, which are soon revealed to be of nonhuman origin. As she and her fellow explorers descend into the "tower," they find that the animal or agent making these inscriptions is ahead of them, with the consequence that the inscriptions at the bottom are the newest. As the other explorers descend into madness and death, our protagonist explores an abandoned lighthouse, this time ascending by stages or steps that lead, paradoxically, to a moldering archive of journals, the journals of all the past, lost expeditions to Area X, secreted at the top of the lighthouse. The hypothesis put forward toward the end of the novel is that the nonhuman creature making the inscriptions is a quasi-evolutionary descendant of the original lighthouse-keeper, and/or an entity from the outside that has assimilated this man, along with many subsequent explorers, into itself.

This novel effectively thematizes the epistemological problem of steps or stages in deep-time reckoning. Their vertical position in a geological column or schematic time scale (*scala*, ladder) is theoretically independent of any form of progressive development, and yet the connotations of "higher" and "lower" steps or stages are inescapable. In the same way, Darwinian theory acknowledges the risk of labeling ancestral forms as "primitive" while nonetheless triumphing in the clarity that makes possible the retrieval of these forms from the deep past. In Chapter 4, I compared Darwin's epiphany concerning human evolutionary descent from the lancelet to Lubbock's progressive descent from Iron Age to Paleolithic humans. In *Annihilation*, as in many science-fiction novels, the reversion or regression of the modern human characters happens very quickly; in this instance it reflects a drastic and unprecedented shock to the ecosystem of Area X, taken by some readers as an allegory of the Deepwater Horizon oil spill and its impact on the Gulf Coast.[25]

By focalizing a group of scientist-explorers, VanderMeer takes up another venerable sci-fi trope, but the ecology that they engage with is so disorienting as to warrant comparison with the Cook voyage narratives examined in Chapter 1. In that case, George Forster finds himself confronted at the end of a long voyage by a culture and a soil so different from those encountered earlier that he is moved to conclude that "the history of the human species . . . cannot yet be unravelled" (*V* II.563). The inscriptions so central to VanderMeer's novel might be compared with the monuments put forward by Buffon to mark the epochs of earth's history (see Ch. 2): although they are organized stepwise, they remain shifting and palimpsestic in part because they are the work of other species. Blake and the ballad revivalists (see Ch. 3) tend to celebrate the recovery of a deep ancestral aptitude for song. In *Annihilation*, the keening song that seems to be the only expression remaining to the posthuman species strikes the narrator as haunting and deeply uncanny, but it is the first sign of a sentient quality in this life-form that appears in the novel. This "crawler" also carries hints of the convergent evolution that fascinates Darwin in *The Descent of Man* (Ch. 4), which tracks human-like traits in "lower" animals and vice versa. The prospect of a descent into barbarism haunts all of the eighteenth- and nineteenth-century authors I have discussed and remains an essential trope for VanderMeer as well.

Whether we find ourselves submerged in the dark abyss of time or surveying its vastness from the "summit of the time scale," the language of deep time has become an indispensable idiom for the paradoxical history of a prehuman world. Deep time is never "just" geological time, to counter another skeptical remark from the same set of course evaluations. It is a language for the scale of difference, a Western paradigm that requires non-Western inputs and preserves a record of the human differences and inequalities that helped trigger the recognition of a difference between two orders of time. For this reason, the texts of Hinduism, the Kanak, the Chukchi, the Yaghan and other "Fuegians," Aboriginal Australians, and many other non-European informants have appeared consistently in these pages along with British and European authors and their texts. Deep time is both an Enlightenment legacy and a wider imaginative space that allows for reflection on common human descent. As a space of reflection, the abyss of time invites comparison with the visual convention known as *mise-en-abîme*, "placed in the abyss." I would like to offer two parallels between gazing into the abyss of time and gazing into the indefinite regress evoked by a picture within a picture (as in Velázquez's *Las Meninas*).[26] In a geological column with any fossil content, the fixed point marked by that content corresponds to, or reflects, the moment in the history of life occupied by the gazer. This reciprocal effect between the observer and the material witness of deep time accounts for the importance I have placed on the presence of fossils in Buffon and Darwin; their absence in the Pacific islands, as noted by the Forsters; and their figurative presence in the form of aural artifacts in the ballad revival. The second parallel has to do with scale: the indefinite regress in each case comes with a scale, measurable epochs in the case of deep time and the ratio of the inset picture to the larger picture in the case of *Las Meninas* and its ilk. These visualizations allow us to approach what cannot be visualized. Part of the power of deep time is that it conjures up a prehuman history in which the history of our species is nonetheless reflected. "You have often / Begun to tell me what I am," says Miranda, and Prospero replies with a question: "what see'st thou else / In the dark backward and abysm of time?"

In the stratigraphy of language, the more portentous and unwieldy "abyss of time" underlies "deep time." "Fathoming deep time" may well be "geology's single greatest contribution to humanity," but it is important to remember that the abyss of time preceding it is literally fathomless.[27] The *OED* gives both "bottomless pit" and "primal chaos" as glosses for the Latin *abyssus*, and the scientific act of populating deep time via the history of life carries inescapable echoes of ancient creation myths, which may account for some of the conflict occasioned by the science of deep time. Our newly discovered proximity to the prehuman world can be coopted to secure the dominant pose of the human animal, as in Buffon's Age of Man or the evolutionary nostalgia of the Anthropocene. A more faithful reckoning, in my view, will produce a bittersweet and sobering recognition like that achieved by Byron's *Cain*, who hails the "small blue circle" of his "scarcely-yet-shaped planet," or the kind of inspired geological description incited by Alexander von Humboldt and carried on today by "temporally literate" writers such as Bjornerud (whose phrase this is), VanderMeer, Robert McFarlane, and many others: "May the eagerly curious human spirit be permitted, from time to time, to swerve from the present into the darkness of prehistory to intuit what may not yet be clearly recognized, and so to delight itself with the ancient myths of geognosy, which return in many forms."[28]

NOTES

Introduction. Deep Time: A Counterhistory

1. See, among others, Rhoda Rappaport, *When Geologists Were Historians,* esp. ch. 7.

2. Davies, *Birth of the Anthropocene,* ch. 2.

3. Kloetzli, "Myriad Concerns"; see also Nayak, "Cycles of Great Time."

4. Archer, *Long Thaw,* 2; Smail and Shryock, *Deep History,* 5–6, 23–24.

5. On the expanded time horizon associated with modernity, see Koselleck, *Zeitschichten,*10, and Blumenberg, *Lebenszeit und Weltzeit,* 218–24.

6. Ballard, *Drowned World,* 133; Gould, *Time's Arrow, Time's Cycle,* 2.

7. Kolbert, *Sixth Extinction,* 265, 45. Martin Rudwick, whose translations and scholarship have been instrumental in rehabilitating Cuvier's reputation, also summarizes the disfavor into which he fell among twentieth-century scholars. See *Georges Cuvier,* x, 258, as well as Gould, *Time's Arrow, Time's Cycle,* 113–14.

8. Z 4. This translation by Zalasiewicz, Milon, and Zalasiewicz is the first complete English translation of Buffon's book in the 230 years since publication. The first English translation by William Smellie (1785) was heavily abridged. Kolbert's narrative also incorporates *Epochs of Nature.* See *Sixth Extinction,* 27.

9. Zalasiewicz, unpublished MS version of introduction to *Epochs of Nature* (2016), 14. My thanks to the author for sharing this early draft with me.

10. For these criticisms, see Finney and Edwards, "'Anthropocene' Epoch"; and Autin and Holbrook, "Is the Anthropocene an Issue of Stratigraphy or Popular Culture?"

11. The best-known instance of "the abyss of time" in English occurs in Playfair, "Biographical Notice of James Hutton," 73. For accounts of earlier sources, see Rossi, *Dark Abyss of Time;* and Albritton, *Abyss of Time.*

12. Chakrabarty, "Climate of History," 201.

13. Smail and Shryock refer to this convergence as "the time revolution." Smail and Shryock, *Deep History,* 5–6, 23–24. See also Gamble and Kruszynski, "John Evans, Joseph Prestwich, and the Stone That Shattered the Time Barrier."

14. Blumenberg, *Lebenszeit und Weltzeit,* 116–29, 171–72, 221.

15. Nayak, "Cycles of Great Time"; Kloetzli, "Myriad Concerns." See also Chakrabarti, *Inscriptions of Nature.*

16. Cuvier, *Recherches,* I.101.

17. Scholars following Gould's attribution of "deep time" to McPhee include Zielinski, *Deep Time of the Media,* 5; McGurl, "Posthuman Comedy," 538n9; Kugler, "Die Tiefenzeit von Dingen

und Menschen," 411n3; Heringman, "Deep Time at the Dawn of the Anthropocene," 57; and Chakrabarti, *Inscriptions of Nature*, 192. Chakrabarti offers his own very worthwhile riposte to "mainstream accounts of deep time."

18. Dimock, *Through Other Continents*; Zielinski, *Deep Time of the Media*; McGurl, "Posthuman Comedy," 538. Jussi Parikka has retraced the path from Gould back to Huttonian "deep time" (via Zielinski) in *A Geology of Media*.

19. See, among others, Dean, *James Hutton and the History of Geology*; and Jack Repcheck, *Man Who Found Time*. Representative textbook treatments of Hutton may be found in Larson and Birkeland, *Putnam's Geology*, 4th ed., 4–5, 124; and Marshak, *Essentials of Geology*, 6th ed., 217 ("father of geology"), 343, and 369 ("no vestige of a beginning").

20. Rudwick, *Bursting the Limits of Time*, 170. Hutton challenged his colleagues' consensus concerning the greater age of crystalline rocks such as granite in "The Supposition of Primitive Mountains Refuted" in *Theory of the Earth*, I.311.

21. "Internalist" histories of geology provide by far the richest source material to date on the history of deep time as a concept. To the works already cited by Rossi, Albritton, and Gould, I would add the essays collected in Lewis and Knell, eds., *Age of the Earth*; Wyse Jackson, *Chronologers' Quest*; and Bjornerud, *Timefulness*. Toulmin and Goodfield's classic *The Discovery of Time* integrates the other sciences into its history, as does Richet, *Natural History of Time*. Rudwick's *Bursting the Limits of Time* has been particularly important for the present study.

22. Chakrabarti, *Inscriptions of Nature*, 6–8 and passim.

23. Recalling his visit to the unconformity at Siccar Point with James Hutton, Playfair writes: "The mind seemed to grow giddy by looking so far back into the abyss of time." "Biographical Notice," 73. See also McPhee, *Annals of the Former World*, 95. Similarly, Buffon wrote of the vast "space of time" (*l'espace de durée*) in which "the human spirit loses itself" (*EN* 40–41), and Abraham Gottlob Werner wrote of the "vast spaces of time [*ungeheure Zeiträume*] of the earth's existence." Werner, *Kurze Klassifikation*, 5. Koselleck makes the point that any discourse about time is reliant on spatial metaphors. Koselleck, *Zeitschichten*, 9.

24. In *Milton* and *Jerusalem*, Blake describes the production of history through the "extension" of a single "moment" (*Milton* 29:1ff. / *E* 127; *Jerusalem* 48:30ff. / *E* 197). Other instances in Romantic literature include conjectures on the early history of the species by theorists of the ballad revival. See Chapter 3.

25. *Drowned World*, 148. "Descent into Deep Time" is the title of Ballard's fifth chapter.

26. Koselleck, "*Neuzeit*."

27. Blumenberg, *Lebenszeit und Weltzeit*, 224. Taking Voltaire's speculations on human antiquity as an example, Blumenberg notes that "world time" initially has the anthropocentric motive of purchasing time (*Zeitgewinn*) for the past and future development of the human species. Wolf Lepenies makes a related point about the refashioning of Western civilization into the most advanced result of a universal process of development. Lepenies, *Ende der Naturgeschichte*, 77.

28. The Industrial Revolution, Jonsson argues, "constituted not a conclusive escape from material limits but a temporary reprieve bought with finite fossil fuel stock." "Industrial Revolution," 680–81. In "History of the Species" (470), citing Curtis Stager's *Deep Future* (2011), Jonsson traces the "cornucopian ideology" of future abundance to "millennial theology"—though it should be noted that the millennial time of sacred history also has its "shallow" side, the ever-imminent "Day of the Lord."

29. The figures of the distant-future reader and human species actor are discussed in Heringman, "Anthropocene Reads Buffon." See also Hamilton and Grinevald, "Was the Anthropocene Anticipated?"

30. "Tiefenzeit," Korpusbelege DWDS-Kernkorpus (1900–1999), available at www.dwds.de/r?corpus=kern;q=Tiefenzeit.

31. "Sediments of time" was coined by Sean Franzel and Stefan-Ludwig Hoffmann as the title of their volume of translations from Koselleck. This volume includes the essay "Zeitschichten" but not the introduction to Koselleck's *Zeitschichten* volume, my source for the "new temporal horizon" (*der neue temporale Horizont*) defined by Kant and Buffon. Koselleck, *Zeitschichten,* 10.

32. Darwin, *Journal of Researches,* 220, 180. On the ethnographic and ethnocentric aspects of "the geological imagination," see Buckland, "'Inhabitants of the Same World.'"

33. Fabian, *Time and the Other,* 16–17. Using the literature on excavations in nineteenth-century India, Chakrabarti makes a compelling case that geological and "human" time remained imbricated with each other as late as the late nineteenth century. See *Inscriptions of Nature.*

34. Davy, *Humphry Davy on Geology,* 69; Cuvier, *Rapport historique,* 181. Rachel Laudan discusses German mining in the mid-eighteenth century as a context for Lehmann and the primitive/secondary division. *From Mineralogy to Geology,* 58.

35. Georg Forster, "Noch etwas über die Menschenrassen," *Werke in vier Bänden,* II.96.

36. Darwin, *Journal of Researches,* 220, 180; *DM* 689. The largely still-current German word for "primitive" peoples, *Naturvölker,* is apposite here.

37. Schnyder, "Einleitung," *Erdgeschichten,* 10.

38. Koselleck, "Einleitung," *Geschichtliche Grundbegriffe,* I.xv.

39. Rudwick, *Bursting,* 127.

40. Humboldt, *Ansichten der Natur,* 110–11. For further discussion of deep time as *mise-en-abîme,* see the afterword. Geognosy (*Geognosie*) is the form of earth science popularized by Werner, who trained Humboldt and many other German intellectuals. It had an enthusiastic following in Britain as well. See further Rudwick, *Bursting,* ch. 2.

41. Lehmann, *Versuch einer Geschichte,* 118; d'Holbach, *Essai,* 245.

42. Koselleck, *Zeitschichten,* 9; d'Holbach, *Essai,* xxi–xxii.

43. Foucault, *Order of Things,* 275. For *Sattelzeit,* see Koselleck, "Einleitung," xv. For a twenty-first-century revaluation of Foucault's paradigm, see Calè and Craciun, "Disorder of Things."

44. "Einleitung," xv; *Zeitschichten,* 11; "Einleitung," xxv, xxi.

45. Foucault relies heavily on Cuvier's comparative anatomy to make this claim, partly because he must touch on many scientific areas, especially in the human sciences, to document the broad epistemic shift for which he is arguing. Lepenies attributes the transition "from natural history to the history of nature" instead to a methodological collapse of Enlightenment natural history in the face of a data explosion that forced new kinds to be classified as potentially "transitional," hence historicized, in relation to known forms. *Ende der Naturgeschichte,* 61. Koselleck, acknowledging both these approaches, places more emphasis on the division of geological time as a critical stage in the production of historical modernity. If this is so, then geology might not be completely "metahistorical," as Koselleck himself later argues: we might see geological time instead as coexisting in tension with historical time, rather than as being transferred from history to nature and back again. See *Zeitschichten,* 10, 84, 11.

46. Rudwick, *Bursting,* 127.

47. Rein, "Georg Christian Füchsel," 23. Rein points out that Füchsel's sections extend beneath the surface, which allows geological mapping in four dimensions.

48. Smellie's English rendering of *siècles* in *FA* is the more ambiguous "ages," which ironically is more amenable to a twenty-first-century reading as well.

49. Darwin, *Annotated Origin*, 282.

50. Schnyder, "Geologisch-Meteorologische Phantasien," 103.

51. Deluc, Review of *Theory of the Earth*; J. R. Forster, "Dr. Forster an Prof. Lichtenberg."

52. See, among others, Cutler, "Nicolaus Steno and the Problem of Deep Time." A further nine essays in this volume (*Revolution in Geology*, ed. Rosenberg) are devoted to Steno.

53. Playfair, qtd. in Porter, *Making of Geology*, 1. On the role of comets and other seventeenth-century influences on Buffon, see *Z* xx–xv and *EN* xlix–li. For a detailed discussion of the contemporary criticisms of Buffon's *Epochs*, see Chapter 2.

54. Gould, *Time's Arrow, Time's Cycle*, 51–59, 13. In his account of the "Burnet controversy," Gould also makes the point that Whiston's theory concerning the geological effects of comets seemed less deranged by the late 1980s, when the Alvarez hypothesis of mass extinction caused by an asteroid was gaining traction. See 176–77.

55. Rudwick, *Bursting*, 118, 170.

56. Marshak, *Essentials of Geology*, 217, 343.

57. Barnett, *After the Flood*.

58. Cuvier, *Rapport historique*, 181.

59. Cuvier, *Recherches*, I.34–35.

60. Cuvier, 116. Cuvier's diligent reading of the fossil archive breaks "the supposed continuity of time" assumed by the "classical" natural histories of the Enlightenment, according to Foucault. See *Order of Things*, 270.

61. Cuvier, 101.

62. I am juxtaposing Fabian's view of natural time as an artifact against Rudwick's view of its discovery in the rock record. See Fabian, *Time and the Other*, 16–17; Rudwick, *Bursting the Limits of Time*, 348.

63. Davy, *Humphry Davy on Geology*, 123.

64. In the revised and expanded German version of his *Voyage Round the World*, George Forster uses the term *Stubenphilosophen*, "armchair philosophers" (*R* 904). The Forsters' narratives do feature several attempts to integrate Pacific peoples into the frameworks of conjectural history, but these attempts become increasingly reflexive in the course of the voyage; repeated experiences of first contact eventually lead them (George especially) to eschew these frameworks.

65. On the expansion and development of "the history of the human species," or conjectural history, in the nineteenth century, see Palmeri, *State of Nature, Stages of Society*.

66. Georg Forster, "Noch etwas über die Menschenrassen," 96.

67. Nyong'o, *Afro-Fabulations*, 21, 103–4, 113. In a somewhat similar vein, Yusoff critiques the "racial blindness of the Anthropocene." *Billion Black Anthropocenes, xiii*.

68. Rossi, *Dark Abyss of Time*, 109–11. Cf. Erasmus Darwin, *Botanic Garden, Additional Note XX, p. 51*.

69. Rossi, *Dark Abyss of Time*, 107–8; Laurent, *Dark Abyss of Time*, xviii; Lyle, *Abyss of Time*, 50.

70. Thomas, *Ode sur le temps*, 3–4; Dryden, *The Tempest*, 5; cf. "time's abyss, the common grave of all," in Juvenal, Satire X.145, trans. Dryden; and Dryden, *All for Love*.

71. Albritton, *Abyss of Time*, 10.

72. Herder, "Extracts from a Correspondence," 156–57; Gidal, *Ossianic Unconformities*.

73. Smail and Shryock, *Deep History*, 5–6, 23–24; Foucault, *Order of Things*.

74. See also Palmeri, *State of Nature, Stages of Society*, chs. 4–5; Buckland, "'Inhabitants of the Same World.'"

75. Ballard, *Drowned World*, 114.

Chapter 1. Primitive Rocks and Primitive Customs

1. See, respectively, *Journal of Researches*, 220, and "On the Distribution of the Erratic Boulders," 417.

2. Foucault, *Order of Things*, 275; Rossi, *Dark Abyss of Time*; Rudwick, *Bursting the Limits of Time*, 182 and passim.

3. Hutton, *Theory of the Earth*, vol. 1, ch. 4, esp. 311–12; Fabian, *Time and the Other*, 13ff.

4. Chakrabarti, *Inscriptions of Nature*, 5.

5. Davy, *Humphry Davy on Geology*, 75. Davy suggests that Jean-André Deluc's childhood in Geneva lends credibility to his geological theory, inspired by "the contemplation of the grandest and most elevated of the mountain chains of Europe" (53). The history of primitive rocks is discussed more fully in the introduction to this book.

6. Davy, *Humphry Davy on Geology*, 69; Cuvier, "Historical Report," 118; d'Holbach, *Essai d'une histoire naturelle*, 217; Lehmann, *Versuch einer Geschichte*, 99.

7. Forster, *Introduction to Mineralogy*, 58.

8. Commerson's "Post-Script on the Island of New Cytherea or Tahiti" (1768) is translated in full in Richard Lansdown, *Strangers in the South Seas*, 84. It is worth noting that Commerson also uses the term in a skeptical way, as Hutton would in geology: what is ancient is primitive enough.

9. As noted by the editors (*V* I.439n31), the "annihilating future" (*die vernichtende Zukunft*) was an afterthought added to George's German text (*R* 180–81). On the motif of green branches as a sign of peace, see also *O* 43.

10. Previous scholarship on the Cook voyages has addressed the influence of Enlightenment theories, and especially Scottish theories, of the stadial development of civil society on the Forsters' voyage narratives (see note 15 below). There needs to be more emphasis on the way their experience in the field leads them to depart from these models—both Scottish and European—a departure heralded by Reinhold in this passage. European culture is actually the first "degenerated" society to which Reinhold introduces us, in the same passage (*O* 9–10, cf. 144). George provides a good example of this refocusing of the stages of social development in his concluding discussion of Tanna, in which he analyzes the circumstances that require a different model for the Pacific Islands (*V* II.554).

11. On the antiquarian study of customs and manners, see Alain Schnapp, "Antiquarian Studies in Naples." Reinhold Forster published papers in the Society of Antiquaries' journal, *Archaeologia*, and in other antiquarian venues.

12. Foucault, "Of Other Spaces," 354; Fabian, *Time and the Other*, 32.

13. For a critical reflection on the comparison to ancient Greece specifically, see George Forster's comment on J. K. Sherwin's version of Hodges's painting of the landing at 'Eua

(Middleburgh), Tonga, engraved for the official Cook voyage publication (*V* I.232). Reinhold Forster, more deeply versed in the classics, often compares indigenous peoples to the barbarians depicted by Tacitus and other ancient authors, but this practice is one step removed from the conflation of Tahitians and Arcadians put forward by Joseph Banks during Cook's first voyage. For Banks's comparison, see John Hawkesworth, *Account of the Voyages,* II.120.

14. Cf. Pierre Hugues d'Hancarville, *Collection of Etruscan, Greek, and Roman Antiquities,* I.177.

15. Thomas, *In Oceania,* 86; Guest, *Empire, Barbarism, and Civilisation,* 56; Bindman, *Ape to Apollo,* 149–50.

16. Lansdown, *Strangers in the South Seas,* 69–70; Thomas, *In Oceania,* 71. I am conflating passages from the very useful introduction to Lansdown's book, an anthology of Pacific writings, in which he discusses the classical legacy (*Strangers,* 11–12) and introduces his concept of bipolar vision (16), with the introduction to his section on the "noble savage," where he develops his distinction between cultural and chronological primitivism (65) and his reading of Rousseau. Bindman and Lansdown offer a larger European framework for understanding the voyages, which is just as important—especially in the case of continental intellectuals such as the Forsters—as the Scottish Enlightenment framework emphasized by Thomas and Guest.

17. Commerson, "Post-Script," trans. Lansdown, *Strangers in the South Seas,* 84.

18. The word occurs in Hamilton's published remarks on a specimen and description of a stone by William Anderson, who was on board the same ship as the Forsters, serving as surgeon's mate. See Hamilton, "Account of a Large Stone."

19. Hamilton, *Campi Phlegraei,* I.84n. The *Resolution* did not sail by the obviously volcanic parts of New Zealand (such as the east coast of the South Island or Auckland in the north), or Forster would have revised his judgment.

20. Davy, *Humphry Davy on Geology,* 118. On "Plutonic," see Rudwick, *Bursting the Limits of Time,* 316. The thermometer gives the dates of later printings of Whiston (originally 1696) and Leibniz (originally 1693).

21. Cf. Hamilton, *Campi Phlegraei,* I.5. The disavowal of theory became increasingly commonplace and was adopted as a kind of ideology by early Geological Society members (Davy himself was one of the founders in 1807). Rudwick cites a complaint by Nicolas Desmarest in 1795 that there were already too many theories of the earth. *Bursting the Limits of Time,* 304, 339–40.

22. Davy, *Humphry Davy on Geology,* 123, 60, 130, 96, 135.

23. Davy, *Humphry Davy on Geology,* 68; Hutton, *Theory of the Earth,* I.318, I.311–12.

24. For this language, see the title of John Whitehurst's *Inquiry into the Original State and Formation of the Earth* (1778), which explained the "toadstone" intrusions into Derbyshire limestone strata more or less correctly via the action of "subterraneous fire."

25. Forster, *Ansichten vom Niederrhein,* part 1, ch. 3, *Werke in Vier Bänden,* II.398; Humboldt, *Ansichten der Natur,* 92.

26. Hamilton, *Campi Phlegraei,* I.4. Martin Guntau cites Reinhold Forster's work as a typical example of a bourgeois Enlightenment worldview that sought to capture the totality of Nature. See his *Genesis der Geologie als Wissenschaft,* 25.

27. Davy, *Humphry Davy on Geology,* 138–39. If New Zealand is ultimately improved not by volcanoes but by another agent of revolution, it may retain the "metals and petrefactions"

that, as Forster argues, are unlikely to occur on any island altered by volcanism (*O* 41). Forster's last publication before the voyage was *An Easy Method of Assaying and Classing Minerals* (London, 1772).

28. At times the Forsters' arrogance or prejudice impeded these exchanges, but it is notable that they (George especially) also gained some critical distance from their ethnocentrism during the stay on Malekula. See Margaret Jolly, "'Ill-Natured Comparisons.'"

29. This is a case in which George's German translation of his *Voyage* (which I translate or re-translate in my text) is at the same time a revision, expressing more conviction than the original English: "The volcano, which burns on the island, doubtless works a great change in its mineral productions, and might perhaps have afforded some new observations, if the jealousy of the natives had not continually prevented our examining it" (*V* II.552).

30. Ferber, *Travels through Italy,* 165. Ferber does not name names himself, but Forster (*O* 33n.) cites Hamilton's letter to Paul Henry Maty, published in the *Philosophical Transactions* for 1771, as guilty of this mistake. There's more than a little schadenfreude here, and it's hard not to imagine that Forster, as well as Ferber's translator Raspe, if not Ferber himself, are responding implicitly to a priority dispute that Hamilton seems to have started by charging Ferber with plagiarism in his *Campi Phlegraei* (II.xliii). *The Tactless Philosopher* is the title of Michael Hoare's biography of Reinhold Forster.

31. Ferber, *Travels through Italy,* xxxi, xiv, xvii.

32. Ferber, *Travels through Italy,* 60, 134.

33. As often, George's own German version of his English text is more pointed and extravagant (cf. *V* II.552). Michael Hoare's valuable annotations on the corresponding pages of Reinhold's *Resolution* Journal describe his generally antagonistic attitude toward Buffon and point out that Cook borrowed Forster's extensive notes on Yasur for the Tannese portion of his published narrative of this voyage. See Hoare, ed., Resolution *Journal of John Reinhold Forster,* IV.614–18.

34. Unlike Ferber's, Buffon's natural history was also an important source on human nature for the Forsters, and both of them cite him frequently—not always critically—on matters of ethnicity and geographic distribution of human populations.

35. This is another passage that George revised and expanded considerably for his German version (cf. *V* I.341–42).

36. Cuvier, "Preliminary Discourse," 183, 185.

37. Cuvier, "Preliminary Discourse," 224, 234; "Historical Report," 126.

38. Cuvier, "Preliminary Discourse," 207, 236, 246, 235.

39. Cuvier, "Preliminary Discourse," 228, 252. If humans now are *less* ancient, then secondary formations are more so: "The chalk, which had been thought so modern, thus finds itself pushed far back into the centuries of the penultimate age" (250).

40. See further Thomas, *In Oceania,* ch. 3; and Guest, *Empire, Barbarism, and Civilisation,* ch. 2.

41. Montesquieu, *Spirit of the Laws,* 231–45 (Book 14).

42. Guntau, *Genesis der Geologie als Wissenschaft,* 22. In Humboldt, see, for example, the preface to *Essay on the Geography of Plants,* 61.

43. Forster cites Buffon as his authority for the view that nature becomes "deformed by being left to itself" (*O* 99n.). Ameliorative human intervention happens on Tahiti "for a short time" (100), and although it is recurring, this language also suggests that the remediation is only temporary, implicitly situating a shorter human time scale within a longer geological one, in the

manner of Buffon. The idea of a fallen state evoked by this narrative in its more theological register anticipates one of the objections raised much later to Darwin's efforts to integrate human and geological scales of time. In the second edition of *The Descent of Man*, Darwin himself quotes a reviewer's charge that he had "reintroduce[d] a new doctrine of the fall of man" (*DM* 66n62).

44. See also *O* 214. The influence of Kames is limited here, since Forster could only have consulted Kames's *Sketches* (1774) after the voyage, and Forster departs from stadial theory in general with the thesis of migratory degeneration developed in this passage. He argues that the Fuegians retain "little or nothing" of the "original system of education" still preserved by the Eskimos in North America, and by the Maya and the Inca, who he believes were sent to South America later by Kublai Khan. By contrast to these "happy tribes," the Fuegians are "descended from a degenerated race" (206).

45. Palmeri, *State of Nature, Stages of Society*.

46. Resolution *Journal of Johann Reinhold Forster*, ed. Hoare, III.66; cf. *V* II.541.

47. On the ethnographic difficulties posed by these encounters with the Kanak and the Vanuatuans, see Douglas, "Art as Ethno-historical Text," 70–73.

48. He emphasizes their peace-loving character and credits them for the absence of conflict that makes them "the only people in the South Seas who have not had reason to complain of our arrival among them" (*V* II.592).

49. On the sophistication of Reinhold Forster's ethnography relative to some aspects of twentieth-century anthropology, see Thomas's introduction to the *Observations* (xv–xvi, xxx).

50. George comments that "self-preservation is doubtless the first law of nature, and the passions are subservient to its purposes" (*V* II.533). The contrast between self-preservation as the main principle of this "elementary savagery," to borrow the editors' helpful gloss on this passage, and "philanthropy" or society as the principle of more civilized peoples, recalls Edmund Burke's explanation of the difference between the sublime and the beautiful. It is a systematic classification in any case.

51. Confirmation of this incident of cannibalism did not come until after their departure from New Zealand (*V* II.606–11). See further Salmond, *Trial of the Cannibal Dog*, 228–30.

52. The German text introduces the idea of resistance to oppression (*R* 812). See further Guest, "Looking at Women: Forster's Observations in the South Pacific" (*O* xli–liv).

53. George Forster's outlook is considerably more jaded by the time of the third Dusky Sound episode. In this and the subsequent narrative of Tierra del Fuego, George gives vent to his disgust with both sailors and natives, and his ethnocentrism degenerates into outright racism. This shift reflects a certain level of fatigue as well as continuing disappointment over what the Forsters saw as Cook's unwillingness to privilege the agenda of natural history; it also reflects the influence of existing negative portrayals of the Fuegians.

54. Reinhold Forster touts the fieldwork that gives his theory an empirical grounding, as discussed at the beginning of this chapter. The pronounced influence of classical antiquity sets his very long chapter on human variety apart from other conjectural histories in yet another way. Each section of "Remarks on the Human Species in the South Sea Isles" (*O* 143–376) has a Greek or Latin epigraph, and nearly half of these are from Lucretius, whose *De Rerum Natura* fully integrates human and natural sciences. Many more quotations in the body of the text (from Pliny, among others) reinforce the primacy of natural history as a framework, neglected by many Enlightenment practitioners of the "history of mankind."

55. Bindman, *Ape to Apollo,* 129. Bindman also notes that the Forsters differ from Bougainville on this point.

56. I have interpolated my own translation of G. Forster's German version for the second half of this quotation, because his language in the German is stronger and makes the connection more explicit: "Menschen und Sitten als . . . der vornehmste Endzweck eines jeden philosophischen Reisenden" (*R* 675). See also Gascoigne, "The German Enlightenment and the Pacific," 147–48. J. R. Forster's review of *Epoques de la Nature* appeared as "Dr. Forster an Prof. Lichtenberg."

57. Forster, "Noch etwas über die Menschenrassen," *Werke in Vier Bänden,* II.78, 75.

58. Winckelmann, *Gedanken,* 8–9.

59. On this topic see D'Hancarville, *Recherches sur l'origine;* and Knight, *Account of the Remains.*

60. The term *race* occurs more widely in the English texts, particularly in Reinhold Forster's *Observations,* and George's conjecture here that the Tannese and the New Caledonians stem from two different "original races" (*V* II.590–92), Melanesian and Polynesian, owes something to his father's account of "the causes of the difference in the races of men in the south seas" (*O* 172–75). "Race" and "nation" are used more or less synonymously by both Forsters in English, a usage that underscores the ultimately monogenist premise of Reinhold's account. See also Douglas, "Art as Ethno-historical Text."

61. G. Forster, "Noch etwas über die Menschenrassen," 96. George also argues that the word's derivation from *radix* implies only descent ("*Abstammung überhaupt*"), rather than an original condition. *Stamm* in German refers to the trunk or main branches of a tree as well as to distinct human populations or tribes.

62. G. Forster, "Noch etwas über die Menschenrassen," 81–82; *EN* 43/*Z* 37. Forster's choice of a "color scale" as his metric casts doubt on his claim that he is pursuing a value-neutral analysis.

63. Thomas makes the point that for J. R. Forster, environmentally specific adaptations of population groups attested to ethnic mutability, and hence monogenesis rather than racial hierarchy (*O* xxiii–xxvii). George Forster, however, notes in his later work that monogenesis too could be used as an alibi for slavery. "Noch etwas über die Menschenrassen," 100.

64. On racism and relativism in this context, see Jolly, "'Ill-Natured Comparisons.'"

65. "Fuegians" is a vague and now superseded term for multiple ethnic groups, as explained fully in Ch. 4.

66. Darwin, *Expression of Emotions,* 257; *DM* 689.

67. Meillassoux, *After Finitude,* ch. 1.

68. Guest, *Empire, Barbarism, and Civilisation,* 95–105. On the tension at Malekula, see Thomas at *V* II.815n36. See also George's Forster's analysis of the picturesque qualities of a georgic landscape on Tanna (*V* II.548).

69. Jolly, "'Ill-Natured Comparisons'"; Douglas, "Art as Ethno-historical Text."

70. See Coote, Gathercole, and Meister, "'Curiosities Sent to Oxford.'"

71. Cuvier, "Preliminary Discourse," 193.

72. The quotation is taken from the "Advertisement" to *Essay on the Theory of the Earth by M. Cuvier,* which appeared while Byron was writing this canto.

73. *Don Juan,* IX.xxxvii–xl, ll. 289–98, 320, *Complete Poetical Works of Byron,* 901.

Chapter 2. The "Profoundest Depths of Time" in Buffon's *Epochs of Nature*

1. "Dr. Forster an Prof. Lichtenberg," 148.

2. For both references, see the "Table of Epochs" from William Smellie's 1785 translation, given as Fig. 1a in the introduction. In what follows, I will highlight manuscript evidence of Buffon's revision process showing that he ultimately refused to decide on a clearly demarcated point of human origins or ascendancy.

3. On racism and colonialism in Buffon, see Sloan, "Idea of Racial Degeneracy"; and Salih, "Filling Up the Space." On the differences between Enlightenment ethnocentrism and modern racism, see Palmeri, *State of Nature, Stages of Society,* 281–86.

4. Published in 1749 in vol. 1 of Buffon's *Histoire naturelle.* As the Forsters were to learn after their voyage, Buffon attributes a more central role to volcanoes in the more nuanced account of *Epochs of Nature,* though he locates their main effects in prehuman time. See also *The* Resolution *Journal of John Reinhold Forster,* IV.614–18.

5. *Epochs of Nature* was the fifth in a series of supplementary volumes begun in 1774. The first two revisit and update the theory of the earth put forward by Buffon twenty-five years earlier in the first volume of the *Natural History,* and the fourth similarly revisits his natural history of man, also first published in 1749. The first supplement also serves as an introduction to the history of minerals, a sequence of five volumes that occupied Buffon in the last years of his life (1783–88). *Epochs of Nature* is therefore only one product of Buffon's extensive reconsideration of both geology and anthropology in the latter part of his career.

6. G. Forster, *Werke in Vier Bänden,* II.87, 81–82.

7. Rudwick, *Bursting the Limits of Time,* 150. On the Forsters and Linnaeus, see Dettelbach, "'A Kind of Linnaean Being': Forster and Eighteenth-Century Natural History" (*O* lv–lxxiv).

8. "Dr. Forster an Prof. Lichtenberg," 154–55.

9. De Baere, *La pensée cosmogonique de Buffon,* 77–78, 147–52; Duncan, *Human Forms,* 39.

10. Stalnaker, "Buffon on Death and Fossils," 21; Rudwick, *Bursting the Limits of Time,* 142; Toulmin and Goodfield, *Discovery of Time,* 149.

11. Wyse Jackson, *Chronologers' Quest,*118; Zalasiewicz, Sverker Sörlin, Libby Robin, and Jacques Grinevald, "Introduction: Buffon and the History of the Earth" (*Z* xvi). Richet's formulation is also helpful: "Buffon delivered, in *Epochs of Nature,* the 'great views' that contradicted both the brevity of Mosaic times and the eternity of the Greeks to which Maillet had returned." Richet, *Natural History of Time,* 122. Both Richet and Wyse Jackson point out that Buffon's experiments to determine the earth's cooling were taken up again much later by physicists including Joseph Fourier and William Thomson, Lord Kelvin.

12. "Dr. Forster an Prof. Lichtenberg," 140; Meister, "Avril," 169.

13. "Dr. Forster an Prof. Lichtenberg," 148–49.

14. Dimock, *Through Other Continents;* Zielinski, *Deep Time of the Media.* For more detailed discussion of Dimock, see the conclusion of this chapter.

15. Autin and Holbrook, "Is the Anthropocene an Issue of Stratigraphy or Popular Culture?"

16. See Chakrabarty, "Climate of History," 201; Smail and Shryock, *Deep History.*

17. Rossi, *Dark Abyss of Time,* 107–9 and passim. On the history of this figure and its attribution to Buffon, see further the last section of this chapter.

18. Latour, *We Have Never Been Modern,* 13–48; Foucault, *Order of Things,* 275. In Foucault's version, Lamarck "introduced a radical discontinuity into the Classical scale of beings," and this "discontinuity of living forms made it possible to conceive of a great temporal current" or "'history' of nature." If the "historicity proper to life itself" is modeled on human history, then temporal distinction becomes specific to the short (human) time scale and is not, or not yet, transscalar, as we find it in Buffon. Like other contemporary scholars, I am here exploring and resisting the restrictive binarism that arises from Foucault's seminal paradigm of epistemic change.

19. Industrial humanity's signature emission of CO_2—ironically a "colourless, odourless gas," as Steffen et al. point out—took many thousands of years, on most accounts, to begin making a mark that now appears permanent and significant on the planetary "scale of the Earth as a . . . system." Steffen et al., "Anthropocene," 842–43.

20. Rudwick, *Bursting the Limits of Time,* 181–82, 118.

21. For detailed discussion of Koselleck and conceptual history, see the introduction.

22. See, among others, Zalasiewicz et al., "Introduction: Buffon and the History of the Earth" (*Z* xiii–xxxiv); Kolbert, *Sixth Extinction,* 27; and Hamilton and Grinevald, "Was the Anthropocene Anticipated?"

23. Sullivan and Shinkle, "Dark Green in the Early Anthropocene"; Menely, "Late Holocene Poetics"; Reno, *Early Anthropocene Literature in Britain.*

24. Zalasiewicz et al. review the historiography and designate Buffon's earth as "a different planet" from ours (*Z* xiv–xvi). See also Hamilton and Grinevald, "Was the Anthropocene Anticipated?"

25. On precursors, including Antonio Stoppani and Vladimir Vernadsky, see Steffen et al., "Anthropocene" 843–44; on narrative and scale, see Nixon, *Slow Violence,* 216.

26. On early industrial air, see Menely, "'Present Obfuscation,'" 484–85.

27. I am giving my own version of the French: "nous tâcherons . . . de former une chaîne qui, du sommet de l'échelle du tems, descendra jusqu'à nous" (*EN* 5). I believe that Zalasiewicz et al.'s "ladder" (*Z* 5) is unnecessarily literal, and they do use "scale" in another instance (*Z* 116). Smellie offers the freer translation "commencement of time" (*FA* 259). The Table of Epochs appears in the above form in Smellie (*FA* 305–6) and is excerpted in the *Critical Review* (Fig. 1a above).

28. *EN* xxxv. Strictly speaking, there are only six days of Creation, and although *Epochs of Nature* was originally announced as comprising six epochs, the symmetry was at least partly an accident of Buffon's late division of Epoch II into two parts, as Roger reports here. Epoch Seventh, he adds, was an afterthought, based on Buffon's reading of Nicolas-Antoine Boulanger's manuscript, *Anecdotes de la Nature* (*EN* xxxv). Thus, while Buffon's Age of Man may appear to be an allegory of the seventh day (God's day of rest after the creation) or perhaps an ironic riposte to Genesis, this appearance seems more accidental than otherwise.

29. Buffon attributes the change to environmental and cultural factors operating during this long interval of time, rather than to evolution in a strict sense. Buffon did explore the possibility of species change over time, and Darwin acknowledges his role in the history of evolutionary thought in a note added to the fifth edition of *Origin of Species,* xv. Roger points out some of the crucial differences between Buffon's views and later evolutionary thought in *Buffon: A Life,* 322–23.

30. I am interpolating "great increase in population" and "men just emerging from the state of nature" from Smellie (*FA* 382, 384).

31. *Z* 124, emphasis added. Inspired by the *Antiquities* of Flavius Josephus and J.-S. Bailly's commentary on Josephus, Buffon dates the first high civilization (in present-day Siberia and Tartary) between 7000 and 4000 BCE and places "thirty centuries of ignorance" between this period and recorded history (*Z* 123, 175–76).

32. This is Smellie's translation (*FA* 383), overly free in this case, but it captures the tenor of Buffon's argument well. Zalasiewicz et al.'s more faithful translation reads, "[man] is barely reassured today by the passage of time" (*Z* 120; cf. *EN* 206). On the flint implements and Buffon's use of antiquarian scholarship, see Roger's note on the passage (*EN* 306n2) as well as Taylor, "*Époques de la Nature* and Geology," 377–78.

33. Smail, *On Deep History and the Brain*, 59–66, 194.

34. In addition to J.-S. Bailly's commentary on Josephus (*EN* 306n7), the relevant antiquarian literature includes John Whitehurst's *Inquiry into the Original State and Formation of the Earth*, contemporary with the *Époques*, which offers a spirited vindication of the "great antiquity of arts and civilization" (272).

35. Sloan, "Idea of Racial Degeneracy," 306–8.

36. "Was die Menschen von der Natur lernen wollen, ist, sie anzuwenden, um sie und die Menschen vollends zu beherrschen." Adorno and Horkheimer, *Dialektik der Aufklärung*, 10. The "recursive moment" (*das rückläufige Moment*) marks Enlightenment's inexorable regression into myth (3). Salih's identification of racism with speciesism speaks to the same point. Salih, "Filling up the Space," 99.

37. Bewell, "Jefferson's Thermometer," 128, 132; Smail and Shryock, *Deep History*, 247. See also Bewell, *Natures in Translation*, 15–33.

38. For Shelley's comment, see his journal-letter to Thomas Love Peacock included in Mary Shelley, *History of a Six Weeks' Tour*, 161. Christopher Cokinos describes the Holocene as "this awesome precious interglacial surge." *Bodies, of the Holocene*, 14.

39. Steffen et al., "Anthropocene," 859. See also O'Donnell, "Buffering the Sun."

40. I am adapting Blake's proverb, "What is now proved was once, only imagin'd." *Marriage of Heaven and Hell*, 8:16/E 36.

41. From a postcolonial vantage point, this preemptive act of Indian Removal signals the ideological baggage sometimes attached to the idea of deep time, which becomes more explicit after Darwin (see Ch. 4). In this context, Buffon's exaggeration of the lower population density of the Americas, influenced particularly by contemporary accounts of the Guyanese interior, is presented as negative evidence for the "epochal" scale of European environmental impact, the main subject of Epoch VII.

42. Ironically, Buffon's attempt to make his system *more* orthodox by postdating the species backfired, and it was this portion of the text that the Sorbonne condemned.

43. Anderson, "More about Some Possible Sources." This is the last of a series of four articles on the subject, in which Anderson concludes, after surveying several candidates, that the military officer Sonnini de Manoncourt was the most important of Buffon's sources on Guyana (76–79). On scientific "centers of calculation," see Latour, *Science in Action*; for an eighteenth-century application, see David Philip Miller, "Joseph Banks, Empire, and 'Centers of Calculation'"; on the Jardin du Roi in this historiographic context, see Easterby-Smith, "Recalcitrant Seeds."

44. Rudwick, ed., *Georges Cuvier*, 124–25, 125n34.

45. Whitehurst, *Inquiry*, 203.

46. Bataille, *Accursed Share*, 22.

47. By this time, Cuvier's much more sophisticated account of extinction was the one that Lamarck was contesting. See Rudwick, *Bursting the Limits of Time*, 452.

48. In his excursus note on fossil wood and impressions of plants (as well as fish) in slate, Buffon makes an appeal for documentation of these monuments by engravers and close study by botanists, in part to confirm the number of extinctions (*Z* 158).

49. *Critique of Judgment*, 109; cf. *Kritik der Urteilskraft*, 341–42 (B 92–93). Buffon's sentence presents at least one difficulty: "La seule chose qui pourroit être difficile à concevoir, c'est l'immense quantité de debris de végétaux que la composition des ces mines de charbon suppose" (*EN* 93). "Seule" in the sense of "only" may refer to the preceding explanation, which points to the obvious impressions of wood, leaves, etc., associated with coal—hence its origin is easy to grasp, but not the quantity. In an editorial note, Roger points out that this passage is heavily indebted both to Lehmann and to Lehmann's translator and editor, d'Holbach, who introduces the analogy to present-day South American tropical forests (*EN* 293–94n40).

50. This passage is inserted in the left margin of a fair copy in Trécourt's hand identified as "Cahier 3" in MS 883, fol. 22–23 (82–83), Musée national d'histoire naturelle, Paris. On Buffon's information concerning Guyana and the timing of these reports, see Anderson, "More about Some Possible Sources." Roger's edition includes facsimile pages showing the differing hands of Buffon and his secretary, Jacques Trécourt, which both appear in the manuscript. My analysis here is based on a careful study of the original manuscript, but Roger's invaluable critical edition (cited throughout as *EN*) reproduces almost all the unpublished manuscript text in the bottom margins.

51. Flourens, ed., *Des manuscrits de Buffon*, 66–67. Roger outlines this revision in a note to the text of Epoch III (101n.) and then includes the altered and deleted MS text in his notes to the corresponding passage in Epoch I (39–43n.)

52. The opening question was heavily revised (cf. *EN* 35), and other portions of the passage reproduced here (Fig. 2.1) were deleted. A comparable passage, showing the same series of dates, is reproduced in the introduction (Fig. 3). The latter passage was revised more lightly and can be recognized more readily in the printed text (*EN* 60).

53. Playfair, "Biographical Notice of James Hutton," 73. See also Darwin, *Annotated Origin*, 282.

54. Since this date MS 883 includes two sections labeled "cahier 4." The one documented by Roger is actually an earlier draft of Epoch V, however, whereas the more recently discovered *cahier* (found in the municipal archives or *mairie* of the village Buffon) is very close to the published Epoch IV in content and is paginated 1–6, 1–4, and 7–28. The two may be further distinguished by their headings: "4ème Epoque" for the former and "Quatrième Epoque" for the latter.

55. Latour, *Reassembling the Social*, 80; Cohen, *Fate of the Mammoth*, 96.

56. In this note, Buffon extensively quotes his American correspondence and other recent literature on mammoth and mastodon finds, including the journal of George Croghan, who discovered the famous "Ohio animal" in 1765 (*Z* 137–42). Neither of the modern editions reproduces the engravings.

57. Playfair, qtd. in Porter, *Making of Geology*, 1. This contrast is discussed more fully in the introduction.

58. For an overview of these theories, see de Luca, *Words of Eternity*, 180–87.

59. The manuscript entitled "Quatrième époque," as noted above, was acquired by the Musée national d'histoire naturelle in 2000 and integrated with MS 883. The last page of this section reads, in part: "Nous remercierons le Créateur de n'avoir pas rendu l'homme témoin de ces désastres" (fol. 28). In print, "ces désastres" become "ces scènes effrayantes et terribles, qui ont précédé, et pour ainsi dire annoncé la naissance de la Nature intelligente & sensible" (*EN* 138).

60. Here Smellie's English seems to me more idiomatic than the modern translation (*Z* 129). Cf. "ces vingt espèces . . . font plus de bien sur la Terre" (*EN* 216).

61. "Dr. Forster an Prof. Lichtenberg," 144. A substantial part of Forster's review is occupied with a reinterpretation of both ancient and modern sources on the timing and scope of flooding events around the Caspian and Black Seas, offering a learned riposte to Buffon's discussion of some of the same authors and issues (*Z* 88, 104–8).

62. "Dr. Forster an Prof. Lichtenberg," 149.

63. On deep time and epic, see McGurl, "Posthuman Comedy"; Dimock, "Low Epic"; and the concluding discussion of this chapter.

64. McPhee, *Annals,* 50. B. P. Reardon notes that "the qualities of vision (or 'imagination'), generous scale, and sustained description" distinguished romance from epic from a quite early period. Reardon, *Form of Greek Romance,* 129. See also Dean, "*Odyssey* as Romance."

65. Fielding, *Tom Jones,* 414–15 (IX.1); Godwin, "Of History and Romance" (1797), *Philosophical and Political Writings,* V.301. I am indebted for the last point to de Baere's discussion of the "vraisemblable" in *La pensée cosmogonique,* 237–46.

66. This remark appears in his preliminary discourse to the *Researches on Fossil Bones* (1812), as translated by Martin Rudwick in *Georges Cuvier,* 228.

67. Roger transcribes Guettard's manuscript letter in its entirety in the notes to his critical edition of Buffon (*EN* cxxxix–cxln7). Jeff Loveland notes that Buffon replies to such "empirical critiques" of the *Epochs* in his *Natural History of Minerals,* "berating strict empiricists for lacking in genius." Loveland, *Rhetoric and Natural History,* 115.

68. Buffon's *Natural History* was both prestigious and immensely popular, and for this reason its literary reach could be construed as detrimental to its scientific authority, even at a time when there was no strict binary division between literature and science. See de Baere, *La pensée cosmogonique,* 14.

69. Kolbert, *Sixth Extinction,* 46.

70. Kolbert, *Sixth Extinction,* 27. Cuvier's theory is the leitfmotif of Kolbert's "history of the science of extinction" (93). After her sustained initial discussion of Cuvier's theory (28–46), which includes Balzac's praise of Cuvier as a "poet" (38), Kolbert returns many times to the idea that Cuvier was right (94, 226, 265, and passim).

71. On interspecies recognition, see Kolbert, *Sixth Extinction,* 224; for the Herculaneum analogy, see Cuvier, *Essay on the Theory of the Earth,* i.

72. Loveland, *Rhetoric and Natural History,* 95–96. De Baere discusses Buffon's gloss on the creation account of Genesis in the *Epochs* in *La pensée cosmogonique* (224–34) and sums up his literary influences (252). "High romance" is from John Keats, "When I Have Fears That I May Cease to Be" (1818). From a historical perspective, it is worth noting that even Jacques Boucher de Perthes, who made his prehistoric discoveries in the 1840s, was still widely disbelieved because of the same lack of strong archaeological (stratigraphic) evidence or systematic dating. On this point, see Ch. 4.

73. Longinus, "On Sublimity," 9.9, in *Classical Literary Criticism*, ed. Russell and Winterbottom, 152. Kant also refers to the Hebrew Bible, specifically the prohibition on graven images (second commandment) in Exodus 20:4. *Critique of Judgment*, 135 (B 124).

74. Buffon is indebted here to secondary sources including Boulanger (see further below) and possibly Spinoza (*EN* 276n66). The main verb in Genesis 1:2 (הָיְתָה) is in the Qal perfect. See Seow, *Grammar for Biblical Hebrew*, 162. בָּרָא, the verb preceding it, is a famous crux and may be read as a gerund, though Buffon appears not to have known this (cf. *Z* 18/*EN* 20). In short, there is good support for the view that this passage is open-ended with respect to duration.

75. Richet, *Natural History of Time*, 131. Richet, however, provides a valuable chapter on the history of Biblical exegesis in its relation to the age of the world up to the eighteenth century (24–53).

76. Lyell, *Principles of Geology*, 1.5–7, 20. In this chapter, Lyell examines the "Great Year" of the Greek philosophers also discussed in "Dr. Forster an Prof. Lichtenberg," 152.

77. Thomas, *Ode sur le temps*. Cited by line number in the text.

78. *Complete Poetical Works*, 438 (l. 61), 161 (l. 778). On Thomas, see the brief but useful entry by Peter France.

79. Boullanger, *Anecdotes physiques*, 228, 232–34, 213. The extended gloss on Genesis includes a detailed account of each of the six "days" of creation (232–47). Boutin also discusses the plagiarism charge, quoting the primary sources extensively (18–20). Boutin prefers the spelling "Boullanger," but I have retained "Boulanger" in my text because this spelling is typically used in English-language sources. Both spellings were current during his lifetime.

80. Boullanger, *Anecdotes physiques*, 369; Desmarest, "Anecdotes de la nature," 536–37. I owe the reference to Rudwick, *Bursting the Limits of Time*, 451. Rudwick also discusses Boulanger's influence here. Here is Desmarest's French: "La profondeur de l'abîme des temps où notre esprit est par-là obligé de se plonger, paroit si immense, si peu conformé à notre façon de penser qu'il n'est pas étonnant que le plus grand nombre des hommes soit peu disposé à croire à des revolutions, quoique verifiée par plusieurs monumens." See further Anecdote VII on the duration of time (561–62).

81. Richet, *Natural History of Time*, 134, ix.

82. Rossi, *Dark Abyss*, 109–12. Compared to geology, astronomy is the site of much more extensive speculation about the age of the world from a much earlier date, and Hans Blumenberg concentrates on this science in his initial account of "world time" in *Lebenszeit und Weltzeit*. Richet incorporates physics and astronomy with geology and myth in his synoptic account.

83. McGurl, "Posthuman Comedy," 537–38.

84. Dimock, "Low Epic," esp. 615. The journal also published a reply to this reply, which develops further the issue of scale and literary representation: McGurl, "'Neither Indeed.'"

85. Herodotus, *Histories*, 107, 106. Boulanger is the only one of the eighteenth-century figures studied here to take note of this passage, though curiously it is of interest to him for its reference to fossil shells in the hills rather than the timescale. See Boullanger, *Anecdotes*, 216 and 434n196.

Chapter 3. William Blake, the Ballad Revival, and the Deep Past of Poetry

1. See also Genesis 3:8–9, in which the voice of God calls to Adam and Eve among the trees in Eden after they have eaten the forbidden fruit, and the echoes of this passage in John 1:1–4.

2. A somewhat similar argument is made in Jetha and Ryan, *Sex at Dawn*.

3. Friedman, *Ballad Revival*, 3. In Wordsworth this survival takes the form of the recurring "burr" of the nonverbal hero of the ballad "The Idiot Boy" (1798).

4. Burns, *The Best-Laid Schemes*, ed. Crawford and McLachlan, 22. Cited parenthetically hereafter as *BLS*. On Ritson and Burns, see Friedman, *Ballad Revival*, 264.

5. Ritson, "Historical Essay" (hereafter *HE*), i.

6. McLane, *Balladeering*, 75, 132.

7. Philips, ed., *Collection of Old Ballads* (hereafter *COB*), I.iii, I.67.

8. Addison, "Chance Readings" (*Spectator* 85); Aikin, *Essays on Song-Writing*, 26; Friedman, *Ballad Revival*, 5; Green, "Introduction," viii.

9. Friedman, *Ballad Revival*, 270, 264.

10. Ritson's *Select Collection* was published in 1783 and the first three poems in *Songs of Innocence* appeared together in Blake's burlesque manuscript "An Island in the Moon" (1784). *Songs of Innocence* first appeared in 1789, and the complete *Songs of Innocence and Experience* in 1794, the same year as the fourth edition of Percy's *Reliques of Ancient English Poetry*. See also Bentley, "Blake and Percy's *Reliques*."

11. Examples of a tetrameter line (most often the third) being split in early printed ballads include "A Good Throw for Three Maidenheads" (1629), Pepys Ballads I.314–15 (Cambridge University), EBBA 20149. English Broadside Ballad Archive, online at http://ebba.english.ucsb.edu/ballad/20149; "A Servant's Sorrow" (1618), in Philips, ed., *COB*, III.139–44; and "The Greenland Voyage" (curiously georgic but undated), in Philips, *COB*, III.172–75. The implied stanza form in Blake's two poems is a quatrain consisting of one trimeter followed by three tetrameter lines. The dimeter couplets obtained here by splitting the third line constitute the entire stanza in two poems from *Innocence*, "The Ecchoing Green" and "Spring." Blake also uses the typical ballad stanza (4-3-4-3) throughout both *Innocence* ("The Little Boy Lost" and "Found"; "The Divine Image"; "Nurse's Song") and *Experience* ("Nurse's Song" and "The School Boy"). On the musical character of the *Songs*, see Hilton, "What Has *Songs* To Do with Hymns?"; and Hutchings, "William Blake and the Music of the *Songs*."

12. See, among others, *Blake's Poetry and Designs*, 29n7.

13. The word *seed* is included but then canceled in the draft of "Earth's Answer" (19). See *The Notebook of William Blake*, ed. Erdman. I am following Erdman's numbering of the pages in question—N109, N11, and N113, all reversed; the drafts mentioned here are numbered by Erdman as Poem 15 through Poem 20. In his commentary, Erdman argues that the line "free Love with bondage bound" links Poems 17 through 19 and inspired "in a mirtle shade," a lengthy (but ultimately rejected) continuation of "Infant Sorrow," which appears in *Songs of Experience* as a short poem of eight lines. See also Poem 33, "To My Mirtle." In the context of this cluster of draft material, Blake's Earth emerges as an avatar of the maiden in "A Little Girl Lost," a later *Song of Experience* that nods explicitly toward conjectural history: "Children of the future age, / Reading this indignant page, / Know that in a former time, / Love! Sweet love! was thought a crime" (E 29).

14. E. D. Hirsch's argument that "Earth's Answer" is a parody of "Introduction" depends on the premise that the latter was written first. Hirsch, *Innocence and Experience*, 213–14. Since it is neither in the Notebook nor otherwise preserved in draft, this view remains speculative, if intuitively convincing. Much earlier Wicksteed argued for reading this poem as an answer to the

preceding poems in the notebook, and Erdman (citing Wicksteed) believes that "Introduction" may have been composed later, after the Notebook drafts (E 714).

15. In Hirsch's reading, the Bard's anxiety about Earth's "turning" paves the way for outright parody in "Earth's Answer," which reveals his prophecy of Earth rising as an illusion. Hirsch, *Innocence and Experience,* 212–14. See also Lüdeke, *Zur Schreibkunst,* 183–84.

16. Lüdeke, *Zur Schreibkunst,* 187–88, 160, 181.

17. Lüdeke, *Zur Schreibkunst,* 163, 49–50, 163.

18. See, for example, the frontispieces to both *Innocence* and *Experience* and the first several poems in *Innocence* (especially "The Shepherd" and "The Lamb").

19. "The Children in the Wood. Or, The Norfolk Gentleman's Last Will and Testament." National Library of Scotland, Crawford.EB.898. EBBA 33763. English Broadside Ballad Archive, online at http://ebba.english.ucsb.edu/ballad/33763/image.

20. Friedman, *Ballad Revival,* 265

21. Ritson, "Preface," *Select Collection,* I.ii.

22. On Blake and Macpherson, see Folkenflik, "Macpherson, Chatterton, Blake"; on Macpherson's remediation of oral traditional ballads in his Ossian poems, see Gidal, *Ossianic Unconformities,* 19 and passim.

23. Percy, "Essay on the Ancient Minstrels in England," *Reliques of Ancient English Poetry* (hereafter *Reliques*), I.345–48.

24. McLane, *Balladeering,* 119, 123.

25. McLane, *Balladeering,* 123, 124.

26. McLane, *Balladeering,* 71.

27. Burns here uses his preferred mock-heroic stanza, the Habbie stanza, rather than the ballad. It also turns out in the next stanza that Grose has theories about Adam and Abel.

28. Wordsworth, *Lyrical Ballads,* 263, 267.

29. Scott, *Minstrelsy of the Scottish Border* (hereafter *MSB*), 326.

30. Stenhouse, ed., *The Scots Musical Museum,* V.370; Child, *English and Scottish Popular Ballads,* I.335. Child, citing an 1803 letter indicating that Scott himself "suspected that they might be the work of some poetical clergyman or schoolmaster," declares these added stanzas "as unlike popular verse as anything can be." Child identifies eleven stanzas in Scott's version as spurious but does not seem to consider the possibility that Scott may have composed these stanzas himself.

31. Herd, *Ancient and Modern Scots Songs,* iv, 300–302.

32. Child, *English and Scottish Popular Ballads,* I.336; Stenhouse and Johnson, *Scots Musical Museum* V.368, iii.

33. Thus it seems to me that McLane oversimplifies when she refers to a single "ballad chronotope." McLane, *Balladeering,* 71. See also Bakhtin, *Dialogic Imagination,* 84; on "areas of uncertainty," see Latour, *Reassembling the Social,* 21–22.

34. Conard, Maline, and Münzel, "New Flutes."

35. Shelley M. Bennett states that the design for this frontispiece was attributed to Stothard by Samuel Boddington, a patron and early collector of his works. *Thomas Stothard,* 67. (Though this plate is unsigned, Stothard is credited as delineator on many of the others.) On artifacts and deep history, see Smail, *On Deep History,* 59–66.

36. Ritson, *Scotish Song,* I.lxxxi–lxxxii. In the preface to *Select Collection,* Ritson notes that all the songs in that work (unlike his later ones) have been published before (viii–ix). Ritson's

skepticism contrasts forcefully with Child's willingness to entertain more than 2,400 years of unbroken oral transmission in his "Tam Lin" headnote quoted above. When Ritson does turn to oral tradition, he seems deliberately to choose examples that support this degeneration theory, as witnessed by Scott's complaint in his "Remarks on Popular Poetry."

37. On the comparative method, see Stocking, *Victorian Anthropology*, 127, 164–74. One defining Victorian example is Henry Maine's "comparative jurisprudence." Maine, *Ancient Law*, 118. For a broad survey of the many senses of this term, see Griffiths, "Comparative Method."

38. For an extensive discussion of this "Death-Song," see McLane, *Balladeering*, 104–12. "We could say, too, that the invention of tradition emerges precisely as an anthropological, comparative discourse: not only Native Americans, but Scots border-raiders, Morayshire dairy-maids, Highlanders, and African 'Negroes' all appear by 1800 as objects of an elaborate English and Scottish discourse on tradition and mediation—a transnational vernacular cultural imaginary" (112).

39. Herder, "Extracts from a Correspondence," 156–57.

40. Herder, "Extracts from a Correspondence," 157. Herder opposes this valorized "barbarism" to a scientific worldview: "Know, then, that the more barbarous a people is—that is, the more alive, the more freely acting (for that is what the word means)—the more barbarous, that is, the more alive, the more free, the closer to the senses, the more lyrically dynamic its songs will be. . . . The more remote a people is from an artificial, scientific manner of thinking, speaking, and writing, the less its songs are made for paper and print, the less its verses are written for the *dead letter*" (155–56, emphasis added). Herder proceeded to choose the same epigraph from Persius for his *Ideas* that Rousseau had used for his second *Discourse* a generation earlier.

41. On Herder and Enlightenment anthropology, see Zammito, *Kant, Herder, and the Birth of Anthropology*; and Wolff and Cipolloni, eds., *Anthropology of the Enlightenment*.

42. On the long consensus around degeneration theory, see Ian Duncan, "Our Mutual Friend," 290–92.

43. Herder, "Von Ähnlichkeit," 44. The rejection of European travel narrative as a reliable source on indigenous culture provides the anticolonial moment in this passage (43–44). On Herder's critique of colonialism, see Zhang, *Transculturality*, 119–60; and Spencer, *Herder's Political Thought*, 185–214. Scott comments that the German translation of Percy's version of Sir Patrick Spens is one of the best things in "the Volks-Lieder of Professor Herder, in which it is only to be regretted that the actual popular songs of the Germans form so trifling a proportion" (*MSB* 98).

44. Herder, "Von Ähnlichkeit," 34.

45. Herder, "Von Ähnlichkeit," 44–45.

46. Herder, "Von Ähnlichkeit," 44.

47. Herder, "Von Ähnlichkeit," 43–44; McLane, *Balladeering*, 112.

48. Inasmuch as the earth speaks in biota, the animal population of this area—here roughly the area between the Black Sea and the Caspian Sea, though elsewhere Herder seems to refer to areas farther to the east—also speaks of human origins, according to a zoological account by Peter Simon Pallas, which asserts that all the usual domestic animals are found in their wild state "in the mountain chains of central Asia" (qtd. in *I* 387).

49. *I* 387–88, 392. Herder may be alluding to the idea that the Mediterranean was produced by a megaflood originating in the Black Sea basin, attributed to Strabo and other ancient geographers by J. R. Forster in "Dr. Forster an Prof. Lichtenberg."

50. I am using *kind* to translate Herder's *Gattung* at the end of this sentence. *Gattung* tends to be associated with genus rather than species, but since Herder is explicitly arguing for

monogenesis here, *kind* seems appropriate. *Pesserai* is a term (typically pejorative) used by Europeans to describe the peoples of Tierra del Fuego, based on their misunderstanding of a native word.

51. See, among others, Wolff, "Discovering Cultural Perspective."

52. Macpherson, *Fragments,* iii.

53. The term is Philip Connell's. See his "British Identities," 164.

54. According to the *Digitales Wörterbuch der deutschen Sprache* (dwds.de), *Vorzeit* occurs occasionally before the eighteenth century but only becomes frequent after 1770. German is richer than English in synonyms for prehistory, including both *Vor-* and *Urzeit* and *Vor-* and *Urgeschichte* as well as the adjective *prähistorisch,* which became current in the 1870s. On the evolving distinction between *Vor-* and *Urzeit* and its disciplinary ramifications, see Kasper, "Urwelt und Altertum."

55. Macpherson, *Temora,* 132n. Ironically, this particular work is entirely without a basis in tradition. See Gidal, *Ossianic Unconformities,* on the trope of "proleptic" or "projective self-eulogy" (109, 183) and on Herder's instantiation of it (25–27).

56. On this connection, see, among others, Duncan, "Pathos of Abstraction." Gidal notes that even Adam Smith himself "took Ossianic poetry as evidence of the very stadial theories that had informed Macpherson's project." Gidal, *Ossianic Unconformities,* 52.

57. Leask, "Sir Walter Scott's *The Antiquary,*" 197. For Percy's view, see the preface to his *Five Pieces of Runic Poetry,* vi.

58. *Herders Werke,* II.289; cf. Herder, *Volkslieder,* 226–27.

59. David Hume's similar interest in the empirical verification of ballad texts stems from the opposite motive of skepticism about the authenticity of the Ossian poems. See Grobman, "David Hume."

60. *Lyrical Ballads,* 172.

61. *Lyrical Ballads,* 425–26.

62. *Lyrical Ballads,* 174; Chapais, *Primeval Kinship.*

63. McLane, *Balladeering,* 79, 75, 72–73. McLane also recognizes the imprint of conjectural history in Wordsworth's account of "shepherd sociability" in the Lake District, similarly threatened by extinction (204–5), but her account of Wordsworth's own poetic practice as foregrounding "speech in the now" (168) may be usefully contrasted with the ambivalent practice of Scott as a "native" and "fieldworker."

64. Here the scale shifts again from the prehistorian's perspective to that of the journalist (Philips was the founding editor of the Whig paper *The Free-Thinker*), and the apparent historical distance of the object (as in parallax) shifts with it.

65. The "genuine antique poem" nonetheless turns out to be no older than the 1440s (by Percy's own admission).

66. Cohen, *Social Lives of Poems,* 159, 163.

67. Ritson, "Preface," *Select Collection,* I.ii; Aikin, *Essays on Song-Writing,* 27, 33. Ritson makes similar claims in *HE* lxx–lxxi.

68. *Herders Werke,* II.97; Herder, "Von Ähnlichkeit" 35.

69. *Herders Werke,* II.296. In the note to a Lithuanian folksong included in *Volkslieder,* Herder quotes his mentor Hamann's ethnographic explanation for the meter of the Homeric epics, based on his encounters with ethnic Lithuanian farm workers in the Prussian countryside. Hamann reported an epiphany that occurred to him while listening to the Lithuanian work songs: the

"repetitive cadence" of these songs resembled Homer's "monotonous meter," suggesting the proximity of the Homeric meter to oral tradition (279–80).

70. Gidal, *Ossianic Unconformities*, 34–36. Gidal traces the evolution of this approach well into the nineteenth century (ch. 3). Mathias notes Wood's particular influence on the passage quoted previously from Herder. *Herders Werke*, II.522.

71. Addison, "On Popular Poetry" (*Spectator* 70); cf. *Spectator* 74, *Spectator* 85.

72. Wood, *Essay on the Original Genius and Writings of Homer* (1769; 1775), cited in Gidal, *Ossianic Unconformities*, 36. On Wood's ethnographic approach, see also Sachs, "On the Road."

73. "Editing Robert Burns for the 21st Century," available at https://burnsc21.glasgow.ac.uk/the-jolly-beggars/; *Best-Laid Schemes*, 86n. The poem (or cantata) remained unpublished during Burns's lifetime, and the earliest printed versions omitted this note. The publication history is fully explained by author of the website, Kirsteen McCue, in her edition of *Robert Burns's Songs for George Thomson*, 573–80.

74. Wordsworth, *Prelude*, 162 (1805: V.203ff.).

75. Koselleck, "Does History Accelerate?"; Blumenberg, *Lebenszeit und Weltzeit*.

76. Bakhtin, *Dialogic Imagination*, 16.

77. Mayhew, *London Labour*, 285.

78. Friedman, *Ballad Revival*, 6; McLane, *Balladeering*, 43.

79. For animals, see also "Bonny Birdy" (Child #82) as well as "The Birds Harmony," discussed in Palmer, "Picturing Song Across Species." For inorganic speakers, compare "The Twa Sisters" (Child #10) and "The Jew's Daughter" (Child #155), among others.

80. Dimock, *Through Other Continents*; Puchner, *Written World*. McGurl invokes the "original geological meaning" of deep time in "The Posthuman Comedy," 538.

81. Cuvier, *Recherches*, I.101.

82. In the long middle section of *Lebenszeit und Weltzeit* (71–312), his genealogy of the divergence of lived historical time (*Lebenszeit*) from cosmic time (*Weltzeit*) during the Enlightenment, Blumenberg shows how this expansion occurs simultaneously on multiple scales.

83. The quotations are from the "Introduction" to *Songs of Experience* (E 18). Nonetheless, Blake's view of the bard comes closer to Wordsworth's sense of poems as "powers, / for ever to be hallowed" (*Prelude* V.219–20) than to Burns's or Scott's views of the matter.

84. *Complete Poetry and Prose*, ed. Erdman, 127. *Milton* (*M*) and *Jerusalem* (*J*) are cited parenthetically hereafter by plate and line as well as by Erdman page number (here *M* 28:62–29:3/E 127).

85. Buffon, operating in another rhetorical register, relies on figures of compression in attempting to persuade readers to consider the history of the earth on a scale of tens of thousands rather than thousands of years.

86. Locke, *Essay*, 181 [II.xiv].

87. See Whitehead, "'Went to See Blake.'"

88. *Jerusalem*, 272; De Luca, *Words of Eternity*, 179–91.

89. Whitehurst, *Inquiry*, 273, 257–58.

90. Whitehurst, *Inquiry*, 272. Whitehurst, then, had technical expertise that might have supported his refusal to separate human and natural time. For Blake, natural time remains sharply distinguished from eternity as the domain of the human spirit. He depends nevertheless on vivid geological figures, such as the sinking of Atlantis, to articulate this distinction; he also depends on rhythmic actions and affections of the body, such as walking and pulsation, to negotiate the location of time in eternity.

91. The main burden of Erin's long speech is the distinction between states and individuals that is so central to the moral and temporal universe of *Jerusalem*. "Learn therefore," she exhorts her listeners, "to distinguish the Eternal Human / . . . / . . . from those States or Worlds in which the Spirit travels" (*J* 49:72–74/E 199). If a state is an Eternal's manifestation in time, then even a moment—like the more famous grain of sand in "Auguries of Innocence"—can extend indefinitely.

92. Cf. Amanda J. Goldstein's account of the production of human anatomy as "bilateral figuration" in *Milton*. Goldstein, *Sweet Science*, 89.

93. The youth and the maiden in "Ah, Sunflower! Weary of Time," from *Songs of Experience*, mistakenly regard time as a barrier to the metaphysical beyond to which they aspire. Los's act of walking here dissolves that barrier. Rallying a misguided Christianity in "To the Christians," the *Jerusalem* narrator observes that "We are told to abstain from fleshly desires that we may lose no time from the work of the Lord. Every moment lost, is a moment that cannot be redeemed" (*J* 77: 11–12). But even publicans and harlots have "mental gifts," Blake argues in this provocative passage, and it is fallacious to claim that their spiritual life in eternity can be diminished either by acts or by moments. Blake's concern is not so much to uphold the Christian dualism of time and eternity as to show that the progress through time is an eternal act.

94. Likewise, the scale of the earth's body is no longer commensurate with the early modern or "classical" paradigm of the macrocosm; the scale of modern geology is too large for Athanasius Kircher's or Thomas Burnet's analogies to stand (with the possible exception of Erasmus Darwin's graphic "Section of the Earth" in *The Botanic Garden*).

95. Darwin, *Annotated Origin*, 310, 282.

96. Flourens refers to a single sentence in the manuscript for *Epochs of Nature* in which the original figure of four or five thousand years is crossed out and replaced with one million years, which in turn is replaced by the equivalent ten thousand centuries, and then with one thousand centuries; 75,000 years is the figure that finally appears in print (*EN* 41). See Flourens, *Des manuscrits de Buffon*, 69. See also his *Examen du livre du M. Darwin*.

97. On *chronos* and *kairos* specifically, see Tannenbaum, *Biblical Tradition*, 86–87.

98. Dimock, "Nonbiological Clock," 167; Goldsmith, *Blake's Agitation*, 232–34. I owe the Dimock reference to Goldsmith as well. Cooper makes the helpful point that Blake "situates Eternity at the interface of every passing moment as an infinitely deep recession arising out of chronological time's self-contradictory indefiniteness and instability." Cooper, *William Blake and the Productions of Time*, 19.

99. Klancher, *Transfiguring the Arts and Sciences*, 13. On the inseparability of "invention" and "execution," see Blake's "Public Address to the Chalcographic Society" (E 560–71) and important discussions of this issue in Barrell, *Political Theory of Painting*, ch. 3; and Eaves, *Counter-Arts Conspiracy*, ch. 4. On Blake and the body of knowledge, see further Heringman, "Primitive Arts and Sciences."

100. Tannenbaum, *Biblical Tradition*, 87–88. Baulch provides a helpful gloss on Blake's use of *kairos*, calling it an "interval of indefinite time." Baulch, "For in This Period," 5.

101. *Blake: The Complete Poems*, ed. Stevenson, 676; Gilpin, "William Blake and the World's Body," 43–44; Whitehurst (citing Bacon), *Inquiry*, 273.

102. Whitehurst, *Inquiry*, 28; Forster, *Observations*, 342. For the importance of the ancient wisdom doctrine in Erasmus Darwin, see Priestman, *Poetry of Erasmus Darwin*, 139–67.

103. Koselleck uses a geological metaphor, *Sattelzeit*, to contextualize this process of bridging the distance between early modern and modern concepts. See his "Introduction" to *Geschichtliche Grundbegriffe* (xv, xxi) and see my introduction.

104. Joppien and Smith, *Art of Captain Cook's Voyages*, II.110. Blake's professional engravings in this vein include his images for John Gabriel Stedman's *Narrative of a Five Years' Expedition* (1796) and Plate 14 of John Hunter, *An Historical Journal of the Transactions at Port Jackson and Norfolk Island* (1793).

105. Goldstein, *Sweet Science*, 86, 89.

106. See especially his account of time "gained" for the past ("Zeitgewinn für die Vergangenheit"). *Lebenszeit und Weltzeit*, 224.

107. Ballard, *Drowned World*, 56, 89. On evolution and the arts, see Dissanayake, *Homo aestheticus*.

Chapter 4. The Descent into Deep Time in Darwin and Lubbock

1. Lubbock was, from 1864, a founding member of the X Club. Barton accentuates the importance of Lubbock's "social position" in the consolidation of power and authority by this group in her study *The X Club*, a valuable source on Lubbock and his milieu. See Barton, *X Club*, 2. For more on the "prehistoric turn," see my introduction.

2. Chakrabarti, *Inscriptions of Nature*, 6–8. Chakrabarti offers a useful overview of the disciplinary origins of prehistory.

3. Green, "Introduction," viii.

4. On the comparative method in Victorian social science, see Stocking, *Victorian Anthropology*, 164–69. Darwin himself might have described his innovation as an "evolutionary synthesis" of natural history and the new social science, but I am avoiding this term here because of its strong association with twentieth-century evolutionary biology.

5. Fabian, *Time and the Other*, 16–18, 32, 74–75. I am indebted to previous scholarship on Darwin for many insights into this fascinating scene of encounter, especially Chapman, *European Encounters with the Yamana People*; Duncan, "Darwin and the Savages"; Schmitt, *Darwin and the Memory of the Human*, ch. 1; and Buckland, "'Inhabitants of the Same World.'" Detailed engagement with these sources follows in the body of the chapter.

6. For the debate, see *Critical Inquiry* 29 (Spring 2013): 614–38, and the more detailed discussion in Ch. 2.

7. Duncan, "Darwin and the Savages," 19, 23, 25; Buckland, "'Inhabitants of the Same World,'" 220. Darwin refers to "geological time" in *The Origin of Species* (e.g., *Annotated Origin*, 365). The *OED* records Thomas Carlyle's strictly historical use of "deep time" in 1832 as the first instance, but scientific use of it dates to the twentieth century. For discussion, see my introduction.

8. For background on Tylor, see Logan, *Victorian Fetishism*; on Lubbock, see, again, Barton, *X Club*.

9. Darwin, *Annotated Origin*, 310.

10. On the powerful afterlife of conjectural history in the nineteenth century, see Palmeri, *State of Nature, Stages of Society*.

11. Smail and Shryock, *Deep History*, 5–6, 23–24. See also the pioneering article by Spears, "Evolution in Context."

12. On Lapeyrère, see Schnapp, *Discovery of the Past*, 222–24, 228–31; on Boulanger, see Roger, *EN* xxiv–xxvi; on Giovene, see Toscano, "Figure of the Naturalist-Antiquary."

13. See, respectively, *Journal of Researches*, 220, and "On the Distribution of the Erratic Boulders," 417. On these boulders, see further Edward B. Evenson et al., "Enigmatic Boulder Trains."

14. Tylor, *Origins of Culture*, 60. See also Tylor, *Researches into the Early History of Mankind* (1865).

15. Desmond and Moore survey these developments in their introduction (*DM* xxxvii–xxxviii). See further van Riper, *Men Among the Mammoths*, 139–43.

16. Desmond and Moore critique the distinction between Darwinism and social Darwinism as anachronistic (*DM* liv–lv).

17. Degérando, qtd. in Fabian, *Time and the Other*, 7; see also 16, 18, 75.

18. Stocking, *Victorian Anthropology*, 171, 302–4. At their most clear-sighted, the Enlightenment travelers departed from both the philosophical idiom of the noble or ignoble savage and from the early explorers' topos of indigenous people as belated ancient Greeks or Celts. The Forsters, as I argued in Chapter 1, were sensitive to the geological context that rendered their fieldwork destabilizing to human time. The comparative fixity of "savage" and "civilized" time in Darwin's *The Descent of Man* and (especially) in Lubbock's *Pre-Historic Times* results in part from their commitment to the nineteenth-century pseudo-science of human "races." On the relationship between conjectural history and racism, see Palmeri, *State of Nature, Stages of Society*, 281–86.

19. Palmeri, *State of Nature, Stages of Society*.

20. Fabian, *Time and the Other*, 75, 12; Darwin, *Journal of Researches*, 220.

21. Schmitt, *Darwin and the Memory of the Human*, 36, 41, 42. Concerning the Fuegians, Schmitt observes that "seeing them causes Darwin to indulge in the quintessentially Victorian speculation that he might be face to face, if not literally with his own ancestors, at least with people very like them" (42). Schmitt's "at least" marks another slippage in meaning.

22. Schmitt, *Darwin and the Memory of the Human*, 46. Darwin recycles this trope in the very last chapter of *The Descent of Man* (676) when he says that anyone who doesn't want to look at nature as a savage does—that is, as a set of disconnected phenomena—should accept evolutionary theory (and human descent from animals).

23. Lawson, *Darwin and Evolution for Kids*, 46–47.

24. Schmitt, *Darwin and the Memory of the Human*, 48.

25. Radick, "Did Darwin Change His Mind about the Fuegians?" 54.

26. Darwin, *Journal of Researches*, 213. The reference to animality (*dem Thiere näher*) is missing from Forster's corresponding English text (*V* II.630–31).

27. Thomas, "Introduction" (*O* xxx); Fabian, *Time and the Other*, 18.

28. Hulme, "Cast Away"; Burnett, "Savage Selection." Most of Darwin's encounters took place with the Yaghan (Yámana) people, who inhabited the southern parts along the Beagle Channel, but other population groups sometimes referred to in his and other voyagers' discussions include the Selk'nam (Ona) in the northeast, the Mannenkenk or Haush on the main island, and the Alakaluf to the west. See McEwan, Borrero, and Prieto, eds., *Patagonia*, 7–8, 147–50.

29. See, respectively, Pigafetta, *First Voyage*, 13; and Columbus, *Four Voyages*, 55. Lubbock cites the voyage of Adolph Decker in 1624 to similar effect (*PT* 525).

30. In the "aborigines" chapter of his *Notes on the State of Virginia*, Thomas Jefferson stresses the linguistic and cultural diversity of the forty tribes recorded at the time of first contact and their political organization in small societies. Their diversity, Jefferson writes, "proves them of a greater antiquity than those of Asia" (102), while their small, self-governing societies may be recoded in republican terms as more rather than less evolved (93). Both points allow him to attack Buffon's thesis that humans and other animals degenerated as they crossed over from the Old into the New World. Darwin's conclusion on this point is strikingly antithetical: "The perfect equality among the individuals composing the Fuegian tribes, must for a long time retard their civilization." Darwin, *Journal of Researches*, 229.

31. Harley, "Animal Encounter," 5; Darwin, *Journal of Researches*, 213. In her *Autobiologies*, Harley argues that sympathy with nonhuman animals is a central motif from the Beagle *Diary* that Darwin revisits in *The Descent of Man* and its companion volume, *On the Expression of the Emotions in Man and Animals* (66–67).

32. Darwin, *Journal of Researches*, 216. Darwin's "ethnological meditation" is positioned near the beginning of his narrative about Tierra del Fuego, but as Chapman points out it is prompted by an encounter at Wollaston Island that actually took place over a year later. Chapman, *European Encounters*, 175. As Chapman shows, he rearranges events in the published text to begin with a retrospective survey comprising some direct observation along with reading and hearsay. The ethnological tone resurfaces in the more chronological narrative that follows, as when he designates the Yaghan as "savages of the lowest grade" in narrating the events of 21 January 1833. Darwin, *Journal of Researches*, 220.

33. Radick, "Did Darwin Change His Mind about the Fuegians?"; Schmitt, *Darwin and the Memory of The Human*, 42.

34. 11 April 1833, "Letter no. 204."

35. See, among others, Gohau, "Darwin the Geologist."

36. Buckland, "'Inhabitants of the Same World,'" 223.

37. Buckland, "'Inhabitants of the Same World,'" 229, citing Lyell, *Principles*, I.165–66 and I.76. My own reference to the latter passage is meant to signal the very rapid movement of Lyell's thought experiment here from ancient Egyptian "superstitions" to those that impeded the seventeenth-century Italian naturalist Falloppio. See Lyell, *Principles of Geology*, I.76–78.

38. Buckland, "'Inhabitants of the Same World,'" 226, 228.

39. Buckland, "'Inhabitants of the Same World,'" 233–34, 236. Buckland's insight here has strong potential for critical application to Anthropocene thought, toward which she orients her historical argument in this essay. "Colonial politics," she argues, "remains embedded at the heart of attempts to comprehend the scale of human significance in relation to the earth" (222).

40. Lyell, *Principles of Geology*, vol. 1, ch. iii–v; Palmeri, *State of Nature, Stages of Society*, 163–78.

41. William Whewell, in the addresses he gave as president of the Geological Society in 1838–39, participated in Lyell's triumphalist rhetoric. In his own philosophy of science, however, he was surprisingly conscious of the gradual and uneven nature of paradigm change in science. See Whewell's "Address[es] to the Geological Society" as well as "On the Transformation of Hypotheses" (1851).

42. Koselleck, "Einleitung," xv. Koselleck here defines the *Sattelzeit* as a period during which terms such as *democracy, republic, revolution,* and *history* went through a discernible "process of

translation." Though the translation of the abyss of time into "deep time" is a more gradual process, Koselleck's figure helps us to see the continuity from mid-eighteenth century geognostic fieldwork (which I linked to *Sattelzeit* in the introduction to this book) up through Darwin's fieldwork on the *Beagle* and beyond.

43. Compare G. B. Cipriani's highly idealized image of the occupants of a similar Fuegian "wigwam" in Hawkesworth, *Account of the Voyages* (1773), and the more faithful rendering of the same comparatively naturalistic drawing by Sydney Parkinson engraved by T. Chambers for Parkinson's *Journal of a Voyage to the South Seas* (1773), reproduced in Heringman, *Sciences of Antiquity*, 39–42. See also Meek, *Social Science and the Ignoble Savage*.

44. Chapman, *European Encounters*, 234–36, xxi.

45. Desai, "Reverse Evolution and Evolutionary Memory." Schmitt outlines a very pervasive nineteenth-century trope of species memory that "permeates the whole body," a sort of race memory that he depicts Darwin as resisting on one level but embracing on others. Schmitt, *Darwin and the Memory*, 23 and passim.

46. Darwin, *Expression of the Emotions*, 257.

47. On Lubbock and Darwin, see Pearn, "The Teacher Taught?"

48. Darwin, *Journal of Researches*, 215–16.

49. Burnett, "Savage Selection," 123.

50. *Poetics* [1452a], 96–97; the "astonishment" produced by the "best" kind of recognition (1455a) resembles the shock of surprise evoked by Darwin.

51. Stocking, *Victorian Anthropology*, 169.

52. Blumenberg, *Lebenszeit und Weltzeit*, 224.

53. Van Riper organizes his detailed study of the Victorian origins of prehistory around this topos. See especially his discussion of the discoveries, debates, and visits to Boucher de Perthes in France that established the consensus presupposed by Lubbock's *Pre-Historic Times*. Van Riper, *Men among the Mammoths*, 74–134. See also Gamble and Kruszynski, "John Evans, Joseph Prestwich, and the Stone That Shattered the Time Barrier."

54. Though not exactly a writer of popular science, Lubbock did see himself as offering a synthesis of archaeological research, and his books contributed to the popularization of a broad range of archaeological and evolutionary ideas. The popularizer William Houghton dedicated his *Gleanings from the Natural History of the Ancients* (1879) to Lubbock. See further Lightman, *Victorian Popularizers*.

55. Mithen, *After the Ice*, 5–6 and passim.

56. Lubbock was created Baron Avebury in 1900, partly in recognition of his efforts to preserve that site in Wiltshire. In *Pre-Historic Times* he describes this monument as older than Stonehenge, assigning it "either to the close of the Stone Age, or to the commencement of the Age of Bronze" (124).

57. Pearn notes that Lubbock added the explicit defense of Darwin here to the second edition of 1869, helping to draw Darwin's attention to the relationship between prehistoric extinctions and human evolution as the latter was writing *The Descent of Man*. Pearn, "Teacher Taught," 10.

58. Cf. Lyell, *Geological Evidences of the Antiquity of Man*, 184–85. Lubbock felt that Lyell had asserted a false priority claim and perhaps even plagiarized his work on human prehistory. See further the Darwin Project, "The Lyell-Lubbock Dispute."

59. He notes that the Geological Society has declined to acknowledge or publish very recent evidence on the subject, suggesting in effect that recent discoveries at Brixham Cave and elsewhere were being suppressed (316–17). Here too the emphasis falls on the very recent turn in the climate of ideas that finally makes these considerations possible.

60. The article is signed only "H.D.R." The title makes the author's bias clear enough: "The Reputed Traces of Primeval Man."

61. Rudwick, "Emergence of a Visual Language." It is notable that Lubbock provides additional quantitative measures through statistics and statistical tables showing the numbers and types of bones found at a variety of sites, the number of game animals needed to support a hunter-gatherer population in a certain area, and other such parameters. Many of the tables were added to the second edition of 1869, which is the one cited here and also the one used by Darwin for *The Descent of Man*.

62. Lubbock promises that his prehistory, unlike others, will not be "written in the spirit of the novelist" (1), but this is easier said than done.

63. Desmond and Moore, *Darwin's Sacred Cause*.

64. Lévi-Strauss dismantles these prejudices, many of which held sway through the mid-twentieth century, in *The Savage Mind*.

65. Wallace, "Origin of Human Races," clxv.

66. Wallace, "Origin of Human Races," clxvi–clxvii, clxix.

67. Schmitt makes a similar point about this passage in *Darwin and the Memory of the Human*, 49.

68. Darwin, *Journal of Researches*, 216. As Buckland points out, these "powers" include the imagination. Buckland, "'Inhabitants of the Same World,'" 231.

69. Pearn, "The Teacher Taught?" 13–14.

70. Pearn, "The Teacher Taught?" 14–15.

71. Pearn, "The Teacher Taught?" 14. Pearn presents evidence of Darwin's worry that Lubbock might have been right after all, but I believe that she overstates Lubbock's influence somewhat, both here and in the case of the Miocene ancestor. Though it was extremely influential in Lubbock's lifetime, *Pre-Historic Times* "had all but disappeared from the canon of archaeological literature" by 1910. See Clark, "Science of John Lubbock," 4.

72. Throughout this section, Darwin is challenging (and/or supplementing) evidence advanced on behalf of original "communal marriage" by Lubbock, McLennan, and Lewis Henry Morgan (e.g., 655n5).

73. For a general account of cultural evolution, see Henrich, *Secret of Our Success*; on counting, Beller and Bender, "Limits of Counting." For a view of *The Descent of Man* in its historical context, see Bowler, *Theories of Human Evolution*.

74. Duncan, "Natural Histories of Form," 60, 52.

75. Palmeri, *State of Nature, Stages of Society*, 166–67. Palmeri gives evidence from the notebooks that Darwin was reading eighteenth-century conjectural histories by Bernard Mandeville and David Hume around 1840, when some of the major ideas for *The Descent of Man* were first drafted, and that he revisited these works along with similar ones by Adam Smith and Lord Monboddo in the years after 1866, when he was planning and writing *The Descent of Man* (169–71). Palmeri's introductory chapters offer useful surveys of common formal properties of conjectural histories, including the reference to what "probably" or "must have" happened (13–14, 34–36, and passim).

76. Menely, *Animal Claim*, 1. For Darwin's views on the universal intelligibility of gestures, facial expressions, and "cries," see also *DM* 208.

77. Duncan, "Natural Histories of Form," 64. For my purposes, music offers the clearer illustration of a "wholly natural origin of art."

78. In her pioneering critique, "Darwin and the Descent of Woman," Evelleen Richards rightly notes that until fairly recently Darwin's sexism was historically relativized by scholars, with "Darwin himself absolved of political and social intent and his theoretical constructs of ideological taint" (58). Richards begins her more recent *Darwin and the Making of Sexual Selection* with a chapter on Darwin and the Fuegians.

79. The most notorious example is in Monboddo, *Origin and Progress of Language*, I.289–90. Palmeri cites notebook evidence showing that Darwin revisited this work while preparing the second edition of *The Descent of Man* and noted Monboddo's reference to the idea that "the first language among men was music." Palmeri, *State of Nature, Stages of Society*, 171; cf. *DM* 639n40.

80. On this issue, see Desmond and Moore, *Darwin's Sacred Cause*, 366–67.

81. It is notable that Darwin uses the "sweetness" of women's voices as indirect evidence for the greater humanity of the "semi-human progenitor" as opposed to the modern savages who "value their women merely as useful slaves" (*DM* 639). Although these discussions are marked by Darwin's characteristic ambivalence, Duncan argues that aesthetics in *The Descent of Man* takes on the role of a core human inheritance that Darwin sees as leveling racial and ethnic differences. Duncan, "Natural Histories of Form," 64.

82. The passage refers to "two or three anthropomorphous apes," and Darwin added a note to the second edition citing the discovery in 1872 of the somewhat younger Miocene biped Oreopithecus (182n17). The *Danuvius guggenmosi* discovered in Swabia in 2019 is the earliest of the three, adding to the ambiguity surrounding the African or European origin of the earliest anthropoid bipeds. See Böhme et al., "New Miocene Ape."

83. For a critical analysis of the deceptive correlation between the categories of ancestral and primitive, or lowest and lowliest, see Crisp and Cook, "Do Early Branching Lineages Signify Ancestral Traits?"

84. Crisp and Cook critique the fallacy of understanding one living organism as more ancient or ancestral than another: "whereas character states can be relatively ancestral (plesiomorphic) or derived (apomorphic), these concepts are nonsensical when applied to whole organisms." Crisp and Cook, "Do Early Branching Lineages Signify Ancestral Traits?" 122.

85. Beer notes the structuring presence of an "awareness of the unfathomable past" that is "one of the traits of Darwin's writing" and points out its particular influence on the fiction of Thomas Hardy. Beer, *Darwin's Plots*, 41.

86. Ballard, *Drowned World*, 56–57, 89.

87. Reproduced in Rudwick, *Scenes from Deep Time*, 48–49.

88. On race and colonialism in *The Drowned World*, see Gandy, "J. G. Ballard and the Politics of Catastrophe"; and Matharoo, "Fata Morgana at the End of the World." Other notable examples of the descent into deep time as a romance motif include Jules Verne's *Journey to the Center of the Earth* (1864) and Arthur Conan Doyle's *The Lost World* (1912), both featuring Mesozoic survivals.

89. Buckland, "'Inhabitants of the Same World.'" On Darwin and race, see the essays collected in Sera-Shriar, ed., *Historicizing Humans: Deep Time, Evolution, and Race,* especially those by Radick and Koditschek.

90. Palmeri, *State of Nature, Stages of Society,* 281–86.

91. The equivalent terms in French and German, *sauvage* and *Wilder,* are never as fully invested with pejorative associations as the English "savage." In *La pensée sauvage,* Lévi-Strauss debunks the ethnographic myth of "savages'" incapacity for abstract thought by positing a "science of the concrete." He also corrects the ethnographic assumption that "savage thought" represents an "anterior" historical or developmental stage, arguing instead that savage and "cultivated" thought "co-exist and interpenetrate." Lévi-Strauss views art, across cultures, as a "zone in which savage thought . . . is relatively protected." Lévi-Strauss, *Savage Mind,* 219. The point of the passage is that there is no such thing as a "savage mind," in light of which the English title seems a bit of a misnomer.

92. Brantlinger, *Dark Vanishings,* 164–88.

93. Buckland, "'Inhabitants of the Same World,'" 234–35.

94. Darwin, *Journal of Researches,* 25. On Darwin's opposition to slavery throughout his life, and as a motive for writing *The Descent of Man,* see, again, Desmond and Moore, *Darwin's Sacred Cause.*

95. To underline this point, Palmeri draws an intriguing connection between Darwin's use of the comparative method and Nietzsche's approach in *The Genealogy of Morals. State of Nature, Stages of Society,* 169–72, 191–92.

Afterword: Evolutionary Nostalgia and the Romance of Origins

1. NSF Award 0928053 (Dix Pettey, Principal Investigator).

2. Rossi, *Dark Abyss of Time,* vii.

3. Gee, *In Search of Deep Time,* 34, 32–33.

4. Gee, *In Search of Deep Time,* 5.

5. Lyell, *Principles of Geology,* I.4.

6. Lyell, *Principles of Geology,* 80, 79, 16.

7. Lyell, *Principles of Geology,* 20, 8.

8. See Bartholomew, "Lyell and Evolution," esp. 296–303. "Letter no. 4106" (Darwin) is the first of many letters to Lyell in which Darwin hints at his disappointment over Lyell's reluctance to apply evolutionary theory to humans. This same letter of 18 April 1863 also touches on the controversy that resulted in Lubbock's plagiarism charge against Lyell in the preface to the first edition of *Pre-Historic Times.* See further Lyell, *Geological Evidences of the Antiquity of Man,* ch. 24, "Bearing of the Doctrine of Transmutation on the Origin of Man and His Place in the Creation." Lyell prefers the term "post-Pliocene" for the earliest hominid fossils (6), those associated for him with the "glacial period."

9. Lyell, *Principles of Geology,* I.8–9.

10. Lyell, *Principles of Geology,* 22–24.

11. Lyell, *Principles of Geology,* 69, 73, 74. Lyell's idea of signs being more approachable than numbers recalls Buffon's rhetorical question concerning the mind's tendency to "lose itself" more readily in the expanse of duration than in other quantitative domains (*Z* 36). In his history

of geological ideas, Lyell gives Buffon a limited amount of credit for offending the Sorbonne by proposing younger "secondary" continents. Lyell, *Principles*, I.48–49.

12. Rudwick, *Scenes from Deep Time*, 255n1.

13. Gould, *Time's Arrow, Time's Cycle*, 3; Freud, "A Difficulty in the Path of Psycho-Analysis" (1917), qtd. in Beer, *Darwin's Plots*, 12–14; Peter Schnyder, "Geologie."

14. Archer, *Long Thaw*, 2.

15. Bjornerud, *Timefulness*, 7, 17. In the latter passage "timefulness" is defined as "a feeling for distances and proximities in the geography of deep time."

16. Dames, *Amnesiac Selves*, 36.

17. Wright, *Swan Book*, 178–79, 187–88.

18. Chatwin, *Songlines*, 23, 15. For the evolutionary excursus, see 264–85.

19. Heinrich, *Why We Run*; McDougall, *Born to Run*. As noted above, the Raràmuri (Tarahumara) of Chihuahua play a role in these accounts analogous to the role of the Pintupi nation in Chatwin. McDougall, *Born to Run*, passim; Heinrich, *Why We Run*, 168–75.

20. Bjornerud, *Timefulness*, 16; McGurl, "Posthuman Comedy" (discussed in Chapter 2); Marshall, "What Are the Novels of the Anthropocene?"

21. Bjornerud, *Timefulness*, 16; Heinrich, *Why We Run*, 9–10. In addition to the books by Chatwin, Heinrich, and McDougall, I am also alluding here to works such as Wolff, *Paleo Solution*; and Ryan and Jetha, *Sex at Dawn*. For a thoughtful critique of these archaizing trends, see Zuk, *Paleofantasy*.

22. Heinrich, *Why We Run*, 163; Bjornerud, *Timefulness*, 17.

23. This proliferation has only intensified since Stephanie LeMenager published her exhaustive reckoning, "Climate Change and the Struggle for Genre" (2017).

24. VanderMeer, *Annihilation*, 14.

25. See Strombeck, "Inhuman Writing."

26. On the painting, see Foucault, *Order of Things*, 3–16; on the trope, see Dällenbach, *Mirror in the Text*. On deep time itself as a painterly theme, see O'Rourke, "Staring into the Abyss of Time."

27. Bjornerud, *Timefulness*, 16.

28. Byron, *Cain* II.i.29, II.ii.304, *Complete Poetical Works* 636, 640; Humboldt, *Ansichten der Natur*, 110–11: "Dem neugierig regsamen Geiste des Menschen sei es erlaubt, bisweilen aus der Gegenwart in das Dunkel der Vorzeit hinüberzuschweifen; zu ahnen, was noch nicht klar erkannt werden kann, und sich so an den alten, unter vielerlei Formen wiederkehrenden Mythen der Geognosie zu ergötzen."

BIBLIOGRAPHY

Primary Sources

Addison, Joseph. "Chance Readings; the 'Children in the Wood'; the beauty of many old English Ballads." 1710. *Spectator* 85. In *Addison: Selections from Addison's Papers*, edited by Thomas Arnold, 430–33. Oxford, UK: Clarendon, 1886.

———. "On Popular Poetry." 1710. *Spectator* 70. In *Addison: Selections from Addison's Papers*, edited by Thomas Arnold, 378–83. Oxford, UK: Clarendon, 1886.

Aikin, John. *Essays on Song-Writing: with a Collection of such English Songs as are most eminent for Poetic Merit*. London: Joseph Johnson, 1772.

Aristotle. *Poetics*. Translated by Richard Janko. In *The Norton Anthology of Theory and Criticism*, 2nd ed., edited by Vincent Leitch et al., 88–115. New York: Norton, 2010.

Ballard, J. G. *The Drowned World*. 1962. Reprint ed. New York: Liveright, 2012.

Barrett, Mike. "Babylons I–XIV." *Samizdat* 1 (Autumn 1998): 14. Also available at https://mikebarrettarchive.com.

Blake, William. *Blake: The Complete Poems*. Edited by W. H. Stevenson. London: Longman, 1972.

———. *Blake's Poetry and Designs*. 2nd ed. Edited by John E. Grant and Mary Lynn Johnson. New York: Norton, 2008.

———. *The Complete Poetry and Prose of William Blake*. Edited by David V. Erdman. New York: Anchor/Doubleday, 1988.

———. *Jerusalem: The Emanation of the Giant Albion*. Edited by Morton D. Paley. Princeton, NJ: Princeton UP, 1991.

———. *The Notebook of William Blake: A Photographic and Typographic Facsimile*. Edited by David V. Erdman. Oxford, UK: Clarendon, 1973.

Boullanger, Nicolas-Antoine. *Anecdotes physiques de l'histoire de la nature*. Vol. 2 of *Oeuvres Complètes*. Edited by Pierre Boutin. Paris: Champion, 2006.

Buffon, Comte de [Georges-Louis Leclerc]. *Des manuscrits de Buffon: avec des facsimile de Buffon et de ses collaborateurs*. Edited by Marie-Jean-Pierre Flourens. Paris: Garnier, 1860.

———. *The Epochs of Nature*. Edited and translated by Jan Zalasiewicz, Anne-Sophie Milon, and Mateusz Zalasiewicz. Chicago: U of Chicago P, 2018.

———. *Les Époques de la Nature*. MS 883, Musée national d'histoire naturelle, Paris, France.

———. *Les Époques de la Nature*. Edited by Jacques Roger. 1962. Reprint ed. Paris: Éditions du Muséum, 1988.

———. *Natural History, General and Particular*. 2nd ed. Translated by William Smellie. 9 vols. London: Strahan and Cadell, 1785.

Burns, Robert. *The Best-Laid Schemes: Selected Prose and Poetry of Robert Burns*. Edited by Robert Crawford and Christopher McLachlan. Princeton, NJ: Princeton UP, 2009.

———. *Robert Burns's Songs for George Thomson*. Vol. 4 of *The Oxford Edition of the Works of Robert Burns*. Edited by Kirsteen McCue. Oxford, UK: Oxford UP, 2021.

Byron, George Gordon, Baron. *The Complete Poetical Works of Byron*. Edited by Paul Elmer More. Cambridge, MA: Houghton Mifflin, 1933.

Chatwin, Bruce. *The Songlines*. London: Picador, 1988.

Child, Francis James. *English and Scottish Popular Ballads*. 5 vols. 1882–98. Reprint ed. New York: Dover, 1962.

Columbus, Christopher. *The Four Voyages*. Edited and translated by J. M. Cohen. Harmondsworth: Penguin, 1969.

Cuvier, Georges. *Essay on the Theory of the Earth by M. Cuvier*. Translated by Robert Jameson. Edinburgh: Blackwood, 1822.

———. "Historical Report on the Progress of Geology Since 1789, and on Its Present State." In *Georges Cuvier, Fossil Bones, and Geological Catastrophes*, edited and translated by Rudwick, 115–126.

———. "Preliminary Discourse." In *Georges Cuvier, Fossil Bones, and Geological Catastrophes*, edited and translated by Rudwick, 183–252.

———. *Rapport historique sur les progrès des sciences naturelles depuis 1789*. Paris: Imprimerie Impériale, 1810.

———. *Recherches sur les ossemens fossiles des quadrupèdes*. 2 vols. Paris: Deterville, 1812.

Darwin, Charles. *The Annotated Origin: A Facsimile of the First Edition of Origin of Species*. Edited by James T. Costa. Cambridge, MA: Belknap Press, 2009.

———. *Darwin*. Edited by Philip Appleman. New York: Norton, 2000.

———. *The Descent of Man and Selection in Relation to Sex*. Edited by James Desmond and Adrian Moore. Harmondsworth, UK: Penguin, 2004.

———. *Expression of the Emotions in Man and Animals*. 1872. In *The Complete Work of Charles Darwin Online*, edited by John van Wyhe, 2002. Available at http://darwin-online.org.uk/.

———. *Journal of Researches into the Natural History and Geology of the Various Countries Visited During the Voyage of HMS* Beagle. 2nd ed. London: John Murray, 1845.

———. Letter to J. S. Henslow, 11 April 1833. Darwin Correspondence Project, "Letter no. 204," available at www.darwinproject.ac.uk/letter/DCP-LETT-204.xml, accessed 18 January 2022.

———. Letter to Charles Lyell, 18 April 1863. Darwin Correspondence Project, "Letter no. 4106," www.darwinproject.ac.uk/letter/?docId=letters/DCP-LETT-4106.xml, accessed 18 January 2022.

———. "On the Distribution of the Erratic Boulders . . . of South America." *Transactions of the Geological Society*, series 2, vol. 6 (1842): 415–31.

———. *On the Origin of Species by Means of Natural Selection, or the Preservation of Favoured Races in the Struggle for Life*. 5th ed. London: John Murray, 1869.

Darwin, Erasmus. *The Botanic Garden; A Poem, in Two Parts*. London: Joseph Johnson, 1791.

Davy, Humphry. *Humphry Davy on Geology: The 1805 Lectures for the General Audience*. Edited by Robert Siegfried and Robert H. Dott. Madison: U of Wisconsin P, 1980.

Deluc, Jean-André. Review of *Theory of the Earth*. *British Critic* 8 (1796): 337–52.

Desmarest, Nicolas. "Anecdotes de la nature." In *Géographie physique*, vol. 2, pp. 532–86. *Encyclopédie methodique*. Vol. 22. Paris: H. Agasse, 1803. https://books.google.com/books?id=eJnMatQwFsAC&newbks=1&newbks_redir=0&pg=PP7#v=onepage&q&f=false.

Dryden, John. *The Tempest, or The Enchanted Island*. London: H. Herringman, 1676.

Ferber, Johann Jakob. *Travels through Italy, in the Years 1771 and 1772*. Translated by R. E. Raspe. London: Printed for L. Davis, 1776.

Fielding, Henry. *The History of Tom Jones, a Foundling*. 1749. Reprint ed. New York: Vintage, 1950.

Flourens, Marie-Jean-Pierre. *Examen du livre du M. Darwin sur l'origine des espèces*. Paris: Garnier, 1864.

Forster, Georg[e]. *Reise um die Welt*. Edited by Gerhard Steiner. Frankfurt: Insel, 1967.

———. *A Voyage Round the World*. Edited by Nicholas Thomas and Oliver Berghof. 2 vols. Honolulu: U of Hawaii P, 2000.

———. *Werke in Vier Bänden*. Edited by Gerhard Steiner. Frankfurt: Insel, 1967.

Forster, Joh[an]n Reinhold. "Dr. Forster and Prof. Lichtenberg." Review of *Epoques de la Nature*. *Göttingisches Magazin der Wissenschaft und Litteratur* 1 (1780): 140–57.

———. *Introduction to Mineralogy*. London: Joseph Johnson, 1768.

———. *Observations Made During a Voyage Round the World*. Edited by Nicholas Thomas, Harriet Guest, and Michael Dettelbach. Honolulu: U of Hawaii P, 1996.

Fumerton, Patricia, dir. English Broadside Ballad Archive. http://ebba.english.ucsb.edu.

Godwin, William. *Political and Philosophical Writings*. Vol. 5. Edited by Pamela Clemit. London: Pickering, 1993.

Hamilton, Sir William. "Account of a Large Stone near Cape Town, in a Letter from Mr. Anderson." *Philosophical Transactions of the Royal Society* 68 (1778): 106.

———. *Campi Phlegraei: Observations on the Volcanos of the Two Sicilies*, 2 vols. Naples: Pietro Fabris, 1776.

Hancarville, Pierre de [P. F. Hugues]. *Collection of Etruscan, Greek, and Roman Antiquities from the Cabinet of Sir William Hamilton*. 4 vols. Naples: F. Morelli, 1766–1767 [1768–1776].

———. *Recherches sur l'origine, l'esprit et les progrès des arts de la Grèce*. Edited by Burton Feldman. 3 vols. 1785. Reprint ed. New York: Garland, 1984.

Hawkesworth, John. *Account of the Voyages undertaken by the order of His present Majesty for making discoveries in the Southern Hemisphere*. 3 vols. London: Printed for W. Strahan and T. Cadell, 1773.

Herd, David. *The Ancient and Modern Scots Songs, Heroic Ballads, &c.* Edinburgh: Martin and Wotherspoon, 1769.

Herder, Johann Gottfried. "Extracts from a Correspondence on Ossian." 1769. In *German Aesthetic and Literary Criticism*, edited by H. B. Nisbet, 151–76. Cambridge, UK: Cambridge UP, 1985.

———. *Ideen zur Philosophie der Geschichte der Menschheit*. Edited by Martin Bollacher. Vol. 6 of *Werke in Zehn Bänden*. Frankfurt: Deutscher Klassiker Verlag, 1989.

———. *Volkslieder*. Edited by Theodor Mathias. Vol. 2 of *Herders Werke*. Leipzig and Vienna: Bibliografisches Institut, 1900.

———. *Volkslieder, Übertragungen, Dichtungen*. Edited by Ulrich Gaier. Vol. 3 of *Werke in Zehn Bänden*. Frankfurt: Deutscher Klassiker Verlag, 1990.

———. "Von Ähnlichkeit der mittleren englischen und deutschen Dichtkunst." 1777. In *Kleinere Aufsätze I*, edited by Karl Breul, 31–46. Cambridge, UK: Cambridge UP, 1941.

Herodotus. *The Histories*. Translated by Aubrey de Selincourt. Baltimore: Penguin, 1954.

Hoare, Michael, ed. *The* Resolution *Journal of John Reinhold Forster, 1772–75*. 4 vols. London: Hakluyt Society, 1982.

Holbach, Baron de (Paul Henri Thiry), ed. and trans. *Essai d'une histoire naturelle de couches de la terre par M. Jean-Gotlob Lehmann*. Vol. 3 of *Traités de physique, d'histoire naturelle, de minéralogie, et de metallurgie*. Paris: J. T. Hérissant, 1759.

Humboldt, Alexander von. *Ansichten der Natur*. Stuttgart: Reclam, 1969.

Humboldt, Alexander von, and Aimé Bonpland. *Essay on the Geography of Plants*. Translated by Sylvie Romanowski. Chicago: U of Chicago P, 2008.

Hutton, James. *Theory of the Earth*. 2 vols. 1795. Weinheim, Germany: Engelmann, 1959.

Jefferson, Thomas. *Notes on the State of Virginia*. Edited by William Peden. New York: Norton, 1972.

Kant, Immanuel. *Critique of Judgment*. Translated by Werner S. Pluhar. Indianapolis: Hackett, 1987.

———. *Kritik der Urteilskraft und Schriften zur Naturphilosophie*. Edited by Wilhelm Weischedel. Darmstadt, Germany: Wissenschaftliche Buchgesellschaft, 1983.

Knight, Richard Payne. *An Account of the Remains of the Worship of Priapus, Lately Existing at Isernia, in the Kingdom of Naples*. London: Printed by T. Spilsbury, 1786.

Lansdown, Richard, ed. *Strangers in the South Seas: The Idea of the Pacific in Western Thought*. Honolulu: U of Hawaii P, 2006.

Lehmann, Johann Gottlob. *Versuch einer Geschichte von Flötz-Gebürgen*. Berlin: Klüter, 1756.

Locke, John. *An Essay Concerning Human Understanding*. Edited by Peter H. Nidditch. Oxford, UK: Clarendon, 1975.

Lubbock, John. *Pre-Historic Times, as Illustrated by Ancient Remains and the Manners and Customs of Modern Savages*. 2nd ed. London: Williams and Norgate, 1869.

Lyell, Charles. *Geological Evidences of the Antiquity of Man*. 1873. 4th ed. Reprint ed. New York: AMS, 1973.

———. *The Principles of Geology*. 3 vols. 1830–33. Reprint ed. Lehre, Germany: J. Cramer, 1970.

Macpherson, James. *Fragments of Ancient Poetry*. 1760. Reprint ed. Los Angeles: William Andrews Clark Memorial Library, 1966.

———. *Temora: An Ancient Epic Poem*. London: Printed for T. Becket and P. A. de Hondt, 1763.

Maine, Henry Sumner. *Ancient Law: Its Connection with the Early History of Society*. 1861. Reprint ed. Tucson: U of Arizona P, 1986.

Mayhew, Henry. *London Labour and the London Poor: The London Street-Folk*. London: Griffin, Bohn, and Co., 1861.

Meister, Jacques-Henri. "Avril." *Correspondance Littéraire* 10 (1779): 169–72.

Monboddo, James Burnett, Lord. *Of the Origin and Progress of Language*. Vol. 1. 2nd ed. Edinburgh: Printed for J. Balfour, 1774.

Montesquieu, Baron de [Charles de Secondat]. *The Spirit of the Laws*. Edited and translated by Anne M. Cohler, Basia Carolyn Miller, and Harold Samuel Stone. Cambridge, UK: Cambridge UP, 1989.

Parkinson, Sydney, and Stanfield Parkinson. *A Journal of a Voyage to the South Seas, in His Majesty's Ship, the* Endeavour. 1773. Adelaide: Libraries Board of South Australia, 1972.

Percy, Thomas. *Five Pieces of Runic Poetry Translated from the Islandic Language*. London: Dodsley, 1763.

———. *Reliques of Ancient English Poetry*. 1765. Edited by Henry B. Wheatley. Reprint ed. Mineola: Dover Publications, 1966.

Philips, Ambrose, ed. *A Collection of Old Ballads, Corrected from the best and most Ancient Copies Extant*. 3 vols. London: Printed for J. Roberts, 1723–25.

Pigafetta, Antonio. *The First Voyage around the World, 1519–22: An Account of Magellan's First Voyage*. Edited by Theodore J. Cachey, Jr. Toronto: U of Toronto P, 2007.

Playfair, John. "Biographical Notice of James Hutton." *Philosophical Transactions of the Royal Society of Edinburgh* 5 (1805), Part III: 39–100.

Ritson, Joseph. "Historical Essay on the Origin and Progress of National Song." In *Select Collection of English Song*, Vol. 1, pp. i–lxxii.

———. *Scotish Song, in Two Volumes*. London: Joseph Johnson, 1794.

———. *Select Collection of English Song*. 3 vols. London: Joseph Johnson, 1783.

Rogers, Henry Darwin [H.D.R.]. "The Reputed Traces of Primeval Man." *Blackwood's Edinburgh Magazine* 88 (October 1860): 422–37.

Rudwick, Martin J. S., ed. and trans. *Georges Cuvier, Fossil Bones, and Geological Catastrophes: New Translations & Interpretations of the Primary Texts*. Chicago: U of Chicago P, 1998.

Scott, Walter. *Minstrelsy of the Scottish Border*. 1830. Edited by Thomas Henderson. London: Harrap, 1931.

Shelley, Mary. *History of a Six Weeks' Tour through a Part of France, Switzerland, Germany, and Holland*. London: Hookham and Ollier, 1817.

Stenhouse, William, ed. *The Scots Musical Museum . . . Originally Published by James Johnson*. 6 vols. Edinburgh: Blackwood, 1839. Available at https://digital.nls.uk/special-collections-of-printed-music/archive/87804170.

Taine, John [Eric Temple Bell]. *The Time Stream*. 1931. Reprint ed. New York: Garland, 1975.

Thomas, Antoine-Léonard. *Ode sur le temps. qui a remporté le prix de l'Académie Françoise*. Paris: V. Brunet, 1762. Available at https://books.google.com/books?id=7YkZxQEACAAJ.

Thomson, James. *The Complete Poetical Works of James Thomson*. Edited by J. Logie Robertson. Oxford, UK: Oxford UP, 1908.

Tylor, Edward Burnett. *The Origins of Culture: Part I of "Primitive Culture."* 1871. Reprint ed. New York: Harper & Brothers, 1958.

———. *Researches into the Early History of Mankind and the Development of Civilisation*. London: John Murray, 1865.

VanderMeer, Jeff. *Annihilation*. New York: Farrar, Straus, and Giroux, 2014.

Werner, Abraham Gottlob. *Kurze Klassifikation der verschiedenen Gebirgsarten*. 1786. Translated by A. M. Ospovat. New York: Hafner, 1971.

Whewell, William. "Address to the Geological Society." [1838]. *Proceedings of the Geological Society of London* 2 (1833–38): 624–49. Available at www.biodiversitylibrary.org/item/96958.

———. "Address to the Geological Society." [1839]. *Proceedings of the Geological Society of London* 3 (1838–42): 61–98.

———. "On the Transformation of Hypotheses." 1851. In *Selected Writings on the History of Science*, edited by Yehudah Elkana, 385–92. Chicago: U of Chicago P, 1984.

Whitehurst, John. *Inquiry into the Original State and Formation of the Earth*. 1778; 1786. New York: Arno Press, 1978.

Winckelmann, Johann Joachim. *Gedanken über die Nachahmung der griechischen Werke*. Stuttgart: Reclam, 1995.

Wordsworth, William. *Lyrical Ballads*. Edited by R. L. Brett and A. R. Jones. London: Methuen, 1968.

———. *The Prelude: 1799, 1805, 1850*. Edited by Jonathan Wordsworth, M. H. Abrams, and Stephen Gill. Reprint ed. New York: Norton, 1979.

Wright, Alexis. *The Swan Book*. 2013. New York: Washington Square Press, 2018.

Secondary Sources

Adorno, Theodor W., and Max Horkheimer. *Dialektik der Aufklärung*. Frankfurt: Fischer, 1969.

Albritton, Claude. *The Abyss of Time: Changing Conceptions of the Earth's Antiquity after the Sixteenth Century*. San Francisco: Freeman, Cooper, 1980.

Anderson, Elizabeth. "More about Some Possible Sources of the Passages on Guiana in Buffon's *Époques de la Nature* (II)." *Trivium* 9 (1974): 70–80.

Archer, David. *The Long Thaw: How Humans Are Changing the Next 100,000 Years of Earth's Climate*. Princeton, NJ: Princeton UP, 2009.

Autin, Whitney J., and John M. Holbrook. "Is the Anthropocene an Issue of Stratigraphy or Popular Culture?" *GSA Today* 22 (2012): 60–61.

Baere, Benoît de. *La pensée cosmogonique de Buffon: Percer la nuit des temps*. Paris: Champion, 2004.

Bakhtin, Mikhail. *The Dialogic Imagination: Four Essays*. Translated by Caryl Emerson and Michael Holquist. Austin: U of Texas P, 1981.

Barnett, Lydia. *After the Flood: Imagining the Global Environment in Early Modern Europe*. Baltimore: Johns Hopkins UP, 2019.

Barrell, John. *The Political Theory of Painting from Reynolds to Hazlitt*. New Haven: Yale UP, 1986.

Bartholomew, Michael. "Lyell and Evolution: An Account of Lyell's Response to the Prospect of an Evolutionary Ancestry for Man." *British Journal for the History of Science* 6.3 (June 1973): 261–303.

Barton, Ruth. *The X Club: Power and Authority in Victorian Science*. Chicago: U of Chicago P, 2018.

Bataille, Georges. *The Accursed Share: An Essay on General Economy*. Vol. 1. Translated by Robert Hurley. New York: Zone Books, 1988.

Baulch, David. "'For in This Period the Poets Work is Done': Kairos as Interval in William Blake's *Milton* and *Jerusalem*." Paper delivered at the North American Society for the Study of Romanticism, UC-Berkeley, 13 August 2016.

Beer, Gillian. *Darwin's Plots: Evolutionary Narrative in Darwin, George Eliot, and Nineteenth-Century Fiction*. London: Routledge and Kegan Paul, 1983.

Beller, Sieghard, and Andrea Bender. "The Limits of Counting: Numerical Cognition Between Evolution and Culture." *Science* 319 (11 January 2008): 213–15. DOI: 10.1126/science.1148345.

Bennett, Shelley M. *The Mechanisms of Art Patronage in England Circa 1800. Columbia: U of Missouri P, 1988.*

Bentley, G. E. "Blake and Percy's *Reliques*." *Notes and Queries* 201 [NS III] (1956): 352–53.

Bewell, Alan. "Jefferson's Thermometer: Colonial Biogeographical Constructions of the Climate of America." In *Romantic Science: The Literary Forms of Natural History*, edited by Noah Heringman, 111–38. Albany: SUNY Press, 2003.

———. *Natures in Translation: Romanticism and Colonial Natural History*. Baltimore: Johns Hopkins UP, 2017.

Bindman, David. *Ape to Apollo: Aesthetics and the Idea of Race in the Eighteenth Century*. Ithaca, NY: Cornell UP, 2002.

Bjornerud, Marcia. *Timefulness: How Thinking Like a Geologist Can Help Save the World*. Princeton, NJ: Princeton UP, 2018.

Blumenberg, Hans. *Lebenszeit und Weltzeit*. Frankfurt a.M.: Suhrkamp, 1986.

Böhme, Madelaine, et al. "A New Miocene Ape and Locomotion in the Ancestor of Great Apes and Humans." *Nature* 575 (6 Nov 2019): 489–93.

Bowler, Peter J. *Theories of Human Evolution: A Century of Debate, 1844–1944*. Baltimore: Johns Hopkins UP, 1986.

Brantlinger, Patrick. *Dark Vanishings: Discourse on the Extinction of Primitive Races, 1800–1930*. Ithaca, NY: Cornell UP, 2003.

Buckland, Adelene. "'Inhabitants of the Same World': The Colonial History of Geological Time." *Philological Quarterly* 97.2 (Spring 2018): 219–40.

Burnett, D. Graham. "Savage Selection: Analogy and Elision in *On the Origin of Species*." *Endeavour* 33.4 (Dec 2009): 120–25.

Calè, Luisa, and Adriana Craciun. "The Disorder of Things." *Eighteenth-Century Studies* 45 (2011): 1–13.

Chakrabarti, Pratik. *Inscriptions of Nature: Geology and the Naturalization of Antiquity*. Baltimore: Johns Hopkins UP, 2020.

Chakrabarty, Dipesh. "The Climate of History: Four Theses." *Critical Inquiry* 35 (2009): 197–222.

Chapais, Bernard. *Primeval Kinship: How Pair-Bonding Gave Birth to Human Society*. Cambridge, MA: Harvard UP, 2008.

Chapman, Ann. *European Encounters with the Yamana People of Cape Horn*. Cambridge, UK: Cambridge UP, 2010.

Clark, J. F. M. "The Science of John Lubbock." *Notes and Records of the Royal Society* 68 (2014): 3–6.

Cohen, Claudine. *The Fate of the Mammoth: Fossils, Myth, and History*. Translated by William Rodarmor. Chicago: U of Chicago P, 2002.

Cohen, Michael C. *The Social Lives of Poems in 19th-Century America*. Philadelphia: Penn Press, 2015.

Cokinos, Christopher. *Bodies, of the Holocene*. Kirksville, MO: Truman State UP, 2013.

Conard, Nicholas J., Maria Maline, and Susanne Münzel. "New Flutes Document the Earliest Musical Tradition in Southwestern Germany." *Nature* 460 (6 August 2009): 737–40. Available at https://doi.org/10.1038/nature08169.

Connell, Philip. "British Identities and the Politics of Ancient Poetry in Later Eighteenth-Century England." *Historical Journal* 49 (2006): 161–92.

Cooper, Andrew M. *William Blake and the Productions of Time*. Farnborough: Ashgate, 2013.

Coote, Jeremy, Peter Gathercole, and Nicolette Meister. "'Curiosities Sent to Oxford': The Original Documentation of the Forster Collection at the Pitt Rivers Museum." *Journal of the History of Collections* 12 (2000): 177–92.

Crisp, Michael D., and Lyn G. Cook. "Do Early Branching Lineages Signify Ancestral Traits?" *Trends in Ecology and Evolution* 20.3 (March 2005): 122–28.

Cutler, Alan. "Nicolaus Steno and the Problem of Deep Time." In *The Revolution in Geology from the Renaissance to the Enlightenment*, edited by Gary D. Rosenberg, 143–48. Boulder, CO: Geological Society of America, 2009.

Dällenbach, Lucien. *The Mirror in the Text*. Trans. Jeremy Whitely with Emma Hughes. Chicago: U of Chicago P, 1989.

Dames, Nicholas. *Amnesiac Selves: Nostalgia, Forgetting, and British Fiction, 1810–1870*. New York: Oxford UP, 2001.

Davies, Jeremy. *The Birth of the Anthropocene*. Berkeley: U of Calfornia P, 2016.

De Luca, Vincent A. *Words of Eternity: Blake and the Poetics of the Sublime*. Princeton, NJ: Princeton UP, 1991.

Dean, Dennis. *James Hutton and the History of Geology*. Ithaca, NY: Cornell UP, 1992.

Dean, John. "The *Odyssey* as Romance." *College Literature* 3 (1976): 228–36.

Desai, Michael M. "Reverse Evolution and Evolutionary Memory." *Nature Genetics* 41 (2009): 142–43. Available at https://doi.org/10.1038/ng0209-142.

Desmond, Adrian, and James Moore. *Darwin's Sacred Cause: How a Hatred of Slavery Shaped Darwin's Views on Human Evolution*. Boston: Houghton Mifflin, 2009.

Dimock, Wai-Chee. "Low Epic." *Critical Inquiry* 39 (2013): 614–31.

———. "Nonbiological Clock: Literary History against Newtonian Mechanics." *South Atlantic Quarterly* 102.1 (Winter 2003): 153–77.

———. *Through Other Continents: American Literature through Deep Time*. Princeton, NJ: Princeton UP, 2006.

Dissanayake, Ellen. *Homo Aestheticus: Where Art Comes From and Why*. New York: Free Press, 1992.

Douglas, Bronwen. "Art as Ethno-historical Text: Science, Representation and Indigenous Presence in Eighteenth- and Nineteenth-century Oceanic Voyage Literature." In *Double Vision: Art Histories and Colonial Histories in the Pacific*, edited by Nicholas Thomas and Diane Losche, 65–99. Cambridge, UK: Cambridge UP, 1999.

Duncan, Ian. "Darwin and the Savages." *Yale Journal of Criticism* 4.2 (Spring 1991): 13–45.

———. *Human Forms: The Novel in the Age of Evolution*. Princeton, NJ: Princeton UP, 2019.

———. "Natural Histories of Form: Charles Darwin's Aesthetic Science." *Representations* 151 (2020): 51–73.

———. "*Our Mutual Friend*." In *Oxford Handbook of Charles Dickens*, edited by Robert L. Patten, John O. Jordan, and Catherine Waters, 285–98. Oxford, UK: Oxford UP, 2018.

———. "The Pathos of Abstraction: Adam Smith, Ossian, and Samuel Johnson." In *Scotland and the Borders of Romanticism*, edited by Ian Duncan, Leith Davis, and Janet Sorenson, 38–56. Cambridge, UK: Cambridge UP, 2004.

Easterby-Smith, Sarah. "Recalcitrant Seeds: Material Culture and the Global History of Science." *Past and Present* 242 (2019): 215–42.

Eaves, Morris. *The Counter-Arts Conspiracy: Art and Industry in the Age of Blake*. Ithaca, NY: Cornell UP, 1992.

Evenson, Edward B., et al. "Enigmatic Boulder Trains, Supraglacial Rock Avalanches, and the Origin of 'Darwin's Boulders,' Tierra del Fuego." *GSA Today* 19.12 (December 2009): 4–10. Available at doi:10.1130/GSATG72A.1.

Fabian, Johannes. *Time and the Other: How Anthropology Makes Its Object*. New York: Columbia UP, 1983.

Finney, Stanley C., and Lucy E. Edwards. "The 'Anthropocene' Epoch: Scientific Decision or Political Statement?" *GSA Today* 26 (2016): 4–10. Available at https//doi.org/10.1130/GSATG270A.1.

Folkenflik, Robert. "Macpherson, Chatterton, Blake, and the Great Age of Literary Forgery." *Centennial Review* 18 (1974): 378–91.

Foucault, Michel. "Of Other Spaces: Utopias and Heterotopias." In *Rethinking Architecture: A Reader in Cultural Theory*, edited by Neil Leach, 350–56. New York: Routledge, 1997.

———. *The Order of Things. An Archaeology of the Human Sciences*. 1966. New York: Random House, 1970.

France, Peter. "Thomas, Antoine-Léonard." *The New Oxford Companion to Literature in French* (online), Oxford, UK: Oxford UP, 1995. www.oxfordreference.com/view/10.1093/acref/9780198661252.001.0001/acref-9780198661252-e-4550.

Friedman, Albert Barron. *The Ballad Revival: Studies in the Influence of Popular on Sophisticated Poetry*. Chicago: U of Chicago P, 1961.

Gamble, Clive, and Robert Kruszynski. "John Evans, Joseph Prestwich, and the Stone That Shattered the Time Barrier." *Antiquity* 83 (2009): 461–75.

Gandy, Matthew. "J. G. Ballard and the Politics of Catastrophe." *Space and Culture* 9.1 (February 2006): 86–88.

Gascoigne, John. "The German Enlightenment and the Pacific." In *Anthropology of the Enlightenment*, ed. Larry Wolff and Marco Cipolloni, 141–71.

Gee, Henry. *In Search of Deep Time: Beyond the Fossil Record to a New History of Life*. New York: Free Press, 1999.

Gidal, Eric. *Ossianic Unconformities: Bardic Poetry in the Industrial Age*. Charlottesville: U of Virginia P, 2015.

Gilpin, George H. "William Blake and the World's Body of Science." *Studies in Romanticism* 43 (2004): 35–56.

Gohau, Gabriel. "Darwin the Geologist: Between Lyell and von Buch." *Comptes Rendus Biologies* 333 (2010): 95–98.

Goldsmith, Steven. *Blake's Agitation: Criticism and the Emotions*. Baltimore: Johns Hopkins UP, 2013.

Goldstein, Amanda J. *Sweet Science: Romantic Materialism and the New Logics of Life*. Chicago: U of Chicago P, 2017.

Gould, Stephen Jay. *Time's Arrow, Time's Cycle: Myth and Metaphor in the Discovery of Geological Time*. Cambridge, MA: Harvard UP, 1987.

Green, Roger Lancelyn. "Introduction." In *A Book of British Ballads*, edited by R. Brimley Johnson, v–xi. 1912. London: Dent, 1966.

Grobman, Neil R. "David Hume and the Earliest Scientific Methodology for Collecting Balladry." *Journal of Western Folklore* 34.1 (January 1975): 16–31.

Guest, Harriet. *Empire, Barbarism, and Civilisation: James Cook, William Hodges, and the Return to the Pacific*. Cambridge, UK: Cambridge UP, 2007.

Guntau, Martin. *Genesis der Geologie als Wissenschaft*. Berlin: Akademie-Verlag, 1984.

Hamilton, Clive, and Jacques Grinevald. "Was the Anthropocene Anticipated?" *Anthropocene Review* 2 (2015): 1–14. Available at https//doi.org/10.1177/2053019614567155.

Harley, Alexis. "Animal Encounter and Pre-Darwinian Darwin." Paper delivered at "Romanticism and Evolution," University of Western Ontario, 14 May 2011.

———. *Autobiologies: Charles Darwin and the Natural History of the Self*. Lewisburg, PA: Bucknell UP, 2015.

Heinrich, Bernd. *Why We Run: A Natural History*. New York: Ecco, 2001.

Henrich, Joseph. *The Secret of Our Success: How Culture is Driving Human Evolution, Domesticating our Species, and Making Us Smarter*. Princeton, NJ: Princeton UP, 2017.

Heringman, Noah. "The Anthropocene Reads Buffon (or, Reading Like Geology)." In *Anthropocene Reading*, edited by Menely and Taylor, 59–77.

———. "Deep Time at the Dawn of the Anthropocene." *Representations* 129 (Winter 2015): 56–85.

———. "Primitive Arts and Sciences and the Body of Knowledge in Blake's Epics." In *William Blake: Modernity and Disaster*, edited by Tilottama Rajan and Joel Faflak, 30–53. Toronto: U of Toronto P, 2020.

———. *Sciences of Antiquity: Romantic Antiquarianism, Natural History, and Knowledge Work*. Oxford, UK: Oxford UP, 2013.

Hilton, Nelson. "What Has *Songs* To Do with Hymns?" In *Blake in the Nineties*, edited by Steve Clark and David Worrall, 96–113. New York: St. Martin's, 1999.

Hirsch, E. D. *Innocence and Experience: An Introduction to Blake*. New Haven: Yale UP, 1964.

Hoare, Michael. *The Tactless Philosopher: Johann Reinhold Forster, 1729–1798*. Melbourne: Hawthorn Press, 1976.

Hulme, Peter. "Cast Away: The Uttermost Parts of the Earth." In *Sea Changes: Historicizing the Ocean*, edited by Bernhard Klein and Gesa Mackenthun, 187–201. New York: Routledge, 2004.

Hutchings, Kevin. "William Blake and the Music of the *Songs*." *Romanticism on the Net* 45 (February 2007). Available at https://ronjournal.org/s/2125.

Jetha, Cacilda, and Christopher Ryan. *Sex at Dawn: The Prehistoric Origins of Modern Sexuality*. New York: Harper, 2010.

Jolly, Margaret. "'Ill-Natured Comparisons': Racism and Relativism in European Representations of ni-Vanuatu from Cook's Second Voyage." *History and Anthropology* 5 (1992): 331–64.

Jonsson, Fredrik Albritton. "A History of the Species." *History and Theory* 52 (2013): 462–72.

———. "The Industrial Revolution in the Anthropocene." *Journal of Modern History* 84 (2012): 679–96.

Joppien, Rüdiger, and Bernard Smith. *The Art of Captain Cook's Voyages*. 3 vols. New Haven: Yale UP, 1985–88.

Kasper, Norman. "Urwelt und Altertum: Zur narrativen Koordination zweier Konzepte im 19. Jahrhundert." In *Erdgeschichten*, ed. Peter Schnyder, 47–70.

Klancher, Jon. *Transfiguring the Arts and Sciences: Knowledge and Cultural Institutions in the Romantic Age*. Cambridge, UK: Cambridge UP, 2013.

Kloetzli, W. Randolph. "Myriad Concerns: Indian Macro-Time Intervals ('Yugas,' 'Sandhyas,' and 'Kalpas') as Systems of Number." *Journal of Indian Philosophy* 41.6 (December 2013): 631–53.

Kolbert, Elizabeth. *The Sixth Extinction: An Unnatural History*. New York: Picador, 2014.

Koselleck, Reinhart. "Does History Accelerate?" In *Sediments of Time*, ed. Sean Franzel and Stefan-Ludwig Hoffmann, 79–99.

———. "Einleitung." In *Geschichtliche Grundbegriffe: Historisches Lexikon zur politisch-sozialen Sprache in Deutschland*, edited by Otto Brunner et al., vol. 1, pp. xiii–xxviii. Stuttgart: Klett, 1975.

———. "*Neuzeit*: On the Semantics of Modern Concepts of Movement." In *Futures Past: On the Semantics of Historical Time*, translated and edited by Keith Tribe, 222–54. New York: Columbia UP, 2004.

———. *Sediments of Time: On Possible Histories*. Translated and Edited by Sean Franzel and Stefan-Ludwig Hoffmann. Stanford: Stanford UP, 2018.

———. *Zeitschichten: Studien zur Historik*. Frankfurt a.M.: Suhrkamp, 2000.

Kugler, Lena. "Die Tiefenzeit von Dingen und Menschen: (Falsche) Fossilien und die 'Bergwerke zu Falun.'" *Weimarer Beiträge* 59 (2013): 397–415.

Larson, Edwin E., and Peter W. Birkeland. *Putnam's Geology*. 4th ed. New York: Oxford UP, 1982.

Latour, Bruno. *Reassembling the Social: An Introduction to Actor-Network Theory*. Oxford, UK: Oxford UP, 2005.

———. *Science in Action: How to Follow Scientists and Engineers through Society*. Cambridge, MA: Harvard UP, 1987.

———. *We Have Never Been Modern*. Translated by Catherine Porter. Cambridge, MA: Harvard UP, 1993.

Laudan, Rachel. *From Mineralogy to Geology: The Foundations of a Science, 1650–1830*. Chicago: U of Chicago P, 1987.

Laurent, Olivier. *The Dark Abyss of Time: Archaeology and Memory*. Translated by Alan Greenspan. Lanham, MD: Altamira, 2011.

Lawson, Kristan. *Darwin and Evolution for Kids: His Life and Ideas with 21 Activities*. Chicago: Chicago Review Press, 2003.

Leask, Nigel. "Sir Walter Scott's *The Antiquary* and the Ossian Controversy." *Yearbook of English Studies* 47 (2017): 189–202.

LeMenager, Stephanie. "Climate Change and the Struggle for Genre." In *Anthropocene Reading*, edited by Menely and Taylor, 220–38.

Lepenies, Wolf. *Das Ende der Naturgeschichte: Wandel kultureller Selbstverständlichkeiten in den Wissenschaften des 18. und 19. Jahrhunderts*. Frankfurt a.M.: Suhrkamp, 1978.

Lévi-Strauss, Claude. *The Savage Mind*. Chicago: U of Chicago P, 1966.

Lewis, Cherry, and Simon J. Knell, eds. *The Age of the Earth: From 4004 BC to AD 2002*. London: Geological Society, 2001.

Lightman, Bernard. *Victorian Popularizers of Science: Designing Nature for New Audiences*. Chicago: U of Chicago P, 2007.

Logan, Peter Melville. *Victorian Fetishism: Intellectuals and Primitives*. Albany: SUNY Press, 2009.

Loveland, Jeff. *Rhetoric and Natural History: Buffon in Polemical and Literary Context*. Oxford, UK: Voltaire Foundation, 2001.

Lüdeke, Roger. *Zur Schreibkunst von William Blake: Ästhetische Souveränität und Politische Imagination*. Munich: Wilhelm Fink, 2013.

Lyle, Paul. *The Abyss of Time: A Study of Geological Time and Earth History*. Edinburgh: Dunedin Academic, 2016.

Marshak, Stephen. *Essentials of Geology*. 6th ed. New York: Norton, 2019.

Marshall, Kate. "What Are the Novels of the Anthropocene? American Fiction in Geological Time." *American Literary History* 27 (2015): 523–38.

Matharoo, Sean. "A Fata Morgana at the End of the World, or, Towards the Eco-Racial Disaster of J. G. Ballard's *The Drowned World*." *Green Letters: Studies in Ecocriticism* 22 (2018): 366–80.

McDougall, Christopher. *Born to Run: A Hidden Tribe, Superathletes, and the Greatest Race the World Has Never Seen*. New York: Alfred A. Knopf, 2009.

McEwan, Colin, Luis A. Borrero, and Alfredo Prieto, eds. *Patagonia: Natural History, Prehistory, and Ethnography at the Uttermost End of the Earth*. Princeton, NJ: Princeton UP, 1997.

McGurl, Mark. "'Neither Indeed Could I Forebear Smiling at Myself': A Reply to Wai-Chee Dimock." *Critical Inquiry* 39 (2013): 632–38.

———. "The Posthuman Comedy." *Critical Inquiry* 28 (Spring 2012): 533–53.

McLane, Maureen. *Balladeering, Minstrelsy, and the Making of British Romantic Poetry*. Cambridge, UK: Cambridge UP, 2008.

McPhee, John. *Annals of the Former World*. New York: Farrar, Straus, and Giroux, 1998.

Meek, Ronald S. *Social Science and the Ignoble Savage*. Cambridge, UK: Cambridge UP, 1976.

Meillassoux, Quentin. *After Finitude: An Essay on the Necessity of Contingency*. Translated by Ray Brassier. London: Continuum, 2008.

Menely, Tobias. *The Animal Claim: Sensibility and the Creaturely Voice*. Chicago: U of Chicago P, 2015.

———. "Late Holocene Poetics: Genre and Geohistory in Charlotte Smith's *Beachy Head*." *European Romantic Review* 28.3 (June 2017): 307–15.

———. "'The Present Obfuscation': Cowper's *Task* and the Time of Climate Change" *PMLA* 127 (2012): 477–92.

Menely, Tobias, and Jesse Oak Taylor, eds. *Anthropocene Reading: Literary History in Geologic Times*. University Park: Penn State UP, 2017.

Miller, David Philip. "Joseph Banks, Empire, and 'Centers of Calculation' in Hanoverian London." In *Visions of Empire: Voyages, Botany, and Representations of Nature*, edited by David Philip Miller and Peter Hanns Reill, 21–27. Cambridge, UK: Cambridge UP, 1996.

Nayak, Meena. "Cycles of Great Time." In *The Blue Lotus: Myths and Folktales of India*, 507–8. New Delhi: Aleph, 2018.

Nixon, Rob. *Slow Violence and the Environmentalism of the Poor*. Cambridge, MA: Harvard UP, 2011.

Nyong'o, Tavia. *Afro-Fabulations: The Queer Drama of Black Life*. New York: NYU Press, 2019.

O'Donnell, Erin. "Buffering the Sun: David Keith and the Question of Climate Engineering." *Harvard Magazine* (July–August 2013): 36–39, 75. Available at www.harvardmagazine.com /2013/07/buffering-the-sun.

Palmer, Megan E. "Picturing Song Across Species: Broadside Ballads in Image and Word." *Huntington Library Quarterly* 79 (2016): 221–44. https://doi.org/10.1353/hlq.2016.0009.

Palmeri, Frank. *State of Nature, Stages of Society: Enlightenment Conjectural History and Modern Social Discourse*. New York: Columbia UP, 2016.

Parikka, Jussi. *A Geology of Media*. Minneapolis: U of Minnesota P, 2015.

Pearn, Alison. "The Teacher Taught? What Charles Darwin Owed to John Lubbock." *Notes and Records of the Royal Society* 68 (2014): 7–19.

Porter, Roy. *The Making of Geology: Earth Science in Britain, 1660–1815*. Cambridge, UK: Cambridge UP, 1977.

Priestman, Martin. *The Poetry of Erasmus Darwin: Enlightened Spaces, Romantic Times*. Farnham, UK: Ashgate, 2013.

Puchner, Martin. *The Written World: The Power of Stories to Shape People, History, Civilization*. New York: Random House, 2017.

Radick, Gregory. "Did Darwin Change His Mind about the Fuegians?" *Endeavour* 34.2 (June 2010): 50–54.

Rappaport, Rhoda. *When Geologists Were Historians, 1665–1750*. Ithaca, NY: Cornell UP, 1997.

Reardon, B. P. *The Form of Greek Romance*. Princeton, NJ: Princeton UP, 1991.

Rein, Siegfried. "Georg Christian Füchsel (1722–1773)—ein Aktualist entdeckt die Tiefenzeit der Erdgeschichte." *Vernate* 28 (2009): 11–30.

Reno, Seth T. *Early Anthropocene Literature in Britain, 1750–1884*. Cham, Switzerland: Palgrave Macmillan, 2020.

Repcheck, Jack. *The Man Who Found Time: James Hutton and the Discovery of the Earth's Antiquity*. New York: Basic Books, 2008.

Richards, Evelleen. "Darwin and the Descent of Woman." In *The Wider Domain of Evolutionary Thought*, edited by David Oldroyd and Ian Langham, 57–111. Dordrecht, Netherlands: Kluwer, 1983.

———. *Darwin and the Making of Sexual Selection*. Chicago: U of Chicago P, 2017.

Richet, Pascal. *A Natural History of Time*. Translated by John Venerella. Chicago: U of Chicago P, 2007.

Roger, Jacques. *Buffon: A Life in Natural History*. Translated by Sarah Lucille Bonnefoi. Ithaca, NY: Cornell UP, 1997.

Rossi, Paolo. *The Dark Abyss of Time: The History of the Earth and the History of Nations from Hooke to Vico*. Translated by Lydia G. Cochrane. Chicago: U of Chicago P, 1984.

Rudwick, Martin. *Bursting the Limits of Time: The Reconstruction of Geohistory in the Age of Revolution*. Chicago: U of Chicago P, 2005.

———. "The Emergence of a Visual Language for Geological Science, 1760–1840." *History of Science* 14 (1976): 149–95.

———. *Scenes from Deep Time: Early Pictorial Representations of the Prehistoric World*. Chicago: U of Chicago P, 1992.

Sachs, Jonathan. "On the Road: Travel, Antiquarianism, Philology." In *Philology and Its Histories*, edited by Sean Gurd, 127–47. Columbus: Ohio State UP, 2010.

Salih, Sarah. "Filling Up the Space between Mankind and Ape: Racism, Speciesism, and the Androphilic Ape." *ARIEL* 38 (2007): 95–111.

Salmond, Anne. *The Trial of the Cannibal Dog*. New Haven, CT: Yale UP, 2003.

Schmitt, Cannon. *Darwin and the Memory of the Human: Evolution, Savages, and South America*. Cambridge, UK: Cambridge UP, 2009.

Schnapp, Alain. "Antiquarian Studies in Naples at the End of the Eighteenth Century: From Comparative Archaeology to Comparative Religion." In *Naples in the Eighteenth Century*, edited by Giovanni Imbruglia, 154–66. Cambridge, UK: Cambridge UP, 2000.

———. *The Discovery of the Past: The Origins of Archaeology*. New York: Harry N. Abrams, 1997.

Schnyder, Peter, ed. *Erdgeschichten: Literatur und Geologie im langen 19. Jahrhundert*. Würzburg, Germany: Königshausen & Neumann, 2020.

———. "Geologie." In *Literatur und Wissen. Ein interdisziplinäres Handbuch*, edited by Roland Borgards et al., 75–79. Stuttgart and Weimar: Metzler, 2013.

———. "Geologisch-Meteorologische Phantasien: Georg Christoph Lichtenberg und die Poetologie moderner Klimawandelnarrative." *Zeitschrift für Kulturwissenschaften* 1 (2016): 103–15.

Seow, C. L. *A Grammar for Biblical Hebrew*. Revised ed. Nashville: Abingdon Press, 1995.

Sera-Shriar, Efram, ed. *Historicizing Humans: Deep Time, Evolution, and Race*. Pittsburgh: U of Pittsburgh P, 2018.

Sloan, Phillip R. "The Idea of Racial Degeneracy in Buffon's *Histoire naturelle*." In *Racism in the Eighteenth Century*, edited by Harold E. Pagliaro and James L. Clifford, 293–321. Cleveland: Case Western Reserve UP, 1973.

Smail, Daniel Lord. *On Deep History and the Brain*. Berkeley: U of California P, 2008.

Smail, Daniel Lord, and Andrew Shryock. *Deep History: The Architecture of Past and Present*. Berkeley: U of California P, 2011.

Spears, André. "Evolution in Context: 'Deep Time,' Archaeology, and the Post-Romantic Paradigm." *Comparative Literature* 48.4 (Fall 1996): 340–58.

Spencer, Vicki. *Herder's Political Thought: A Study of Language, Culture, and Community*. Toronto: U of Toronto P, 2012.

Stalnaker, Joan. "Buffon on Death and Fossils." *Representations* 115 (2011): 20–41.

Steffen, Will, Jacques Grinevald, Paul Crutzen, and John McNeill. "The Anthropocene: Conceptual and Historical Perspectives." *Philosophical Transactions of the Royal Society of London* [Series A] 369 (2011): 842–67.

Stocking, George. *Victorian Anthropology*. New York: Free Press, 1987.

Strombeck, Andrew. "Inhuman Writing in Jeff VanderMeer's Southern Reach Trilogy." *Textual Practice* 34 (2020): 1365–82.

Sullivan, Heather I., and James Shinkle. "The Dark Green in the Early Anthropocene: Goethe's Plants in *Versuch die Metamorphose der Pflanzen zu erklären* and *Triumph der Empfindsamkeit*." *Goethe Yearbook* 26 (2019): 141–62.

Tannenbaum, Leslie. *Biblical Tradition in Blake's Early Prophecies*. Princeton, NJ: Princeton UP, 1982.

Taylor, Kenneth L. "*Les Époques de la Nature* and Geology during Buffon's Later Years." In *Buffon 88: Actes du Colloque International pour le Bicentenaire de la Mort de Buffon*. Edited by J. Gayon, 371–85. Paris: J. Vrin, 1992.

Thomas, Nicholas. *In Oceania: Visions, Artifacts, Histories*. Durham, NC: Duke UP, 1997.

Toscano, Maria. "The Figure of the Naturalist-Antiquary in the Kingdom of Naples: Giuseppe Giovene (1753–1837) and His Contemporaries." *Journal of the History of Collections* 19 (2007): 225–37.

Toulmin, Stephen, and June Goodfield. *The Discovery of Time*. New York: Harper and Row, 1965.

Van Riper, A. Bowdon. *Men Among the Mammoths: Victorian Science and the Discovery of Human Prehistory*. Chicago: U of Chicago P, 1993.

Whitehead, Angus. "'Went to See Blake—also to Surgeons College': Blake and George Cumberland's Pocket Books." In *Blake in Our Time: Essays in Honour of G. E. Bentley, Jr*, edited by Karen Mulhallen, 165–200. Toronto: U of Toronto P, 2010.

Wolff, Larry. "Discovering Cultural Perspective: The Intellectual History of Anthropological Thought in the Age of Enlightenment." In *The Anthropology of the Enlightenment*, ed. Larry Wolff and Marco Cipolloni, 3–35.

Wolff, Larry, and Marco Cipolloni, eds. *The Anthropology of the Enlightenment*. Stanford, CA: Stanford UP, 2007.

Wolf, Robb. *The Paleo Solution: The Original Human Diet*. Las Vegas: Victory Belt, 2010.

Wyse Jackson, Patrick. *The Chronologers' Quest: Episodes in the Search for the Age of the Earth*. Cambridge, UK: Cambridge UP, 2006.

Yusoff, Kathryn. *A Billion Black Anthropocenes or None*. Minneapolis: U of Minnesota P, 2018.

Zammito, John H. *Kant, Herder, and the Birth of Anthropology*. Chicago: U of Chicago P, 2002.

Zhang, Chunjie. *Transculturality and European Discourse in the Age of Colonialism*. Evanston, IL: Northwestern UP, 2017.

Zielinski, Siegfried. *Deep Time of the Media: Toward an Archaeology of Hearing and Seeing by Technical Means*. Translated by Gloria Custance. Cambridge, MA: MIT Press, 2006.

Zuk, Marlene. *Paleofantasy: What Evolution Really Tells Us about Sex, Diet, and How We Live*. New York: Norton, 2013.

INDEX

A NOTE ON THE TYPE

This book has been composed in Arno, an Old-style serif typeface in the classic Venetian tradition, designed by Robert Slimbach at Adobe.

CPSIA information can be obtained
at www.ICGtesting.com
Printed in the USA
JSHW020321041122
32552JS00002B/2